Evidence Contestation

This book examines the practices of contesting evidence in democratically constituted knowledge societies. It provides a multifaceted view of the processes and conditions of evidence criticism and how they determine the dynamics of de- and re-stabilization of evidence.

Evidence is an essential resource for establishing claims of validity, resolving conflicts, and legitimizing decisions. In recent times, however, evidence is being contested with increasing frequency. Such contestations vary in form and severity – from questioning the interpretation of data or the methodological soundness of studies to accusations of evidence fabrication. The contributors to this volume explore which actors, for what reasons and to what effect, question evidence in fields such as the biological, environmental and health sciences. In addition to actors inside academia, they examine the roles of various other players, including citizen scientists, counter-experts, journalists, patients, consumers and activists. The contributors tackle questions of how disagreements are framed and how they are used to promote vested interests. By drawing on methodological and theoretical approaches from a wide range of fields, this book provides a much-needed perspective on how evidence criticism influences the development and state of knowledge societies and their political condition.

Evidence Contestation will appeal to scholars and advanced students working in philosophy of science, epistemology, bioethics, science and technology studies, the history of science and technology and science communication.

Karin Zachmann is a professor of the history of technology at the Technical University of Munich, Germany, and co-speaker of the DFG research group "Practicing Evidence – Evidencing Practice".

Mariacarla Gadebusch Bondio heads the Institute for the Medical Humanities at the University Hospital of Bonn, Germany. She is the co-editor, with Francesco Spöring and John-Stewart Gordon, of *Medical Ethics, Prediction, and Prognosis: Interdisciplinary Perspectives* (New York: Routledge, 2017).

Saana Jukola is a postdoctoral researcher at the Department of Philosophy I, Ruhr University Bochum, Germany, under the DFG-Emmy Noether Research Group "The Return of the Organism in the Biosciences: Theoretical, Historical, and Social Dimensions".

Olga Sparschuh is a postdoctoral researcher in history of technology at the Technical University of Munich, Germany, and the coordinator of the DFG research group "Practicing Evidence – Evidencing Practice".

Routledge Studies in the Philosophy of Science

For more information about this series, please visit:
https://www.routledge.com/Routledge-Studies-in-the-Philosophy-of-
Science/book-series/POS

Evidence Contestation

Dealing with Dissent in
Knowledge Societies

Edited by Karin Zachmann,
Mariacarla Gadebusch Bondio,
Saana Jukola, and Olga Sparschuh

Routledge
Taylor & Francis Group
NEW YORK AND LONDON

First published 2023
by Routledge
605 Third Avenue, New York, NY 10158

and by Routledge
4 Park Square, Milton Park, Abingdon, Oxon, OX14 4RN

*Routledge is an imprint of the Taylor & Francis Group, an
informa business*

ISBN: 978-1-032-21910-3 (hbk)
ISBN: 978-1-032-22644-6 (pbk)
ISBN: 978-1-003-27350-9 (ebk)

DOI: 10.4324/9781003273509

Typeset in Sabon
by KnowledgeWorks Global Ltd.

Contents

Acknowledgments

This publication was funded by the Deutsche Forschungsgemeinschaft (DFG, German Research Foundation) within the framework of the research group 2448 "Practicing Evidence – Evidencing Practice" with the projects BI 838/9-2; DI 1725/2-2; GA 1086/8-2; ZA 302/6-2; RO 2506/15-2; TR 364/29-1; ZA 302/5-2.

Figures

Abbreviations

ACS	American College of Surgeons
AGÖF	Arbeitsgemeinschaft ökologischer Forschungsinstitute (Ecological Research Institutes Working Group)
AORN	Association of periOperative Registered Nurses
AWMF	Association of the Scientific Medical Societies in Germany
BAUZ	Bremer Arbeits- und Umweltschutz-Zentrum (Bremen Centre for Occupational Safety and Environmental Protection)
BBC	British Broadcasting Corporation
BIAS	Basic, Implicit, Assumption in Science
BSSRS	British Society for Social Responsibility in Science
CBD	Convention on Biological Diversity of the United Nations
CBT	Cognitive Behavioral Therapies
CDU	Christlich Demokratische Union (Christian Democratic Union)
DDT	Dichlordiphenyltrichlorethan
DFG	Deutsche Forschungsgemeinschaft (German Research Foundation)
EBM	Evidence-Based Medicine
EC	European Commission
ECSA	European Citizen Science Association
EFSA	European Food Safety Authority
EPA	Environmental Protection Agency
EPPI	Evidence for Policy and Practice Information and Co-ordinating Centre
EU	European Union
FAO	Food and Agriculture Organization of the United Nations
FRG	Federal Republic of Germany
GISP	Global Invasive Species Program
GMOs	Genetically Modified Organisms
GÖK	Gruppe Ökologie (Ecology Group)
GRADE	Grading of Recommendations, Assessment, Development and Evaluations
GRS	Gesellschaft für Reaktorsicherheit (Society for Reactor Safety)
IARC	International Agency for Research on Cancer

IBP	International Biological Program
IFEU	Institut für Energie- und Umweltforschung (Institute for Research on Energy and the Environment)
INRA	Institut national de recherche en agriculture, alimentation et environment (French National Institute for Agriculture)
IPBES	Intergovernmental Science-Policy Platform on Biodiversity and Ecosystem Services
IPS	Industry Association for Crop Protection
IPCC	Intergovernmental Panel on Climate Change
IUCN	International Union for Conservation of Nature
JCAHO	Joint Commission on Accreditation of Healthcare Organizations
KFA	Kernforschungsanstalt Jülich (Nuclear Research Institute)
LGBTQ+	Lesbian, Gay, Bisexual, Transgender, Queer and Other
MFT	Moral Foundations Theory
MI	Motivational Interviews
MPG	Max-Planck-Gesellschaft (Max-Planck-Society)
NGO	Non-Governmental Organization
NICE	National Institute for Health and Care Excellence
NRDC	Natural Resources Defense Council
OECD	Organization for Economic Co-operation and Development
PAN	Pesticide Action Network
RCT	Randomized Control Trial
SCOPE	Scientific Committee on Problems of the Environment
SIM	Social Intuitionism Model
SSIs	Surgical Site Infections
STS	Science and Technology Studies
UCIL	Union Carbide India Limited
UN	United Nations
UNEP	United Nations Environment Programme
WHO	World Health Organization

Contributors

Kevin Altmann is a PhD candidate in sociology at the Bavarian Research Institute for Digital Transformation at the Bavarian Academy of Sciences and Humanities, Germany.

Fredrik Andersen is an associate professor of philosophy of science at Østfold University College and of business ethics at the University of Life Sciences, Norway.

Eva Barlösius is a professor of macrosociology and social structural analysis at the Leibniz University of Hanover and the founder of the Leibniz Center for Science and Society, Hanover, Germany.

Helena Bilandzic is a professor of communication science at the University of Augsburg, Germany.

Anders Dechsling is a PhD candidate in special pedagogics and behavior analysis at Østfold University College, Norway.

Sarah Ehlers is a postdoctoral researcher at the Deutsches Museum, Munich, Germany.

Stefan Esselborn is a postdoctoral researcher in history of technology at the Technical University of Munich, Germany.

Daniel Füger is a PhD candidate in philosophy at Justus Liebig University Giessen, Germany.

Mariacarla Gadebusch Bondio is a professor of history, philosophy and ethics of medicine and heads the Institute for Medical Humanities at the University Hospital Bonn, Germany.

Yngve Herikstad is a lecturer in social education at Østfold University College, Norway.

Saana Jukola is a postdoctoral researcher in philosophy of science at Ruhr University Bochum, Germany.

Martha Kenney is an associate professor of women and gender studies at San Francisco State University, USA.

Susanne Kinnebrock is a professor of communication at the University of Augsburg, Germany.

Christoph Kueffer is a professor of urban ecology at the Institut für Landschaft und Freiraum at Ostschweizer Fachhochschule, Switzerland and a senior scientist at ETH Zurich.

Elif Özmen is a professor of practical philosophy at Justus Liebig University Giessen, Germany.

Edoardo Maria Pelli is a PhD candidate at the chair of marketing and consumer research at the Technical University of Munich, Germany.

Jutta Roosen holds the chair of marketing and consumer research at the Technical University of Munich, Germany.

Eva Ruffing is a professor of German politics in the context of European multi-level governance at the University of Osnabrueck, Germany.

Olga Sparschuh is a postdoctoral researcher in history of technology at the Technical University of Munich, Germany, and the coordinator of the DFG research group "Practicing Evidence – Evidencing Practice".

Stefan Sütterlin holds the chair of psychology at Østfold University College, Norway.

Andreas Wenninger is a postdoctoral researcher in sociology and coordinator at the Bavarian Research Institute for Digital Transformation at the Bavarian Academy of Sciences, Germany.

Karin Zachmann is a professor of the history of technology at the Technical University of Munich, Germany, and co-speaker of the DFG research group "Practicing Evidence – Evidencing Practice".

Introduction

Evidence Critique and Contestation as a Challenge in Academia and Society

Karin Zachmann, Saana Jukola, Olga Sparschuh, and Mariacarla Gadebusch Bondio

On April 4, 2022, the Intergovernmental Panel on Climate Change (IPCC) published a report, stating that the goal of limiting the average global temperature rise to 1.5°C above pre-industrial levels is turning out to be impossible. Without immediate remedial action and a considerable transformation of our consumerist way of life, we cannot hope to stop average temperatures from rising to levels that threaten to disrupt the climate system on a global scale. Why is it that we have not been able to counter this deeply worrying trend sooner, despite the numerous concerned statements issued by individual scientists, activists and scientific associations? Many blame think tanks, whose researchers have questioned climate science evidence for years and raised doubts about the validity of claims about the looming threat. Those critical of established climate science, in contrast, often argue that they are interested in evidence and would be willing to accept the disturbing messages were there better evidence to support it.[1] And then there are the many who, in the face of ambiguous evidence claims, persist in doing nothing.

The debate on climate change, climate research and its use in informing policy-making is just one example of the many hotly contested evidence-related issues of the present day which form the basis of this book. Particularly in moments of crisis, key questions acquire new urgency: What is good evidence for practical decision-making? What are substantive grounds for questioning certain pieces of evidence? Who has the right – or responsibility – to participate in debates about evidence? How do contesters of evidence proceed? What impact do such contestations have in a societal context in which evidence is elevated to the status of a resource? The authors of the book seek to answer these questions.[2]

Before exploring evidence contestation and its impact on society in more detail, we need to establish the premises of our research approach to evidence. The hypothesis of the "paradox of knowledge societies", which we outline in the following section, provides the starting point for anchoring our questions. We continue by discussing both evidence and its contestation as disputed concepts themselves. Glancing over the perspectives on evidence criticism in the humanities and social sciences

DOI: 10.4324/9781003273509-1

reveals the lines of thinking we rely on. We then look at evidence critique as an academic pursuit and discuss what distinguishes it in contrast to contestations and conflicts that unfold when evidence is put to use. Finally, a passage through the contributions of this book helps us assess how the actors in the case studies presented deal with dissent and thereby act on societal conflicts.

The Paradox of Knowledge Societies

A basic feature of contemporary knowledge societies is the paradox that, on the one hand, they are fundamentally dependent on science and scientific expertise, which are becoming increasingly important as a basis for decision-making. On the other hand, scientific knowledge is continually questioned and expertise is increasingly controversial.[3] Critical discussions are no longer confined within the walls of academia but have spread to society at large. Ever wider circles now question the validity of scientifically generated knowledge.[4] Recently, we have witnessed not only ever more nuanced disagreement between experts and their lay audiences but also outright science denialism.[5]

Depending on the context of the dispute, evidence contestations vary in form and severity. To gain the widest possible insight into the practices of evidence critique, we have chosen a broad concept of "evidence contestation"; one which encompasses forms of traditional systematic evidence critique and review within universities and academia as well as radical questioning of the legitimacy of established methods of evidence generation and entire fields of inquiry inside and outside of the academic realm. This extended conception of the term allows us to avoid a strict separation of contexts and actors, and thus to see connections and entanglements between critique and contestations in different societal spheres.

The prime contemporary example of the interaction between academia, media and various publics and actors often operating in several of these domains was of course the COVID-19 pandemic. Due to the urgency of the situation, traditional scientific peer review processes for quality control were dispensed with. Scientists repeatedly corrected their positions, as did decision makers in politics. Evidence – or its quality level – was contested from the outside, by dissatisfied citizens, and from within, by scientists. Certain groups even labeled the COVID-19 pandemic and related research a hoax.[6]

In the wake of this, some long-standing certainties seem no longer to hold true. Previously common, and in some fields institutionalized, evidence practices no longer guarantee credibility or acceptance among the general public. Trust in scientific expertise and even science in general appears to be in jeopardy. Given the role that scientific knowledge plays in public decision-making today, such contestations of evidence may threaten societal cohesion by destroying the basis for science-based agreements and governance.[7] But at the same time, the very questioning of established evidence by a variety of new actors, who also demand a say in key issues, can challenge traditional beliefs in a productive way. It may lead not only to the production of diverse evidence, but also to new compromises, much in the same way that evidence critique has often yielded fruitful results within academia. What emerges from the contributions to this book are the often positive outcomes of leading scientific evidence being challenged, allowing, to varying degrees, for a broader inclusion of other perspectives and new knowledge contributions from laypersons or marginalized groups. If "lay deference to a consensus of experts" is no longer the way knowledge in society is accepted,[8] this entails a much-needed widening of the zone of criticism and a democratization of critique.

Even though these developments have gathered pace in recent years, they are not entirely new. Processes of evidence destabilization have raised awareness of the vital importance of evidence in society and provoked increased efforts to restabilize evidence. At the same time, however, the provisional and contingent nature of evidence has become more prominent. This revolution in our attitudes to and understanding of evidence practices can be traced back to the 1970s, when the Post-World War II optimism regarding the prospects of ongoing prosperity – hitherto taken for granted – began to atrophy, giving rise to uncertainty and increasing ignorance as ineluctable side-effects of the advances in science and technology. The sociologist Ulrich Beck and colleagues identified in this transition into what they saw as a second, reflective modern era, the so-called paradox of late modern knowledge societies.[9] With ever more evidence and knowledge becoming available, criticism and contestation inevitably increased too. Dealing with dissent became the normal state of affairs. The resulting dynamics strongly determine the way in which societies decide on key issues in the late 20th and early 21st centuries; how they deal with sustainability, environmentalism and technological safety; participation of citizens, patients and consumers; food security; media communication; increasing prevalence of non-communicable diseases and other health problems; or medical safety and interventions.

Reviewing different fields and situations in which science and scientific knowledge have been contested, the contributors to this book analyze the processes of destabilization of evidence. They seek to identify the drivers and actors involved and ask not just about the consequences but also about the counterforces involved in such destabilizations. By drawing on methodological and theoretical approaches from history, sociology, philosophy, science and technology studies (STS), communication science and consumer science, the contributions examine, from a variety of perspectives, the matrix in which the certainty of evidence is either contested or ratified. Understanding the practices of contesting evidence illuminates how actors engage with contemporary knowledge society.

The insights gathered here will help us understand how knowledge societies attune to the provisional and contingent nature of evidence.

Evidence and Its Contestation as Disputed Concepts

"Evidence" is a widely used term. Lawyers, scientists and politicians call for evidence, as indeed do laypersons. What exactly evidence means may diverge significantly. However, common to the diverse situations in which the term is used is that it signifies something that is appealed to when we want to convince others to accept or reject a claim: As Thomas Kelly puts it, "evidence, whatever else it is, is the kind of thing which can make a difference to what one is justified in believing or (what is often, but not always, taken to be the same thing) what it is reasonable for one to believe".[10] It should be noted that the meaning of the term varies between languages. For example, the meaning assigned to evidence in English differs subtly from that of the German word *Evidenz*. While the English term has a narrower meaning of "one or more reasons for believing that something is or is not true",[11] *Evidenz* also means "immediate and complete insight, clarity, certainty" (*unmittelbare und vollständige Einsichtigkeit, Deutlichkeit, Gewissheit*).[12] According to Immanuel Kant, evidence is therefore presuppositionless insight or "demonstrative certainty" (*anschauliche Gewissheit*), an insight without methodological mediation.[13] Inconsistency and fluctuations in the usage of the concept of evidence are mainly due to the fact that evidence is understood and used both as a subjective form of truth recognition (seeing, recognizing a fact as evident) and as an objective form of truth discovery (evidence as showing, manifesting a fact).

Depending on the context then, what is taken to be evidence varies. Evidence can refer, for example, to an observation, a physical object, a heard statement or research findings. In everyday life, we often interpret an observed state of affairs as evidence for believing that something is the case. The fact that I cannot see Mary's jacket hanging on the rack serves as evidence for my belief that she has already left the office for the day. In the context of a forensic investigation, a single physical object found at the crime scene, say a bullet, a bloodstain or a hair, can provide the decisive piece of evidence that helps to solve the case. Journalists interview sources to collect evidence to support an account of the events they wish to report on. And scientific research can provide us with results that either increase or decrease our belief that, for example, a treatment for a disease is really working, or that a chemical is environmentally safe.[14]

Scientific disciplines have their own criteria and assumptions about what serves as sufficient or adequate evidence to falsify or support a particular hypothesis. Some, such as anthropology, rely on observations, yet others base their findings predominantly on archival material, as historians do. In medicine, clinical trials have for many decades been of great importance and occupy a high position in the hierarchy of evidence-based medicine (EBM); however, the limitations of randomization in large-scale clinical trials that exclude confounding variability are now well known. As a result, there is increasing talk of reality-based evidence, evidence that is less perfectly generated but arises in the real world in which patients live.[15]

But even within disciplines, there are differences in the production of what counts as reliable evidence. For example, different fields of biomedicine pose diverse questions concerning the biological bases of diseases, their frequency in distinct populations and the effectiveness of treatment and prevention methods. In order to find answers to these questions, various methods of producing evidence are used: While laboratory studies are useful for learning more about the etiology of a disease, they are less fruitful when a researcher tries to determine the prevalence of the disease in a given population.[16]

In recent years, the concept of evidence has become particularly central through the calls for evidence-based practice, that is, practical decision-making based on the best available evidence.[17] Even though there currently are different forms of evidence-based practice, e.g. evidence-based policy (EBP), evidence-based nursing, evidence-based pedagogy and many others, this framework for decision-making originates from the context of clinical medicine.[18] Since the 1990s, EBM has become a movement aimed at improving the treatment of patients by basing it on best possible evidence.[19] A central element of EBM is the hierarchy of evidence that ranks different types of evidence according to their assumed strength. In particular, randomized controlled trials (RCTs) are believed to provide irrefutable evidence. The tenets of EBM, especially the emblematic emphasis put on RCTs, have since spread to other fields. Proponents of EBP, for example, hold that public policy should wherever possible be based on RCTs or other experimental evidence.[20] Such a constricted understanding of evidence has been both rigidified on the one hand and slightly diversified on the other. When the burgeoning call for evidence gave rise to national and international institutions, such as the Cochrane Collaboration or the Evidence for Policy and Practice Information and Co-ordinating Centre (EPPI) at University College London, these organizations institutionalized a second-order knowledge production. By conducting systematic reviews of scientific studies and research papers, they either subscribed to the narrow concept of RCT-based evidence or remained more open and included results from differently designed studies.[21] Either way, the practice of generating evidence through systematic review has a power dimension, authorizing certain types of evidence and knowledge at the expense of others.

In fact, such procedures of systematic review and weighting of evidence according to certain rules are part and parcel of a society's politics of evidence, which can be interpreted with the help of Michel Foucault's concept of a regime of truth. Foucault understands truth as "the ensemble of rules according to which true and false are separated and specific effects of power attached to the true".[22] Analogously, an evidence regime consists of procedures that distinguish what counts as evidence from what does not and what privileges the former at the expense of the latter. The power effect attached to evidence lies in its potential to justify decisions and to enable action in democratically constituted knowledge societies. Thus, whatever it is that counts as evidence in a certain context and moment, it may lend power to those who successfully claim it. Because of this, the contestation of evidence is not only an integral part of a society's evidence regime, but also a form of counter-power. Within Foucault's theory of power, the political art of resistance, the art of "not being governed", is both a gesture of determinedly rejecting unjust laws and the more general art of resisting all processes that shape our subjectivities in accordance with a given "dispositive", or a regime of power.[23] This would include processes of knowledge creation and evidence production. In their reading of Foucault's concept of critique, Judith Butler underlines its virtue-ethical and epistemic function: "Significantly, for Foucault, this exposure of the limit of the epistemological field is linked with the practice of virtue, as if virtue is counter to regulation and order, as if virtue itself is to be found in the risking of established order".[24]

That the concept of evidence and its critique are not clear-cut and unambiguous is mirrored in the contributions to this book. The contributors charge the term with diverse meanings, ranging from single data and observations to whole studies or systematic reviews of such studies. Some go even further and interpret evidence as any result of epistemic practices – both academic and extra-academic – which can be deployed to work as justifications for a belief that is claimed or a decision to be taken.

Despite all its ambiguity, evidence has become the most important resource in democratically constituted societies, premised on the understanding of "reason as a capacity to explain and to justify".[25] Being able to justify our views through persuasion by reason and appeal to evidence distinguishes us as members of a democratic polity and underpins the liberal democratic principle of treating each other with equal respect. The rising mobilization of evidence as a means of persuasion, however, exposes it to increased contestation. The arenas of these contestations are as manifold as the actors involved. A closer look at these arenas, actors and practices of evidence contestation can help to understand its dynamics and its societal effects.

Perspectives on Evidence and Criticism in the Humanities and Social Sciences

Evidence and its uses have been topics of scholarship for decades.[26] The phenomenon of evidence criticism has so far largely been investigated in terms of the critique of science. In the 1970s, at the very moment when the fundamental dependence of modern societies on science became a matter of concern, and gave rise to the concept of the knowledge society, critique and contestation of science was taken up as a subject of inquiry too. Scholars, who in one way or another participated in the newly emergent counter-movements in the sciences, reflected on this development and sought to understand its origins, patterns and consequences.[27] They traced the critical traditions back to early 19th-century Romanticism and investigated the different forms of critique from within and from the outside of science. An important insight from this strand of scholarship was the ambiguity and malleability of science criticism. Whether such criticism was considered to be pseudo-science, anti-science or frontier-science could change over time.[28] Also around this period, research into controversies in the sciences began. Here, the goal was to better understand how scientific controversies were entangled with politics and ethics, and what this entanglement meant for scientific reasoning and for the authority of science. For example, sociologist Dorothy Nelkin, an established authority in the study of controversies, provided grand overviews and interpretations of the development and typologies of scientific disputes.[29]

Since the 1970s, reflective social scientists and STS scholars in particular have not only studied science criticism but also contributed to the questioning of scientific evidence themselves. Many studies have criticized the sciences for being prone to a naïve belief in positivistic claims and decontextualized facts.[30] The social constructivists have busied themselves in enlightening both their colleagues in science and the public at large that all facts are man-made things charged with values, and that scientific certainty is a chimera. The extent to which scientists on the receiving end felt attacked was evident in the declaration of the so-called science wars.[31]

Over the course of time, however, hostilities have gradually dissipated, and the exact sciences (mathematics, optics, astronomy and physics) have increasingly accepted the need for a reflective perspective. However, there was a longer lasting and more dangerous knock-on effect of the constructivists' agenda. As the constructivists denounced claims of scientific certainty as unjustified, they may have unwittingly contributed to a shrinking trust in science and so played into the hands of those who – for political or financial ends – strategically discredited science. This was becoming worryingly apparent at the beginning of the new millennium. In 2003, two years after the terrorist attacks on the twin towers of the World Trade Center in New York, the influential STS scholar Bruno Latour expressed this concern in a lecture at Stanford. He warned that criticism of science was out of control. After decades of critical work on "hastily naturalized, objectified facts", he held that "dangerous extremists" were destroying "hard-earned evidence", evidence that could save lives.[32] Latour summed up this self-reflexive critical gesture as follows:

"Have we behaved like mad scientists who have let the virus of criticism out of the confines of their laboratories and now can do nothing to limit its harmful effects; it is now mutating and gnawing away at everything, even the vessels in which it is enclosed?"[33]

Since the mid-1990s, studies on the criticism of science have focused on the attempted discrediting of science within the context of political or economic conflicts of interests.[34] In this field, a strand of research has developed under the heading of "agnotology". First coined by Robert N. Proctor, agnotology refers to the study of ignorance.[35] Researchers in this field are particularly interested in how ignorance is deliberately constructed to raise doubts about the validity of established views. Research on the strategies adopted by agents such as pharmaceutical companies, chemical industry and political think tanks have shown how scientific evidence is intentionally corroded by funding research to produce counter-evidence.[36]

The authors of this book, however, are less interested in such forms of strategic de-construction of certainties than in furnishing a more comprehensive perspective on evidence criticism and contestation of scientific findings. In other words, they also consider the positive and productive sides of evidence contestation. How, for example, can questioning previously accepted certainties open up new paths for research on nuclear security and safety? They ask what else and who else has caused and is causing the crisis of certainty. They aim to explore whether and how the contestation of evidence actually provides spaces for the development of new forms of evidence practices. And they investigate how the legitimacy of both critique and contestation is negotiated.

Critical accounts of EBM and EBP have assessed the application of pre-determined evidence evaluation criteria in decision-making.[37] This scholarship has challenged the assumption of evidence as a neutral arbiter in political controversies or when difficult treatment decisions have to be made.[38] In the context of medicine and healthcare, some have argued that in addition to quantitative evidence produced in clinical studies, so-called subjective evidence, for example patient diaries and narratives, should be considered.[39] As Trisha Greenhalgh puts it, "narrative, phenomenological, and ethnographic research designs should be viewed as complementary rather than inferior to epidemiological evidence – though qualitative, like quantitative, research must be appraised for rigor and relevance".[40] In particular, patients can help researchers gain new insights, via their understanding and experience of their illness, which would otherwise be unaccounted for by more traditional methodological approaches.

The distinction between objectivity and subjectivity – or, to be more precise, what is taken to be objective and subjective – has long been of interest to philosophers of science who have investigated the value-ladenness of knowledge production and argued that values influence the generation and interpretation of empirical evidence. For instance, according to feminist scholars such as Helen Longino and Alison Wylie, value-laden background assumptions influence what data is taken as evidence.[41] Sabina Leonelli's work on big data supports this point.[42] Raw data in itself is not evidence. Activities involved in collecting, archiving and organizing data are essential if they are to be used as evidence for or against a claim.

So, if evidence is not a neutral, value-free object, is the objectivity of science an unachievable ideal? No, some argue. What is required is a broader understanding of objectivity. Lorraine Daston and Peter Galison have made us aware of the fact that the meaning of objectivity has both changed over time and reflects the respective societal context of scientific practice.[43] Proctor states that the "objectivity of science depends upon certain aspects of contextuality".[44] Such aspects of contextuality are for him the fact that knowledge may serve political interests and that knowledge is necessarily contingent. This requires us to ask new questions, but not to jettison the ideal of scientific objectivity and evidence. Objectivity can be preserved when it is not determined as antithetical to subjectivity. Inter-subjectively accessible evidence may count as objective because it allows justifications from a common point of view. For Ian Hacking, standards of objectivity are shaped via styles of reasoning and they rest on intersubjectivity.[45]

Scholars in the field of STS and history of science have observed how evidence conflicts become proxy battles for political conflicts and have asked what consequences this has for the fate of democracies.[46] When expert contestations of facts and data supplant societal negotiations on values and worldviews, the culture of public debate vital to healthy democracies is lost. A culture of constructive controversy, however, needs a common ground to make sense of disputes. Alexander Bogner finds this common ground in an abstract idea of truth and valid facts as necessary fictions.[47] We argue that our focus on practices of evidence and evidence contestation can lead to a more concrete understanding of how such common ground can be gained and how the de- and re-stabilization of evidence play out.

Evidence Contestation as an Academic and a Societal Issue

One could tentatively arrange the variants of what we call evidence contestation on a continuum: Starting from the academic sphere, in which evidence is critically examined, but in itself non-negotiable in its ontological magnitude, it extends to the social sphere, where what is considered relevant and reliable evidence is increasingly questioned by multiple actors and from conflicting perspectives. By differentiating these practices of evidence contestation according to the results they yield, it becomes clear that they range from (in the main) constructive evidence critique among peers in academia to challenging evidence for use as a basis for decision-making. At the extreme end of the spectrum are practices that are deliberately destructive of what is widely considered evidence. This form of contestation often has harmful effects on democratic societies.

Academia is – besides the legal sphere – the arena in modern societies in which evidence critique has, from early on, been cast in an institutionalized form.[48] As organized skepticism, its function is to work as a guarantor of the reliability and trustworthiness of scientific findings. Peer-review procedures, conferences or intra-scientific debates as forms of exposing new findings to the criticism of colleagues have been established for precisely this purpose: The aim is to make sure that possible mistakes and biases by individual researchers are discovered and corrected. Scientists as actors of evidence critique within academia have internalized organized skepticism as an ethical norm that requires a commitment to solid evidence for validity claims. Thus, when scientists are asked to provide unambiguous and certain evidence, it is a virtue rather than a vice that they are reluctant to do so. However, when users need clear advice to support decisions, which scientists are hesitant to deliver, tensions are provoked. For example, in the context of policy-making, evidence is expected to provide a firm basis for grounding debates. As the COVID-19 crisis clearly demonstrated, extreme time constraints and the need for rapid response and action may override the pre-established principles of scientific validation. During the pandemic, studies were occasionally published before passing the usual peer review process as the crisis necessitated immediate choices.[49] When institutionalized evidence criticism was suspended as new evidence was published prior to peer review, widespread public questioning of decisions made on this basis became even more intense.[50]

This is not confined to current events. The research and development of the first nuclear bombs during World War II, known as the Manhattan project, revealed a similar dynamic. The military required immediately available results to quickly build the bomb and determine the outcome of the war. One of the bottlenecks of the project was the production of fissionable material. The nuclear scientists and engineers explored various methods but remained skeptical and hesitant when asked to decide on a solution. This upset the military. Because of the shortage of sufficient evidence pointing to the one best way, Lieutenant General Leslie Groves, the military head of the Manhattan project, decided to try all possible routes regardless of costs. Time was more precious to him than money.[51] These examples clearly demonstrate that the scope and procedures of evidence critique in the academic realm are influenced by the factual and temporal constraints of real-world evidence application.

The firm anchoring of systematic skepticism and evidence criticism within the sciences also determines their innovative potential, which unfolds in the remarkable changeability of scientific disciplines and research fields. The development of the sciences over the last two centuries bears ample witness to this. Paradigm shifts, the turn to new perspectives, new interpretations of facts, the use of new epistemic practices and technologies and the split or the merger of scientific fields, are all processes that rest on evidence contestation. They are, however, anything but smooth developments, implying serious conflicts, grinding controversies and confrontations that are fought on various fronts. Several contributions in this collection refer to such conflicts, controversies and confrontations.

Evidence contestation, however, seldom stays within the confines of academia, and more often than not, it is initiated by problems and crises that emerge outside of academia. Whether it is changing ideas about gender roles or altered perceptions of what is considered autochthonous and what is alien, such major societal paradigm shifts have had important consequences on the way formerly established evidence is criticized within science. Capturing these blurred lines and connections emanating at the borders between academia and society became our focus during the work on this book. As diverse as the origins of quite different controversies may be, when the main actors fighting them are scientists or practitioners of academic professions, they have in common a shared belief that conflicting claims can be resolved. This belief legitimizes evidence contestations and confers on them the status of productive dissent. As the eminent sociologist Georg Simmel states, each conflict must be founded on a basic agreement. In academia, the striving for a scientific truth provides the common ground on which fierce disputes can be fought out.[52]

Evidence is crucial not only for the justification of validity claims within the sciences, but also as a warrantor for success when scientific claims are put to use in the diverse contexts outside of scientific practice *per se*. Over the course of the 20th century and especially toward its close, promoting science was increasingly seen as an investment that promised high returns, such as technological innovations, competitive advantages and political advice for statist and private investors alike. Thus, the practical applicability of science became more important than ever before.[53] The flipside of growing expectations of the immediate utility of scientific findings was the disturbing experience that new scientific evidence uncovered new fields of ignorance and uncertainty. Furthermore, application-induced contestations of evidence increased, because practicing evidence in practical contexts is different from practicing evidence within the sterile context of science. Nancy Cartwright puts it this way: "What a claim means in the context in which it is first justified may be very different from what it means in the different contexts in which it will be put to use". She further states: "What justifies a claim depends on what we are going to do with that claim, and evidence for one use may provide no support for others".[54]

What kind of evidence is appropriate for which use is not just a difficult, but also a heavily contested issue. Many actors with a myriad of motives and diverse priorities and preferences become involved in making choices based on evidence that is applicable for a certain use. The first group of actors constitutes the primary generators of evidence. These include researchers in universities and other public or privately funded academic institutions, as well as research staff members of industrial and commercial enterprises. Depending on their contracting authority, the researchers determine the usability of the evidence they have provided either more broadly or narrowly. Whereas university researchers may have more leeway when they commit to a research topic, industrial researchers are predominantly determined with regard to a concrete field of application. The researchers' closeness to the application of evidence also depends on how far they have or actively seek access to users or people affected by the applied evidence.

When evidence is generated far away from the places of its application and without concern for the context of use, skepticism and contestation of decontextualized evidence is highly probable. Thus, the need arises to bridge the gap between production and application. New actors become involved either to facilitate the transition of evidence to application or to prevent its use. In these cases, we see that the transition from the generation to the application of evidence is a critical process that takes time and occupies space. Here, various societal actors, ranging from certified or self-proclaimed experts via organized activists and engaged citizens to affected laypeople such as patients, consumers or local residents, to name just a few, become involved and participate in debating the validity of the claims in question. It is in this interspace where controversies unfold and procedures to resolve the conflicts must be found.

It needs to be stressed that the focus on evidence for use implies a shift of priorities from knowing to doing. This has an impact on evidence contestations. Now, acceptance of evidence or its rejection no longer hinges on truth-like criteria alone, but on its potential to facilitate the achievement of goals. This again influences evidence conflicts and contestations as it impacts on what is at stake, what needs to be taken into account and who becomes involved in stating a case, advising on it and making a decision.

Thus, when we focus our attention on evidence for use we gain a more comprehensive understanding of evidence contestations. Looking closely at what evidence is required and questioned, misuse of contestations and controversies for vested interests becomes an issue. Historians of science who have studied, for example, the tobacco industry's intrigues against evidence on the carcinogenicity of smoking have detailed such strategic deployment of evidence critique and contestation.[55] Here, the belief in resolution so crucial for productive criticism is compromised by staged controversies that distort evidence from ordered and manipulated studies.

In contrast to the scientific field, between science and society, the channels and tools of critique in the debate about valid knowledge have been established and consolidated in recent decades. Many of them have become institutionalized, and thus societally accepted as legitimate forms of public participation and building blocks of a deliberative democracy.[56] Public hearings, consensus conferences and stakeholder meetings have become integral to decision-making processes at many levels.[57] One recent example of such institutionalized forms of contestation are citizen assemblies. Appointed to enable more meaningful dialogue among citizens, experts and governments, they provide more room for productive critique and deliberative participation than fierce contestation at the interface among science, politics and the public, particularly regarding pressing issues such as climate change.[58] In some cases, these measures extend to the world level, as international bodies increasingly discuss environmental policy issues at global round tables.[59]

In several of the case studies presented in this book, we can see attempts by actors to establish such channels for criticizing what is presented to them as evidence. Some of the groups that doubt evidence emulate the methods of science by coming up with their own publications and pamphlets that take center stage in the context of counter science. Others have chosen a different path by using official data for their own ends. More recent contesters of evidence make use of digital communication channels to question evidence. What we are seeing in all of these instances is the development of hybrid arenas between science and society that are populated by quite different actors, ranging from scientists, experts and journalists to decision makers and citizens in their capacity as patients, consumers, local residents and many others. In these arenas, patterns of evidence critique and contestation have evolved that can pave the way toward a more inclusive and deliberative democracy. There is, however, also the possibility of misguided critique and contestation of evidence that in turn may pose a threat to democracy. Rising concerns about the advent of the post-truth era point in this direction.[60]

Dealing with Dissent in Science and Society

In a variety of case studies, the contributors to this book explore what counts as evidence for different actors and under what circumstances dissent may emerge and grow. Do they find productive ways to include hitherto unconsidered data, facts, sources or observations for justifying different knowledge claims or decisions? Can new evidence ever coexist with results of evidence practices that were previously accepted? Or does novel evidence necessarily trigger conflicts? And, if this is the case, how can such conflicts be resolved?

Beginning with a theoretical view of evidence critique and contestation in the first section, one way to analyze and systematize increasing controversies and dwindling consensus regarding evidence is through the lens of philosophy. According to Daniel Füger and Elif Özmen, critical debate on evidence is an essential precondition for the trustworthiness of research. By reviewing the normative functions of criticism, they show that questioning the evidential basis of claims enables possible biases that might nudge the results of scientific research in one direction or the other to be uncovered. The authors treat evidence as a "pre-rational choice" of scientists and argue this is both the basis and the consequence of the normative setting, which they call scientific ethos. Dissent and criticism are possible, but only as long as scientists do not reject the institutionalized scientific ethos as a whole. They show that controversy has been a characteristic of modern science since its emergence in the 18th century and continues to be beneficial to this day, both epistemically and ethically.

Another theoretically informed way to consider growing dissent in regard to evidence is through the perspective of the sociology of science. In their contribution, Eva Barlösius and Eva Ruffing develop a heuristic to distinguish between three modes of contestation of scientific evidence, knowledge and expertise in contemporary knowledge societies. These three forms of contestation affect the recognition of science and can have destabilizing consequences at different levels. If the first mode problematizes scientific evidence *per se*, the second mode questions its political use. The third mode challenges the relevance of scientific evidence for political decision-making. This heuristic is used to show how important the tension between trust and distrust in scientific knowledge can be, especially in situations of social destabilization and increasing dissent, as seen in the COVID-19 pandemic. With reference to the third form of contestation, the authors suggest that escalating disputes can be expressions of socio-structural and political struggles. In these conflicts, it is not evidence that is contested, but the way political solutions are arrived at.

The second section hones in on examples of dissent within evolutionary biology and ecology. In these cases, we observe that conflicts taking place between academics may have originated outside the academic sphere. The actors who challenge established evidence here have faced backlashes not only from within their disciplines, but also from beyond – and responded to them in various ways. Starting from changing societal norms, the evolutionary biologist Joan Roughgarden, whose case is delineated by Martha Kenney, contested Darwin's theory of sexual selection by presenting new evidence combined with storytelling in the early 2000s. The majority of her peers in evolutionary biology outright rejected Roughgarden's claims about sexual selection for two reasons: First for challenging Darwin's master narrative from a marginalized subject position and second for referring to contemporary US gender politics in the context of science. Roughgarden responded by calling on her colleagues to test hypotheses beyond the neo-Darwinian orthodoxy. She argued for the coexistence of various hypotheses, each confirmed with results from evidence practices that the respective researchers considered appropriate. Roughgarden made the case for accepting dissent and pluralism in the sciences. Why this was a reasonable decision can be understood with the help of the philosopher Nicholas Rescher. He argues for scientific pluralism for the following reasons:

> The diversity of persons, cultures and experiences makes the goal of actually realizing consensus in cognitive or evaluative practical matters effectively impracticable and unrealistic. Only by abstracting from the physical and social realities – shifting to the level of idealization – can we require or expect a valid consensus.[61]

The case study on invasive species research traces the shift from an ecological expert culture largely immune to critique to one that needs to embrace pluralism and disagreement in the face of growing dissent in the rapidly changing field of ecology. Christoph Kueffer shows how a small and homogeneous group of experts successfully framed the socioecological problem in the postwar period in such a way that their advice was adopted by decision makers and consequentially engendered policies on all levels without much contestation. The generalized rules they formulated about ecology often trumped growing real-world evidence. Since the 1990s, however, scientific expertise and policy-making related to biological invasions have increasingly been exposed to dissent, forcing scientists to respond to such contestations of evidence by diverse actors within the sciences and in society and making it more difficult to reach a consensus on actions. As one way to embrace this new pluralism and disagreement, Kueffer discusses the honest broker strategy. Contemporary invasion scientists attempt to maintain their central role in providing information while also allowing for more diversity and participation in research and decision-making processes in order to render the valuation of alien species impacts more socially robust. While it is questionable whether such a strategy will suffice to contain evidence critique, it functions as one approach for achieving consensus on how to react toward biological invasions as one of the many challenges of the Anthropocene.

The third section on the health sector focuses on tensions and conflicts between intra-scientific evidence practices on the one hand, and evidence generation in application contexts on the other. In their contribution, Saana Jukola and Mariacarla Gadebusch Bondio analyze the recent US debate concerning surgical headwear and guidelines for surgical wear as preventive hygiene measures. In this dispute, the Association of periOperative Registered Nurses (AORN) and their critics dealt with dissent by referring to evidence and different criteria for assessing evidence. Eventually, AORN guidelines were rephrased. In their analysis of the argumentation strategies used in different international contexts (the US and Germany) by diverse professional groups (surgeons and perioperative nurses), Jukola and Gadebusch Bondio argue that this case illustrates the importance of considering the role of authoritarian structures in how arguments and positions are formed. Unlike the US, in Germany, there has been no debate about the guidelines for what kind of headwear should be worn in operating theaters. This implies that it is important to consider how non-epistemic concerns and epistemic values are intertwined in evidence disputes.

Sometimes experts aiming at evidence-based decision-making are faced with evidence discordance. In these cases, decision makers have to find a means of first assessing the available evidence, and, second, deciding which evidence to act upon. The task becomes more complicated when there is no clear agreement as to what criteria to use for assessing evidence. The contribution by Fredrik Andersen, Yngve Herikstad, Anders Dechsling, and Stefan Sütterlin addresses this problem. The case they analyze is the development of the *Norwegian National Guidelines for the Treatment of Substance Use Disorders and Addiction*. As the authors note, in this instance, "there is contestation concerning a wide variety of issues such as legality, treatment, the nature of the disease, the quality of evidence and the general ideology of treatment". Yet all evidence evaluation methods and criteria can be contested: For example, it is not clear if certain methods of producing evidence (e.g. RCTs) should be favored over others or whether local evidence should be preferred. In their chapter, Andersen et al. discuss different evidence amalgamation methods that could help expert groups reach consensus in light of diverse evidence. Reaching consensus is necessary in this case, as the aim of the evidence evaluation is to produce guidelines for the treatment of addiction issues. Andersen et al. suggest that the so-called explanatory approach to evaluating evidence could help to solve the problem of how to deal with evidence discordance.[62]

The fourth section focuses on challenges to evidence from established science by non-academic actors. In Stefan Esselborn and Karin Zachmann's case study on the development of counter science in Germany since the 1970s, we see a propagation of evidence contestation from the socio-epistemic to the institutional level. The protagonists of counter science not only set up alternative institutions such as the Öko-Institut. They also emulated the methods of science by coming up with their own publications and pamphlets, in order to provide evidence that was needed by the anti-nuclear movement in court cases on nuclear power, or by citizens concerned about further environmental and societal issues. Seconded by critical but established voices within academia, counter researchers wrestled with the question of how to produce good evidence and for whom. Responses from not only conventional academic institutions and their actors, but also industry and governmental institutions, on the evidence claims of the dissenters at first ranged from irritation to strict denunciation. Over time however, the hostility on both sides ran dry and the counter researchers managed to fill a niche in the market of knowledge and expertise. Two drivers pushed the integration of counter science into the evidence regime that was thereby evolving more plurality: First, structural changes connected to increasing demands for participation and reflexivity in (mainstream) science; and second the wish for intellectual acceptance and the need for funding on the part of the counter researchers.

Exploring the confrontations on pesticide use, Sarah Ehlers highlights what practices of evidence contestation environmental activists developed to achieve a ban on pesticide exports from the Global North to the Global South during the 1970s and 1980s. While one strategy of the activists was to use official data for their own ends, another was to dismantle the results of cost-benefit analyses provided by the chemical industry and other beneficiaries of pesticide exports by way of taking real-world evidence into account. By appropriating and using evidence from diverse sources, the environmental activists challenged the hitherto established experts and political decision makers. Resolving the conflict required a renegotiation of what is accepted as appropriate evidence and exposing hidden inequalities that had precluded the application of certain facts, data and observations as relevant evidence for decisions in advance. Here, dealing with dissent was not a question of pluralism and liberty, but of justice and equality.

Citizen science, which has advanced as a publicly supported movement in Western states since the turn of the millennium, is first and foremost an activity that tries to extend the range of participatory democracy. Laypersons, who gain access to the realm of established science as contributors and participants, challenge the social exclusiveness of academia and its validity claims based on certification. Kevin Altmann and Andreas Wenninger investigate whether selected citizen science projects in Austria and Germany transform professional science not just in a social sense, but also in the factual dimension. In their view, the latter implies potential for evidence contestation when citizens manage to introduce critical questions and content into science. The authors show that professional scientists try to offset the impact of dissenting citizen perspectives on science by securing professional control of citizen science projects. In the interplay of differently institutionalized modes of citizen science, current knowledge societies must balance critical and constructive impacts.

Section five, on conveying academic evidence, highlights a further expansion of contestation. Edoardo Maria Pelli and Jutta Roosen show that a young field of research and application can provoke uncertainty or contradictory positions. Nutrition science, for example, is frequently criticized for methodological weaknesses and for being influenced by the food industry. The decision-making process for consumers seeking a healthy diet is often complex, as action has to be taken, even though there is no clear evidence indicating a distinct direction to follow. The authors analyze one such case, namely consumers' attitudes toward so-called superfoods, marketed as having an exceptionally high nutritional value, when there is actually little scientific evidence for the health benefits of these foods. This raises the question how consumers decide whether to purchase and consume superfoods. Applying moral foundations theory (MFT), the authors explain that consumers' food attitudes are formed in the absence of strong evidence. According to Pelli and Roosen's application of MFT, attitudes are guided by intuitive moral judgments concerning a particular object or phenomenon, rather than only by rational processes, as suggested by traditional neoclassical economics. Consequently, attitudes toward nutritional aspects of food may be moderated by moral values. This helps understand how laypersons circumvent uncertainty or dissent concerning fuzzy evidence and make choices in their everyday life.

The case study of evidence criticism in journalism highlights a particular way of voicing dissent. Using the example of the societally controversial field of genomic research, Helena Bilandzic and Susanne Kinnebrock take a close look at journalistic strategies for contesting scientific evidence. While science journalists do not produce evidence themselves, they communicate scientific evidence to wider audiences. In the majority of cases, they present scientific findings in a neutral or supportive way. Confrontation or rather dissent, by contrast, appear as an exception and do not target scientific results *per se*, but their ethical implications on society. In these rare cases, journalists use narrative strategies to criticize the research. While references to mythical accounts of human hubris have become classic, they also use the personalized topos of the "mad scientist", or references to authorities, experts, institutions or academic journals, to question the trustworthiness of the scientists and/or institutions producing the evidence. By presenting the findings in one way or another, they add to their perception as true or invalid or they "underline or undermine" them. The effect of this kind of narrative dissent on public trust or distrust in science stands in need of further investigation.

What do these many observations and findings from different disciplines and societal contexts offer when dealing with dissent in contemporary knowledge societies? The case studies in this collection showcase polyphonic doubt and debate regarding evidence both within academia and beyond. What we have seen is that the number of people who criticize evidence has multiplied and diversified in recent decades, as have the arenas where contestations take place. Once confined to academia and committed to the search for truth or optimal solutions, respectively, criticism and contestation of evidence have spread into society at large. Here, an essential feature of today's knowledge societies emerges, in which scientific research is advancing to become the dominant model of action. "Systematic and controlled reflection becomes a widespread principle of action in society", notes Peter Weingart.[63] This in turn entails the multiplication of evidence critique beyond academia. Much current criticism and contestation is triggered by one of two things: Either conflicts resulting from putting evidence to use or calls for more appropriate and comprehensive evidence.

But what does such growing dissent mean for society and how can a common ground be reached despite the increasing diversity and legitimacy of various perspectives and the clamoring and often competing challenges? It may help to compare the current situation of late modern knowledge societies with the advent of the enlightened, bourgeois order more than 200 years ago. At that time, Immanuel Kant developed his radical theoretical-philosophical concept of critique, most famously in his *Critique of Pure Reason*.[64] However, Kantian critique did not address the public but was confined to the egalitarian model of the scholarly republic. In the history of political semantics, Kant's radical demand for a freedom of critique is seen as a palliative to the violence and terror of the political revolution in France.[65] In contrast – and this is what the case studies of this book show – current knowledge societies have extended the entitlement to evidence contestation beyond the walls of academia and thus embraced dissent as a generally constructive force for the whole of society. To some extent, however, we are still caught in the Kantian mindset when we frame political dissent and value conflicts as evidence contestation. This is what scholars like Alexander Bogner address when they warn against the dangers of an epistemocracy.[66] Circumventing these dangers requires openly addressing the political and value dimensions that are embedded in all evidence claims. The contributions to this volume show how this can be, and is, done.

Notes

1 Willem Van Rensburg and Brian W. Head, "Climate Change Skepticism: Reconsidering How to Respond to Core Criticisms of Climate Science and Policy", *SAGE Open* 7, no. 4 (October–December 2017): 1–11.
2 The book project goes back to a workshop held (virtually) at Villa Vigoni in Loveno di Menaggio on Lake Como in March 2021. The members of the research group 2448 "Practicing Evidence – Evidencing Practice" funded by the Deutsche Forschungsgemeinschaft (DFG, German Research Foundation) invited a number of international researchers and recognized experts in their respective fields to discuss the topic of evidence contestation with them over three days. All contributions are original research papers and have not been published before. The interdisciplinary research group has existed since 2017 and has provided an active context for discussing evidence practices ever since. This has already resulted in two collected volumes: Karin Zachmann and Sarah Ehlers, eds.,

Wissen und Begründen. Evidenz als umkämpfte Ressource in der Wissensgesellschaft (Baden Baden: Nomos, 2019) as well as Sarah Ehlers and Stefan Esselborn, eds., *Evidence in Action between Science and Society. Constructing, Validating, and Contesting Knowledge* (New York: Routledge, 2022). The editors of this book are particularly grateful to the very valuable comments of all the research group members, who discussed the introduction to the volume with us in June 2022, and to Karl Detering, Stefan Reifberger, and Victoria Woollven for the indispensable (native) proofreading and editing of the volume. We are especially thankful to the DFG, which funded the workshop and also the publication of the results in open access format.

3 Stefan Böschen, *Hybride Wissensregime. Skizze einer soziologischen Feldtheorie* (Baden Baden: Nomos, 2016), 361; Wiebe E. Bijker, Roland Bal and Ruud Hendriks, *The Paradox of Scientific Authority* (Cambridge, MA: MIT Press, 2009), 1.

4 Dorothy Nelkin, "Science Controversies. The Dynamics of Public Disputes in the United States", in *Handbook of Science and Technology Studies*, eds. Sheila Jasanoff, Gerald E. Markle, James C. Peterson and Trevor J. Pinch (Thousand Oaks: Sage Publications, 2001), 444–456; Thomas Brante, "Reasons for Studying Scientific and Science-Based Controversies", in *Controversial Science. From Content to Contention*, eds. Thomas Brante, Steve Fuller and William Lynch (New York: State Univ. of New York Press, 1993), 177–192.

5 Naomi Oreskes and Erik M. Conway, *Merchants of Doubt. How a Handful of Scientists Obscured the Truth on Issues from Tobacco Smoke to Global Warming* (New York: Bloomsbury Press, 2010); Donald Prothero, *Reality Check. How Science Deniers Threaten Our Future* (Bloomington, Indiana: Indiana Univ. Press, 2013); David Harker, *Creating Scientific Controversies. Uncertainty and Bias in Science and Society* (Cambridge: Cambridge Univ. Press, 2015).

6 Tamir Bar-On and Barbara Molas, *Responses to the COVID-19 Pandemic by the Radical Right: Scapegoating, Conspiracy Theories and New Narratives* (Stuttgart: ibidem-Verlag, 2020).

7 The German Querdenker-Bewegung (Lateral Thinkers' Movement) is a case in point here. See for example Sven Reichardt, ed., *Die Misstrauensgemeinschaft der »Querdenker«: Die Corona-Proteste aus kultur- und sozialwissenschaftlicher Perspektive* (Frankfurt: Campus, 2021).

8 Boaz Miller, "The Social Epistemology of Consensus and Dissent", in *The Routledge Handbook of Social Epistemology*, eds. Miranda Fricker, Peter J. Graham, David Henderson and Nikolaj J.L.L. Pedersen (New York: Routledge, 2020), 230.

9 Ulrich Beck, Wolfgang Bonß, and Christoph Lau, "Theorie reflexiver Modernisierung – Fragestellungen, Hypothesen, Forschungsprogramme", in *Die Modernisierung der Moderne*, eds. Ulrich Beck and Wolfgang Bonß (Frankfurt am Main: Suhrkamp Taschenbuch, 2001), 11–59, esp. 36. Beck is not talking about a paradox, but a situation in which "reflexivity subverts reason". Ulrich Beck and Boris Holzer, "Reflexivität und Reflexion", in *Entgrenzung und Entscheidung: Was ist neu an der Theorie reflexiver Modernisierung*, eds. Ulrich Beck and Christoph Lau (Frankfurt am Main: Suhrkamp, 2004), 165–192, esp. 191.

10 Thomas Kelly, "Evidence", in *The Stanford Encyclopedia of Philosophy, Fall 2014*, ed. Edward N. Zalta (Stanford, CA: Metaphysics Research Lab, Stanford Univ., 2014), accessed November 21, 2022. http://plato.stanford.edu/archives/fall2014/entries/evidence/.

11 "Evidence", in *Cambridge Dictionary*, accessed May 20, 2022, https://dictionary.cambridge.org/dictionary/english/evidence.

12 Jürgen Mittelstraß, "Evidenz", in *Enzyklopädie Philosophie und Wissenschaftstheorie*, Vol. 1 (Mannheim: B.I. Wissenschaftsverlag, 1984), 609.

13 Immanuel Kant, *Kritik der reinen Vernunft*, ed. Wilhelm Weischedel (Frankfurt am Main: Suhrkamp, 1968), B762, 627.

14 On the various meanings and uses of the term, see Kelly, "Evidence", 1.

15 Rachel E. Sherman, Steven A. Anderson, Gerald J. Dal Pan, Gerry W. Gray, Thomas Gross, Nina L. Hunter, and Lisa LaVange, "Real-World Evidence – What Is It and What Can It Tell Us?" *The New England Journal of Medicine* 375, no. 23 (2016): 2293–2297; Food and Drug Administration, *Use of Real-World Evidence to Support Regulatory Decision-Making for Medical Devices: Draft Guidance for Industry and Food and Drug Administration Staff* (July 27, 2016): 1–21; Jacqueline Corrigan-Curay, Leonard Sacks and Janet Woodcock, "Real-World Evidence and Real-World Data for Evaluating Drug Safety and Effectiveness", *JAMA* 320, no. 9 (2018): 867–868.

16 Jacob Stegenga, *Care and Cure: An Introduction to Philosophy of Medicine* (Chicago: Univ. of Chicago Press, 2018).

17 Justin O. Parkhurst, *The Politics of Evidence: From Evidence-Based Policy to the Good Governance of Evidence* (Oxfordshire: Routledge 2017); Robert Coe and Stuart Kime, *A (New) Manifesto for Evidence-Based Education: Twenty Years on* (Sunderland UK: Evidence Based Education, 2019), accessed on May 14, 2022, https://evidencebased.education/new-manifesto-evidence-based-education/.

18 April Mackey and Sandra Bassendowski, "The History of Evidence-Based Practice in Nursing Education and Practice", *Journal of Professional Nursing* 33, no. 1 (2017): 51–55.

19 With the programmatic article of the Evidence-Based Medicine Working Group that had formed around the epidemiologists Gordon Guyatt and David L. Sackett, EBM is defined as "[a] NEW paradigm for medical practice [..., which] de-emphasizes intuition, unsystematic clinical experience, and pathophysiologic rationale as sufficient grounds for clinical decision making and stresses the examination of evidence from clinical research." Gordon Guyatt, John Cairns, David Churchill, Deborah Cook, Brian Haynes, Jack Hirsh, Jan Irvine et al., "Evidence-Based Medicine. A New Approach to Teaching the Practice of Medicine", *Journal of the American Medical Association* 268, no. 17 (1992): 2420–2425; see also David L. Sackett, William M.C. Rosenberg, J.A. Muir Gray, R. Brian Haynes and W. Scott Richardson, "Evidence-Based Medicine: What it is and What it isn't", *The BMJ* 312, no. 71 (1996): 71–72.

20 Justin O. Parkhurst and Sudeepa Abeysinghe, "What Constitutes 'Good' Evidence for Public Health and Social Policy-Making? From Hierarchies to Appropriateness", *Social Epistemology* 30, no. 5–6 (2016): 665–679.

21 Hanne Foss Hansen and Olaf Rieper, "The Evidence Movement. The Development and Consequences of Methodologies in Review Practices", *Evaluation* 15, no. 2 (2009): 141–163.

22 Michel Foucault, "Truth and Power, Interviewers: Alessandro Fontana, Pasquale Pasquini", in *Power/Knowledge. Selected Interviews and Other Writings 1972–1977*, ed. Colin Gordon (Brighton: The Harvester Press, 1980), 132.

23 Michel Foucault, "What is Critique?" in *The Politics of Truth*, eds. Sylvère Lotringer and Lysa Hochroth (New York: Semiotext(e), 1997).

24 Judith Butler, "What is Critique? An Essay on Foucault's Virtue", *Transversal*, published May 2001, accessed November 21, 2022, https://transversal. at/transversal/0806/butler/en. See also Judith Butler, "Critique, Dissent, Disciplinarity", *Critical Inquiry* 35, no. 4 (2009): 773–795.

25 Michael P. Lynch, *In Praise of Reason* (Cambridge: MIT Press, 2012), 31. On evidence as a resource, see Zachmann and Ehlers, *Wissen und Begründen.*

26 For a diachronic analysis of the various conceptualizers and theories of evidence from a philosophical perspective, see: Kelly, "Evidence".

27 Helga Nowotny and Hilary Rose, eds., *Counter-Movements in the Sciences: The Sociology of the Alternatives to Big Science* (Dordrecht: Reidel, 1979).

28 As a more recent book on the many and ever context-dependent modes of science criticism with a focus on the US, see Andrew Jewett, *Science under Fire. Challenges to Scientific Authority in Modern America* (Cambridge: Harvard Univ. Press, 2020).

29 Dorothy Nelkin, *Science Textbook Controversies and the Politics of Equal Time* (Cambridge: MIT Press, 1977). Dorothy Nelkin, ed., *Controversy. Politics of Technical Decisions* (London: Sage, 1979); Dorothy Nelkin, *The Creation Controversy: Science or Scripture in the Schools* (New York: Norton, 1982) to name just a few of Dorothy Nelkin's publications. Other authors followed suit, e.g. Hugo Tristram Engelhardt and Arthur L. Caplan, *Scientific Controversies: Case Studies in the Resolution and Closure of Disputes in Science and Technology* (Cambridge: Cambridge Univ. Press, 1987); Thomas Brante, Steve Fuller and William Lynch, eds., *Controversial Science. From Content to Contention* (New York: State Univ. of New York Press, 1993).

30 Harry Collins and Robert Evans, "The Third Wave of Science Studies. Studies of Expertise and Experience", *Social Studies of Science* 32, no. 2 (April 2002): 235–296; Bruno Latour, "Why Has Critique Run out of Steam? From Matters of Fact to Matters of Concern", *Critical Inquiry* 30 (Winter 2004): 225–248.

31 Andrew Ross, "Introduction", in *Science Wars*, ed. Andrew Ross (Durham: Duke Univ. Press, 1996), 1–15.

32 Latour, "Why Has Critique Run out of Steam?", 227.

33 Latour, "Why Has Critique Run out of Steam?", 227.

34 The landmark study in this field is Oreskes and Conway, *Merchants of Doubt.*

35 Robert N. Proctor, "Agnotology: A Missing Term to Describe the Cultural Production of Ignorance (and Its Study)", in *Agnotology. The Making and Unmaking of Ignorance*, eds. Robert N. Proctor and Londa Schiebinger (Stanford: Stanford Univ. Press, 2008), 1–36.

36 Manuela Fernandez Pinto, "Tensions in Agnotology: Normativity in the Studies of Commercially Driven Ignorance", *Social Studies of Science* 45, no. 2 (2015): 294–315; Janet Kourany and Martin Carrier, eds., *Science and the Production of Ignorance: When the Quest for Knowledge is Thwarted* (Cambridge, MA: MIT Press, 2020).

37 Rani Lill Anjum, Samantha Copeland and Elena Rocca, "Medical Scientists' and Philosophers' Worldwide Appeal to EBM to Expand the Notion of 'Evidence'", *BMJ Evidence-Based Medicine* 25, no. 1 (2020): 6–8.

38 Holger Strassheim and Pekka Kettunen, "When Does Evidence-Based Policy Turn into Policy-Based Evidence? Configurations, Contexts and Mechanisms", *Evidence & Policy: A Journal of Research, Debate and Practice* 10, no. 2 (2014): 259–277; Jacob Stegenga, *Medical Nihilism* (Oxford: Oxford Univ. Press, 2018); Anjum, Copeland and Rocca, "Medical Scientists' and Philosophers' Worldwide Appeal to EBM", 6–8.

39 Trisha Greenhalgh and co-authors distinguish between "objective evidence (e.g. what this patient's test results show), and subjective evidence (e.g. what this patient feels; what matters to him or her)": Trisha Greenhalgh, Rosamund Snow, Sara Ryan, Sian Rees and Helen Salisbury, "Six 'Biases' against Patients and Carers in Evidence-Based Medicine", *BMC Medicine* 13, no. 1 (2015): 1–11; Mariacarla Gadebusch Bondio and Ingo F. Herrmann argue for an extended definition of "subjective evidence" that resembles the notion of "experiential knowledge" and thus emphasizes its epistemic dimension: Mariacarla Gadebusch Bondio and Ingo F. Herrmann, "Cancer and the Life beyond it. Patients Testimony as a Contribution to Subjective Evidence (with an original text by Maria Cristina Montani)", in *Ethical Challenges in Cancer Diagnosis and Therapy*, eds. Axel W. Bauer, Ralf-Dieter Hofheinz and Jochen S. Utikal (Cham: Springer, 2021): 259–273; see also: Sarah Atkinson, Hannah Bradby, Mariacarla Gadebusch Bondio, Anna Hallberg, Jane Macnaugthon and Ylva Söderfeldt, "Seeing the Value of Experiential Knowledge through COVID-19", *History and Philosophy of the Life Sciences (HPLS)* 43, no. 85 (2021), 1–4.

40 Greenhalgh et al., "Six 'Biases'", 3, Table 1: "'Biases against patients and carers in traditional evidence-based medicine (EBM) and how they might be overcome".

41 Helen Longino, *Science as Social Knowledge: Values and Objectivity in Scientific Inquiry* (Princeton: Princeton Univ. Press, 1990); Alison Wylie, "The Constitution of Archaeological Evidence", in *The Archaeology of Identities: A Reader*, eds. Timothy Insoll and Sarah Tarlow (Oxfordshire: Routledge, 2007), 97–18.

42 Sabina Leonelli, *Data-Centric Biology* (Chicago: Univ. of Chicago Press, 2016).

43 Lorraine Daston and Peter Galison, *Objectivity* (New York: Zone Books, 1997).

44 Robert N. Proctor, *Value-Free Science? Purity and Power in Modern Knowledge* (Cambridge, MA: Harvard Univ. Press 1991).

45 Ian Hacking, "Statistical Language, Statistical Truth and Statistical Reason: The Self-Authentification of a Style of Scientific Reasoning", in *The Social Dimension of Science*, ed. Ernan McMullin (Indiana: Notre Dame Univ. Press 1992), 130–157.

46 Alexander Bogner, *Die Epistemisierung des Politischen. Wie die Macht des Wissens die Demokratie gefährdet* (Stuttgart: Reclam, 2021).

47 Bogner, *Die Epistemisierung des Politischen*, 52–57.

48 Ehlers and Esselborn, *Evidence in Action between Science and Society*.

49 Mariacarla Gadebusch Bondio and Maria Marloth, "Clinical Trials in Pandemic Settings: How Corona Unbinds Science", in *The Realm of Corona Normativities. A Momentary Snapshot of a Dynamic Discourse*, ed. Werner Gephard (Frankfurt am Main: Vittorio Klostermann, 2020), 29–40.

50 Saana Jukola and Stefano Canali, "On Evidence Fiascos and Judgments in COVID-19 Policy", *HPLS* 43, no. 61 (2021), 1–4. Phillip Roth and Mariacarla Gadebusch Bondio, "The Contested Meaning of 'Long COVID' – Patients, Doctors, and the Politics of Subjective Evidence", *Social Science & Medicine* 292 (2022): 1–8. More research regarding the production, communication and negotiation of scientific evidence during the COVID-19 crisis is currently undertaken by four subprojects within the DFG-research group 2448: Accessed November 21, 2022, https://www.evidenzpraktiken-dfg.tum.de/en/ap4-die-de-und-restabilisierung-von-evidenz-in-der-coronakrise/. Massimiano Bucchi, "Of Deficits, Deviations

and Dialogues. Theories of Public Communication of Science", in *Handbook of Public Communication of Science and Technology*, eds. Massimiano Bucchi und Brian Trench (London: Routledge, 2008), 57–76.

51 Thomas Hughes, *American Genesis. A Century of Invention and Technological Enthusiasm, 1870–1970* (Chicago: Univ. of Chicago Press, 1989), 392–395.

52 Georg Simmel, "Der Streit" in *Gesamtausgabe*, Vol. 11 (Frankfurt am Main: Suhrkamp, 2016), 307–308.

53 Gibbons et al. highlight the orientation toward application as an important feature of Mode 2 Knowledge production. See Michael Gibbons, Camille Limoges, Peter Scott, Simon Schwartzman and Helga Nowotny, *The New Production of Knowledge: The Dynamics of Science and Research in Contemporary Societies* (London: Sage, 1994). Peter Weingart refers to increased usability expectations within his diagnosis of closer couplings between science and politics and the economy, respectively. See Peter Weingart, *Die Stunde der Wahrheit? Zum Verhältnis der Wissenschaft zu Politik, Wirtschaft und Medien in der Wissensgesellschaft* (Weilerswist: Velbrück Wissenschaft, 2001), 25, chapters 4 and 5.

54 Nancy Cartwright, "Well-Ordered Science: Evidence for Use", *Philosophy of Science* 73 (December 2006): 983.

55 Oreskes and Conway, *Merchants of Doubt*.

56 A comprehensive analysis of deliberative democracy provides Andre Bächtiger, John S. Dryzek, Jane Mansbridge and Mark E. Warren, eds., *The Oxford Handbook of Deliberative Democracy* (Oxford: Oxford Univ. Press, 2018).

57 On consensus conferences see Laszlo Kosolosky and Jeroen Van Bouwel, "Explicating Ways of Consensus-Making in Science and Society: Distinguishing the Academic, the Interface and the Meta-Consensus", in *Experts and Consensus in Social Science*, eds. Carlo Martini and Marcel Boumans (Dordrecht: Springer, 2014), 71–92. See for the history of joint decision-making Andy Blunden, *The Origins of Collective Decision Making* (Haymarket Books: Chicago, 2018). As an early example of Public Hearings Paul Albrecht, *Before it's Too Late. The Challenge of Nuclear Disarmament. The Complete Record of the Public Hearing on Nuclear Weapons and Disarmament* (Geneva: World Council of Churches, 1983); for stakeholder meetings Minu Hemmati-Weber, *Multi-Stakeholder Processes for Governance and Sustainability beyond Deadlock and Conflict* (London: Earthscan Publications, 2002).

58 See for example Rebecca Willis, *Too Hot to Handle? The Democratic Challenge of Climate Change* (Bristol: Bristol Univ. Press, 2020).

59 For example, the United Nations Economic Commission for Europe set up a round table on green genetic engineering: https://unece.org/environmental-policy/events/joint-global-round-table-lmosgmos, accessed November 21, 2022.

60 See for example Lee McIntyre, *Post-Truth* (Cambridge, MA: MIT-Press, 2018).

61 Nicholas Rescher, *Pluralism. Against the Demand for Consensus* (Oxford: Clarendon Press, 1993), 43.

62 Heather E. Douglas, "Weighing Complex Evidence in a Democratic Society", *Kennedy Institute of Ethics Journal* 22, no. 2 (June 2012): 144.

63 Weingart, *Die Stunde der Wahrheit*, 17.

64 Kurt Röttgers, "Kritik", in *Geschichtliche Grundbegriffe: Historisches Lexikon zur politisch-sozialen Sprache in Deutschland*, Vol. 3, eds. Otto Brunner, Werner Conze and Reinhart Koselleck (Stuttgart: Klett-Cotta, 1995), 662–665.

65 Röttgers, "Kritik", 662–665.
66 Bogner, *Die Epistemisierung des Politischen*.

References

Albrecht, Paul. *Before It's Too Late. The Challenge of Nuclear Disarmament. The Complete Record of the Public Hearing on Nuclear Weapons and Disarmament*. Geneva: World Council of Churches, 1983.

Anjum, Rani Lill, Samantha Copeland, and Elena Rocca. "Medical Scientists and Philosophers Worldwide Appeal to EBM to Expand the Notion of 'Evidence'". *BMJ Evidence-Based Medicine* 25, no. 1 (2020): 6–8.

Atkinson, Sarah, Hannah Bradby, Mariacarla Gadebusch Bondio, Anna Hallberg, Jane Macnaugthon, and Ylva Söderfeldt. "Seeing the Value of Experiential Knowledge through COVID-19". *History and Philosophy of the Life Sciences (HPLS)* 43, no. 85 (2021): 1–4.

Bächtiger, Andre, John S. Dryzek, Jane Mansbridge, and Mark E. Warren, eds. *The Oxford Handbook of Deliberative Democracy*. Oxford: Oxford Univ. Press, 2018.

Bar-On, Tamir, and Barbara Molas. *Responses to the COVID-19 Pandemic by the Radical Right: Scapegoating, Conspiracy Theories and New Narratives*. Stuttgart: ibidem-Verlag, 2020.

Beck, Ulrich, Wolfgang Bonß, and Christoph Lau. "Theorie reflexiver Modernisierung – Fragestellungen, Hypothesen, Forschungsprogramme", In *Die Modernisierung der Moderne*, edited by Ulrich Beck, and Wolfgang Bonß, 11–59. Frankfurt am Main: Suhrkamp Taschenbuch, 2001.

Beck, Ulrich, and Boris Holzer. "Reflexivität und Reflexion", In *Entgrenzung und Entscheidung: Was ist neu an der Theorie reflexiver Modernisierung*, edited by Ulrich Beck, and Christoph Lau, 165–192. Frankfurt am Main: Suhrkamp, 2004.

Bijker, Wiebe E., Roland Bal, and Ruud Hendriks. *The Paradox of Scientific Authority*. Cambridge, MA: MIT Press, 2009.

Blunden, Andy. *The Origins of Collective Decision Making*. Chicago: Haymarket Books, 2018.

Bogner, Alexander. *Die Epistemisierung des Politischen. Wie die Macht des Wissens die Demokratie gefährdet*. Stuttgart: Reclam, 2021.

Böschen, Stefan. *Hybride Wissensregime. Skizze einer soziologischen Feldtheorie*. Baden-Baden: Nomos, 2016.

Brante, Thomas. "Reasons for Studying Scientific and Science-Based Controversies", In *Controversial Science. From Content to Contention*, edited by Thomas Brante, Steve Fuller, and William Lynch, 177–192. New York: State Univ. of New York Press, 1993.

Brante, Thomas, Steve Fuller, and William Lynch, eds. *Controversial Science. From Content to Contention*. New York: State Univ. of New York Press, 1993.

Bucchi, Massimiano. "Of Deficits, Deviations and Dialogues. Theories of Public Communication of Science", In *Handbook of Public Communication of Science and Technology*, edited by Massimiano Bucchi, and Brian Trench, 57–76. London: Routledge, 2008.

Butler, Judith. "What Is Critique? An Essay on Foucault's Virtue", *Transversal*. Published May 2001. https://transversal.at/transversal/0806/butler/en.

Butler, Judith. "Critique, Dissent, Disciplinarity". *Critical Inquiry* 35, no. 4 (2009): 773–795.

Cartwright, Nancy. "Well-Ordered Science: Evidence for Use". *Philosophy of Science* 73, no. 5 (December 2006): 981–990.

Coe, Robert, and Stuart Kime. *A (New) Manifesto for Evidence-Based Education: Twenty Years on.* Sunderland, UK: Evidence Based Education, 2019.

Collins, Harry, and Robert Evans. "The Third Wave of Science Studies. Studies of Expertise and Experience". *Social Studies of Science* 32, no. 2 (April 2002): 235–296.

Corrigan-Curay, Jacqueline, Leonard Sacks, and Janet Woodcock. "Real-World Evidence and Real-World Data for Evaluating Drug Safety and Effectiveness". *JAMA* 320, no. 9 (2018): 867–868.

Daston, Lorraine, and Peter Galison. *Objectivity.* New York: Zone Books, 1997.

Douglas, Heather E. "Weighing Complex Evidence in a Democratic Society". *Kennedy Institute of Ethics Journal* 22, no. 2 (June 2012): 139–162.

Ehlers, Sarah, and Stefan Esselborn, eds. *Evidence in Action between Science and Society. Constructing, Validating, and Contesting Knowledge.* New York: Routledge, 2022.

Engelhardt, Hugo Tristram, and Arthur L. Caplan. *Scientific Controversies: Case Studies in the Resolution and Closure of Disputes in Science and Technology.* Cambridge: Cambridge Univ. Press, 1987.

"Evidence". In *Cambridge Dictionary*, https://dictionary.cambridge.org/dictionary/english/evidence.

Fernandez Pinto, Manuela. "Tensions in Agnotology: Normativity in the Studies of Commercially Driven Ignorance". *Social Studies of Science* 45, no. 2 (2015): 294–315.

Food and Drug Administration. *Use of Real-World Evidence to Support Regulatory Decision-Making for Medical Devices: Draft Guidance for Industry and Food and Drug Administration Staff.* July 27, 2016.

Foss Hansen, Hanne, and Olaf Rieper. "The Evidence Movement. The Development and Consequences of Methodologies in Review Practices". *Evaluation* 15, no. 2 (2009): 141–163.

Foucault, Michel. "Truth and Power, Interviewers: Alessandro Fontana, Pasquale Pasquini", In *Power/Knowledge. Selected Interviews and Other Writings 1972–1977. Michel Foucault*, edited by Colin Gordon. Brighton: The Harvester Press, 1980.

Foucault, Michel. "What Is Critique?" In *The Politics of Truth*, edited by Sylvère Lotringer, and Lysa Hochroth. New York: Semiotext(e), 1997.

Gadebusch Bondio, Mariacarla, and Maria Marloth. "Clinical Trials in Pandemic Settings: How Corona Unbinds Science", In *The Realm of Corona Normativities. A Momentary Snapshot of a Dynamic Discourse*, edited by Werner Gephard, 29–40. Frankfurt am Main: Vittorio Klostermann, 2020.

Gadebusch Bondio, Mariacarla, and Ingo F. Herrmann. "Cancer and the Life beyond It. Patients Testimony as a Contribution to Subjective Evidence (with an Original Text by Maria Cristina Montani)". In *Ethical Challenges in Cancer Diagnosis and Therapy*, edited by Axel W. Bauer, Ralf-Dieter Hofheinz, and Jochen S. Utikal. Cham: Springer, 2021.

Gibbons, Michael, Camille Limoges, Peter Scott, Simon Schwartzman, and Helga Nowotny. *The New Production of Knowledge: The Dynamics of Science and Research in Contemporary Societies*. London: Sage, 1994.

Greenhalgh, Trisha, Rosamund Snow, Sara Ryan, Sian Rees, and Helen Salisbury. "Six 'Biases' against Patients and Carers in Evidence-Based Medicine". *BMC Medicine* 13, no. 1 (2015): 1–11.

Guyatt, Gordon, John Cairns, David Churchill, Deborah Cook, Brian Haynes, Jack Hirsh, and Jan Irvine et al. "Evidence-Based Medicine. A New Approach to Teaching the Practice of Medicine". *Journal of the American Medical Association* 268, no. 17 (1992): 2420.

Hacking, Ian. "Statistical Language, Statistical Truth and Statistical Reason: The Self-Authentification of a Style of Scientific Reasoning", In *The Social Dimension of Science*, edited by Ernan McMullin, 130–157. Indiana: Notre Dame Univ. Press, 1992.

Harker, David. *Creating Scientific Controversies. Uncertainty and Bias in Science and Society*. Cambridge: Cambridge Univ. Press, 2015.

Hemmati-Weber, Minu. *Multi-Stakeholder Processes for Governance and Sustainability beyond Deadlock and Conflict*. London: Earthscan Publications, 2002.

Hughes, Thomas. *American Genesis. A Century of Invention and Technological Enthusiasm, 1870–1970*. Chicago: Univ. of Chicago Press, 1989.

Jewett, Andrew. *Science under Fire. Challenges to Scientific Authority in Modern America*. Cambridge, MA: Harvard Univ. Press, 2020.

Jukola, Saana, and Stefano Canali. "On Evidence Fiascos and Judgments in COVID-19 Policy". *HPLS* 43, no. 61 (2021): 1–4.

Kant, Immanuel. *Kritik der reinen Vernunft*, edited by Wilhelm Weischedel. Frankfurt am Main: Suhrkamp, 1968.

Kelly, Thomas. "Evidence", In *The Stanford Encyclopedia of Philosophy, Fall 2014*, Vol. 2, edited by Edward N. Zalta. Stanford, CA: Metaphysics Research Lab, Stanford Univ., 2014.

Kosolosky, Laszlo, and Jeroen Van Bouwel. "Explicating Ways of Consensus-Making in Science and Society: Distinguishing the Academic, the Interface and the Meta-Consensus", In *Experts and Consensus in Social Science*, edited by Carlo Martini, and Marcel Boumans, 71–92. Dordrecht: Springer, 2014.

Kourany, Janet, and Martin Carrier, eds. *Science and the Production of Ignorance: When the Quest for Knowledge Is Thwarted*. Cambridge, MA: MIT Press, 2020.

Latour, Bruno. "Why Has Critique Run Out of Steam? From Matters of Fact to Matters of Concern". *Critical Inquiry* 30 (Winter 2004): 225–248.

Leonelli, Sabina. *Data-Centric Biology*. Chicago: Univ. of Chicago Press, 2016.

Longino, Helen. *Science as Social Knowledge: Values and Objectivity in Scientific Inquiry*. Princeton: Princeton Univ. Press, 1990.

Lynch, Michael P. *In Praise of Reason*. Cambridge, MA: MIT Press, 2012.

Mackey, April, and Sandra Bassendowski. "The History of Evidence-Based Practice in Nursing Education and Practice". *Journal of Professional Nursing* 33, no. 1 (2017): 51–55.

McIntyre, Lee. *Post-Truth*. Cambridge, MA: MIT-Press, 2018.

Miller, Boaz. "The Social Epistemology of Consensus and Dissent", In *The Routledge Handbook of Social Epistemology*, edited by Miranda Fricker, Peter J. Graham, David Henderson, and Nikolaj J.L.L. Pedersen. New York: Routledge, 2020.

Mittelstraß, Jürgen. "Evidenz", In *Enzyklopädie Philosophie und Wissenschaftstheorie*, Vol. 1. Mannheim: B.I. Wissenschaftsverlag, 1984.

Nelkin, Dorothy. *Science Textbook Controversies and the Politics of Equal Time*. Cambridge, MA: MIT Press, 1977.

Nelkin, Dorothy, ed. *Controversy. Politics of Technical Decisions*. London: Sage, 1979.

Nelkin, Dorothy. *The Creation Controversy. Science or Scripture in the Schools*. New York: Norton, 1982.

Nelkin, Dorothy. "Science Controversies. The Dynamics of Public Disputes in the United States", In *Handbook of Science and Technology Studies*, edited by Sheila Jasanoff, Gerald E. Markle, James C. Peterson, and Trevor J. Pinch, 444–456. Thousand Oaks: Sage Publications, 2001.

Nowotny, Helga, and Hilary Rose, eds. *Counter-Movements in the Sciences: The Sociology of the Alternatives to Big Science*. Dordrecht: Reidel, 1979.

Oreskes, Naomi, and Erik M. Conway. *Merchants of Doubt. How a Handful of Scientists Obscured the Truth on Issues from Tobacco Smoke to Global Warming*. New York: Bloomsbury Press, 2010.

Parkhurst, Justin O. *The Politics of Evidence. From Evidence-Based Policy to the Good Governance of Evidence*. London: Routledge, 2017.

Parkhurst, Justin O., and Sudeepa Abeysinghe, "What Constitutes 'Good' Evidence for Public Health and Social Policy-Making? From Hierarchies to Appropriateness". *Social Epistemology* 30, no. 5–6 (2016): 665–679.

Proctor, Robert N. *Value-Free Science? Purity and Power in Modern Knowledge*. Cambridge, MA: Harvard Univ. Press, 1991.

Proctor, Robert N. "Agnotology: A Missing Term to Describe the Cultural Production of Ignorance (And Its Study)", In *Agnotology. The Making and Unmaking of Ignorance*, edited by Robert N. Proctor, and Londa Schiebinger, 1–36. Stanford: Stanford Univ. Press, 2008.

Proctor, Robert N., and Londa Schiebinger, eds. *Agnotology. The Making and Unmaking of Ignorance*. Stanford: Stanford Univ. Press, 2008.

Prothero, Donald. *Reality Check. How Science Deniers Threaten Our Future*. Bloomington: Indiana Univ. Press, 2013.

Reichardt, Sven, ed. *Die Misstrauensgemeinschaft der »Querdenker«: Die Corona-Proteste aus kultur- und sozialwissenschaftlicher Perspektive*. Frankfurt: Campus, 2021.

Rescher, Nicholas. *Pluralism. Against the Demand for Consensus*. Oxford: Clarendon Press, 1993.

Ross, Andrew. "Introduction", In *Science Wars*, edited by Andrew Ross, 1–15. Durham: Duke Univ. Press, 1996.

Roth, Phillip, and Mariacarla Gadebusch Bondio. "The Contested Meaning of 'Long COVID' – Patients, Doctors, and the Politics of Subjective Evidence". *Social Science & Medicine* 292 (2022): 1–8.

Röttgers, Kurt. "Kritik", In *Geschichtliche Grundbegriffe: Historisches Lexikon zur politisch-sozialen Sprache in Deutschland*, Vol. 3, edited by Otto Brunner, Werner Conze, and Reinhart Koselleck, 662–665. Stuttgart: Klett-Cotta, 1995.

Sackett, David L., William M.C. Rosenberg, J.A. Muir Gray, R. Brian Haynes, and W. Scott Richardson. "Evidence-Based Medicine: What It Is and What It Isn't". *The BMJ* 312, no. 71 (1996): 71–72.

Sherman, Rachel E., Steven A. Anderson, Gerald J. Dal Pan, Gerry W. Gray, Thomas Gross, Nina L. Hunter, and Lisa LaVange et al. "Real-World Evidence – What Is It and What Can It Tell Us?". *The New England Journal of Medicine* 375, no. 23 (2016): 2293–2297.

Simmel, Georg. "Der Streit", In *Gesamtausgabe*, Vol. 11, 307–308. Frankfurt am Main: Suhrkamp, 2016.

Stegenga, Jacob. *Medical Nihilism*. Oxford: Oxford Univ. Press, 2018.

Stegenga, Jacob. *Care and Cure: An Introduction to Philosophy of Medicine*. Chicago: Univ. of Chicago Press, 2018.

Strassheim, Holger, and Pekka Kettunen. "When Does Evidence-Based Policy Turn into Policy-Based Evidence? Configurations, Contexts and Mechanisms". *Evidence & Policy: A Journal of Research, Debate and Practice* 10, no. 2 (2014): 259–277.

Van Rensburg, Willem, and Brian W. Head. "Climate Change Skepticism: Reconsidering How to Respond to Core Criticisms of Climate Science and Policy". *SAGE Open* 7, no. 4 (October–December 2017): 1–11.

Weingart, Peter. *Die Stunde der Wahrheit? Zum Verhältnis der Wissenschaft zu Politik, Wirtschaft und Medien in der Wissensgesellschaft*. Weilerswist: Velbrück Wissenschaft, 2001.

Willis, Rebecca. *Too Hot to Handle? The Democratic Challenge of Climate Change*. Bristol: Bristol Univ. Press, 2020.

Wylie, Alison. "The Constitution of Archaeological Evidence", In *The Archaeology of Identities: A Reader*, edited by Timothy Insoll, and Sarah Tarlow, 97–18. Oxfordshire: Routledge, 2007.

Zachmann, Karin, and Sarah Ehlers, eds. *Wissen und Begründen. Evidenz als umkämpfte Ressource in der Wissensgesellschaft*. Baden-Baden: Nomos, 2019.

Part I
Theoretical Framing
Evidence Critique and Contestation

1 What Is Scientific Criticism for? Some Philosophical Reflections on Criticism and Evidence within the Scientific Ethos

Daniel Füger and Elif Özmen

The need for scientific evidence and evidence-informed decision-making increased tremendously during the coronavirus pandemic. In recent years, evidence has therefore become a central concern in public and private debates, policies and far-reaching decisions. At the same time, most people, including politicians, find it difficult to understand and handle the uncertain and changing character of evidence on the development, dissemination and pathogenic effects of the coronavirus. Nonetheless, confidence in science has increased significantly compared to the years before the pandemic.[1] But how is it possible to develop trust when evidence as a core aspect of scientific inquiry and science-based advice is under constant threat of being de-stabilized and revised?

Most activities aimed at achieving knowledge, and scientific activities in particular, require evidence. Producing evidence and referring to evidence ensures the robustness of scientific hypothesis and the quality of scientific theories. Moreover, evidence is necessary – albeit not sufficient – for a scientific activity to be acknowledged both within the scientific community and society in general. Science depends on criticism, which not only includes evaluation, judgments and disapproval, but also the contestation of evidence. While scientific evidence serves to either support or counter a scientific hypothesis, its contestation is only partly a devaluation of previous research findings and mostly a common practice within science. How can these two aspects of evidence generation, which we take for granted, coexist? How in particular are evidence and the contestation and critique of evidence related to the norms of scientific activity? And what are the ties and tensions between the contestation of evidence and criticism of science, between scientific ethos and societal norms, and more broadly speaking between science and the public recognition or refutation of scientific knowledge?

We assume that a philosophical approach will help to address these questions by critically evaluating the underlying concepts. Our main focus lies on the *normative* aspects of evidence and evidence criticism within scientific ethos. On the one hand, it seems perfectly clear that science must (and mostly can) justify why and how it relies on specific

DOI: 10.4324/9781003273509-3

evidence practices. On the other hand, these practices and scientific justifications are reliant not only on inner-scientific but also on non-scientific normative claims, e.g. social, political and moral values. As we will show in the following, these normative claims are very important in dealing with *quid iuris* questions about evidence. We hope that our contribution to this volume will help to understand the normative conditions and functions of evidence and its contestation as part of the normative foundations of science as a whole.

In the first section of this chapter, we clarify what scientific criticism entails and what distinguishes it from general skepticism about science. We examine the historical and systematic meaning of the term criticism in order to contextualize it within science. In the second section, we highlight our understanding of evidence as a normative standard in science which is at the same time contested, dynamic and ambivalent. But there is common ground insofar as scientific criticism in general, and criticism of evidence in particular, is built on a shared conception of evidence. We present this conception with the heuristic of agnosticism of evidence, which allows us to approach the ambivalent functions of evidence. In the third section, we examine the interactions between evidence and criticism within the normative setting of science. In particular, we conclude that the so-called scientific ethos is both a necessary condition for scientific practice and a result of this practice. The fourth section addresses the role of contingencies in science and the openness and ambiguity that they constitute. We suggest that criticism and contingencies, partly driven by the ambivalent character of evidence, nevertheless contribute to the stability of science. In addition, we look at the causes of contemporary evidence crises that arise, among other things, when the legitimation of science within democratic societies fails.

What Is Scientific Criticism for?

In this section, we attempt to clarify the idea of scientific criticism with reference to systematic considerations and historical developments. We are particularly interested in the stabilizing effects of this criticism for the shaping and establishing of modern science. Philosopher Helen Longino distinguishes between two types of criticism with regard to their productive role within science. *Evidential criticism* is considered the basis of experimental and observational research; its main aim is to identify the degree to which a scientific hypothesis is supported by evidence. *Conceptual criticism* deals with more theoretical and metatheoretical problems of science, insofar as the conceptual soundness of an inquiry is questioned, the consistency of a hypothesis with an accepted theory is reviewed and the relevance of evidence in support of a hypothesis is tested.[2] Both types of criticism are key elements of scientific inquiry and scientific progress, and it is important to note that these practices of criticism differ from simple or spontaneous expressions of objection, opposition or discomfort. Scientific criticism should be linked to "recognized avenues for the criticism of evidence, of methods, and of assumptions and reasoning" and to "shared standards that critics can invoke".[3] The knowledge and recognition of the rules (or the ethos, as we refer to this later) of science and scientific criticism is required, including an open-mindedness to critical responses from the scientific community and the appreciation of equal intellectual authority between the members of the scientific community. Scientific criticism addresses problems within science and seeks better or new methodologies, explanations and justifications. However, scientific criticism needs to be explained, justified and methodologically specified.

The close ties between criticism and justification, critique and reasonableness that we currently take for granted are in fact the result of a long historical development that culminated in the idea of a dynamic process of criticism during the Enlightenment. The ancient Greek term *krínein* (to judge, recognize, argue) refers to judgments, mostly in philological or medical contexts. The same applies to the Latin term *criticus*. In the 15th and 16th centuries, the logical and methodological dimensions of criticism were established, but the scientific significance of criticism emerged clearly in the 17th century, where criticism was linked to mathematical science and other methodological changes in scientific practice, such as the *ars critica* in linguistics or a methodological criticism in history.[4] Enlightenment philosophy had a particularly significant impact on the scientific imprint of criticism and its public recognition. Broadly speaking, in the 18th century, critique and criticism becomes an application and condition of reasoning itself. In the entry "Critique" of Denis Diderot's *Encyclopédie* (1751), it is understood as the very activity of reason that subjects all kinds of objects to scrutiny – physical, but also social and moral facts, theories, arguments, interpretations, speculations, opinions – by bringing them before a tribunal of truth: "It would be desirable for a determined and enlightened philosopher to dare to put these judgments, made over the centuries by flattery and interest, before the tribunal of truth".[5] The clearest formulation of the idea of an impartial court of reason, which is always and only committed to the truth, can be found in Immanuel Kant's famous *Gerichtshof der Vernunft*:

> One can regard the critique of pure reason as the true court of justice for all controversies of pure reason; for the critique is not involved in these disputes, which pertain immediately to object, but is rather set the task of determining and judging what is lawful in reason in general in accordance with the principles of its primary institution.

Without this, reason is as it were in the state of nature, and it cannot make its assertions and claims valid or secure them except through war. The critique, on the contrary, which derives all decisions from the ground-rules of its own constitution, whose authority no one can doubt, grants us the peace of a state of law in which we should not conduct our controversy except by 'due process'.[6]

To take Kant's metaphor even further, criticism plays three main roles in the *court of reason* at the same time: The accused, the prosecutor and the judge. The same applies to the critical philosopher. At first, criticism is used to object to a position. Then the criticism itself is discussed. And finally, it acts as the judging institution. As a consequence, criticism can be understood as the application of different aspects of reason and reasoning itself.

It is no coincidence that criticism and the establishment of modern scientific practice developed in parallel. In the 17th and 18th century, the "era of evidence",[7] Aristotelian natural philosophy was replaced by cultures and practices of evidence.[8] At the same time, a self-image of scientific inquiry conducted by scientists as a profession was established and institutionalized, primarily in the new scientific academies.[9] Following Wilda C. Anderson, a thus institutionalized self-image leads away from the Aristotelian ideal of the theorizing philosopher, who contemplates about the world, to the scientist, who is in search of the "speaking voice of nature" in experimental data.[10] Precisely this non-contemplative but hands-on and experimental access to nature influenced the conception of scientific evidence drastically.[11] As a result, experimental practices not only had an impact on the status of evidence within science but also determined the self-conception of scientists.[12]

This short historical excursus illuminates our systematic stance, namely that scientific criticism contains a demanding and normative ideal of proper or just critique of science. Hence, good critics should be familiar with the acknowledged avenues of scientific critique since first and foremost their utterance must be compatible with common scientific norms. Scientific criticism that is bound to fundamental structures in this way cannot develop into a general criticism of science as it would then destroy its own foundation. No matter how critical scientific criticism is, it must follow the institutionalized pathways of science. That does not mean that criticism cannot influence these internal norms by commenting on previous scientific standards and procedures. It can confirm different strategies and establish new standards that future scientists can use as guidelines. But even profound criticism cannot call the institution of science into question as a whole without questioning its own claim to be *scientific* criticism.

Ambivalence and Agnosticism of Evidence

First, we need to address the question of what meanings the term evidence can take on – and here the question of language is important. Second, we examine the relation of scientific criticism to evidence claims because evidence critique and evidence contestation are common and acknowledged ways of showing disapproval in science. This suggests that evidence criticism as one option of scientific criticism could have similarly productive effects for scientific inquiry and progress. One reason, as we aim to show in this section, can be found in the epistemic robustness of evidence-claims. Following this, we argue that conceptually ambivalent meanings and functions of evidence are possible and may even be necessary for scientific criticism and progress, as long as they do not violate the overriding (but still ambiguous) idea of scientific evidence. We hope that the heuristic of *evidence agnosticism* will help to ease the tensions between stability and ambiguity of evidence on the one hand and between references to evidence and contestations of evidence on the other.

Starting with a definition of scientific evidence might initially appear fairly straightforward: The available body of facts, data or information supporting or contradicting a proposition, assumption, hypothesis or theory in science. But on further inspection, it soon becomes obvious – and problematic for the task of providing a definition – that the standards for something to count as evidence and the types of evidence practices differ vastly between specific scientific disciplines.[13] It is therefore not viable to talk about just *one* concept of evidence, or to justify or even understand certain evidence practices without reference to their historical background and institutional embedding. Producing and establishing evidence is related to concrete contexts of production and application. Praxeological perspectives on evidence and the analysis of example cases, as examined in the contributions to this volume, are therefore useful and profitable for interdisciplinary research on evidence.

In addition to the interdisciplinary differences, there are general issues concerning the ontological, epistemic and methodological aspects of evidence, which are negotiated in a variety of philosophical investigations on evidence.[14] These theoretical contributions confirm that there is a persistent ambiguity about the concept of evidence that allows for different semantic interpretations. This might at least partially explain the pluralism of the disciplinary standards, criteria and practices of evidence.

The German lexeme *Evidenz* appears at first glance to be more expansive than the English *evidence*. It not only means proof (*Beweis*) but also refers to epistemic processes of certainty (*Gewissheit*). On the one hand,

Evidenz provides a ground for a belief or conviction and on the other hand the term refers to an obviousness. Furthermore, there is a problem in expressing the distinction between an immediate self-evidence and evidence as a tool for the confirmation or justification of a proposition, hypothesis or theory.[15] This diversity of meanings emerges – especially for us as German-speaking philosophers – from a rich history of the term that has also strongly influenced English-language philosophy.[16] In many insightful discussions with English-speaking colleagues, it seemed this was a purely German phenomenon. In contemporary English usage, the term evidence might appear not to carry this ambivalence and be much clearer than its German cognate. Nevertheless, we cannot agree with this assumption. We note that a sole focus on the linguistic denotative and connotative meaning of a term cannot solve conceptual philosophical problems.

Moreover, we also see ambivalences in the english term evidence. Let us assume that the meaning of evidence is sufficiently covered in "reasons or proofs for a conviction that something is true". What could we thereby say about the content or cause of evidence? We could say that it is the object or the phenomenon itself that provides for our conviction without our intervention. Here the path to an attribution of obviousness would be very short, for how should something of itself provide conviction if it is not obvious? Or is it only the process of cognition that turns an object or a phenomenon into evidence? Here, however, one would have to do some work of persuasion in order to answer the question "What makes something evidence?" For the assumption that it is a process of cognition that makes something evidence is circular: The reason for the truth of my conviction is the conviction itself. Even though there may be philosophical exceptions where we can speak of evidence, for example, if there is a minimal change in probability values for truth, at first glance, it is difficult to assume that there can be evidence that has no link to clarity, immediacy or obviousness. But that does not mean that evidence cannot be criticized because it is unquestionably true from certainty. As we have also mentioned, criticism of evidence is a common part of scientific practice. But such interventions mostly refer to concrete evidence themselves and as such do not question the general concept of insightful evidence. Here one might counter that we are unjustifiably distinguishing between concrete evidence and a general concept and that our philosophical problem applies only to the latter. But the singular meaning of evidence cannot answer the question of what exactly makes a phenomenon evidence, nor why it is justified to treat something as evidence. The starting point for our questions about evidence, then, are not linguistic equivocations but conceptual ambivalences. Besides the linguistic peculiarities, the philosophical concept of evidence exhibits persistent ambiguities – at least, if we continue to ask about the philosophical content of the term.

But let us take a closer look at evidence and the possible ambivalences of the term by considering its history. Etymologically, evidence derives from two similar yet different sources: *Evidentia* (obvious) and *energeia* (explicitly visible).[17] The term evidence currently has both an epistemological and an aesthetic meaning: Something that is evident, can be understood immediately, is true, justified or existent beyond doubt. At the same time, evidence is particularly accessible when it is equated with sensual insights or immediate perception.[18] On the one hand, evidence corresponds to a special *form of clarity* and therefore is often used synonymously with concepts such as self-evidence, distinctness, comprehensibility, unambiguousness and obviousness. On the other hand, by systematically providing justification for beliefs and convictions, evidence is connected to a *mode of knowledge* accompanied by demands for certainty, undeniability, objectivity and authenticity.[19]

The controversy over the proper meaning of evidence can also be seen in current philosophical debates. From an epistemological viewpoint, evidence plays a constitutive role in *what we can know with certainty* – in contrast to mere belief, thinking and subjectively holding to be true.[20] As Wolfgang Stegmüller has put it concisely: "All of our arguing, deriving, refuting, reviewing is a continuous plea to evidence".[21] Such pleas are usually put forward with the expectation that it is possible to justify our theoretical and practical beliefs in principle. Throughout the history of philosophy, this demand for reasonableness and rationality has been presented as a connection between *epistéme* (knowledge) and *evidentia* (certainty), which refers to a direct independent source of knowledge that goes along with an irrefutable claim to truth. In epistemology, *evidence* is thus related to the concepts of justified belief and knowledge: "One thing is 'evidence' for another just in case the first tends to enhance the reasonableness or justification of the second".[22] Candidates for such epistemological evidence differ within different philosophical theories and also from historical epoch to epoch. Thus, a systematic overview shows a broad range of positions from empiricist concepts of evidence (e.g. sense data, direct perception, sensual intuitions, protocol sentences) to rationalist positions (e.g. direct intuition, intuitive insight, staunch belief, innate ideas, necessary ideas, law of thought).[23] Although philosophical epistemology may not help to define evidence, it does help to illustrate a complex relationship between evidence and other basic epistemological categories, such as insight, knowledge, truth, justification, which are very powerful in our image of how to reason about the world.

Empirical and inductive science on the one hand and evidence practices on the other, despite their potential to make explanatory statements about the world, face a conceptual problem: There is necessarily a gap between evidence and general scientific statements. Evidence does not by itself determine which scientific statements are true or not. In a case of conflict between competing statements, one has to decide whether the available evidence is sufficient or not.[24] How can such a decision be justified and legitimated?

Various philosophers see *values* coming into play here. Such values can be both ethical and science-specific, especially since a uniform separation between these categories is not tenable. Feminist scholars call for plural and diverse science in order to be able to close gaps by deploying a variety of viewpoints. Moreover, such values can have a direct impact, i.e. they directly influence decisions made by scientists. But they can also have an indirect effect by being taken into account in the evaluation of justifications. For the philosopher Heather E. Douglas values in their indirect role:

> help determine whether the evidence or reasons one has for one's choice are sufficient [...] but do not contribute to the evidence or reasons themselves. In the indirect role, values assess whether evidence is sufficient by examining whether the uncertainty remaining is acceptable. A scientist would assess whether a level of uncertainty is acceptable by considering what the possible consequences of error would be [...] by using values to weigh those consequences.[25]

If evidence is therefore insufficient and, as we have argued above, there are other conceptual problems as well, should we better refrain from a philosophical approach to evidence altogether? Such defeatism would not do justice to the fact that evidence, despite its inherent ontological and epistemological problems, is a central and practice-guiding concept in science. Therefore, although the philosophical debates about the existence and validity of evidence do demonstrate the complexity of the concept, these notional confusions may still have no destructive effect for evidence-claims and evidence-practices in science. Remarkably, quite a few philosophers of the 20th century proposed that the conceptual ambiguity of evidence should be interpreted as a general impossibility of citing proofs (or rather evidence) for the existence of evidence.[26] After all, arguing for evidence requires embracing the very concept that is being argued for. From this perspective, the reasoning about and the arguing for evidence are ineluctably circular. This would mean that we could not give a convincing answer to the question of what evidence is and that we (as philosophers of science) should stop trying. But this is not meant as a farewell to the concept of evidence or to a philosophical investigation of evidence. Rather than defeatism we propose agnosticism as a philosophical approach to dealing with the problems of evidence.

In this regard, we would like to suggest a heuristic to make the ambiguity of evidence comprehensible without undermining the epistemic and scientific relevance of this concept, nor denying its persistent epistemic problems or neglecting its scientific functions. The heuristic figure of an *agnosticism toward evidence* regards it from the outset as a pluralistic, ambiguous and highly contested concept with inherent contingencies – but still as a concept with an important normative function in terms of the authority and recognition of science. Accordingly, evidence is an abstract, qualitative and evaluative notion with a variety of meanings. Such essentially contested concepts "inevitably involve endless disputes about their proper uses on the part of their users" which "cannot be settled by appeal to, linguistic usage, or the canons of logic alone".[27] Thus, evidence can still be regarded as one of the basic concepts of philosophy, because it addresses the question of how knowledge is possible or beliefs are justified. It illustrates the focus and main motives of epistemology and the philosophy of science. This makes evidence part of the conceptual inventory of the "disenchantment of the world" by processes of objectification, rationalization and mechanization brought about by, among other things, modern science.[28]

But at the same time, we can face up to the presented epistemic and ontological problems of evidence by shifting from epistemology to practical philosophy of science. This entails a change of perspective: Evidence is no longer just seen as a crucial concept in epistemology and scientific inquiry, but as a matter of collective convention or even individual belief with no rational foundation. Wolfgang Stegmüller, who invented the idea of *Evidenz-Agnostizismus*, claims that scientific practice itself can neither justify nor produce evidence by itself. This means that one needs criteria outside of the purely epistemic practice of science to label or understand something as evidence (criteria such as values, heuristics or normative agreements on thresholds). Hence, for Stegmüller, every single concept and scientific use of evidence should be seen as the result of a *vorrationale Urentscheidung* (pre-rational original decision): "[I]n scientific practice we cannot do without the assumption of evidence or insight. But the fact that we cannot do otherwise is no justification".[29] Stegmüller points out that a justification of evidence could not rely on rational insights, "pure thinking" or decisions absent of reason. Instead, we have to assume a *practical belief in evidence* ("Evidenz-Glaube"), which is not justifiable and therefore not based on a rational foundation but is normatively effective nevertheless.[30] In other words, even if criticism of individual evidence is possible (usually by referring to other evidence), it is important to consider belief in the general concept of evidence as a basic decision. Criticism of evidence does not mean rejection of all evidence. The heuristic of agnosticism allows us to establish a new starting point for investigating why scientists are justified in believing in evidence, although there is no foundational justification of evidence. Hence, we can ask for the *normative requirements, functions and consequences* of scientific evidence, as we now do in the next section.

Functions of Evidence within the Scientific Ethos

One can reasonably assume that good scientific practice yields reliable knowledge about the world due to the standards and requirements that are necessary for a scientific outcome. But what are the normative reasons and conditions for good science? In this section, we address this question by examining the functions of evidence in the normative setting of science – the *scientific ethos*. In this way, it should become clear that the normative function of evidence has important implications for the general understanding of science and scientific criticism.

"What aims should this discipline pursue?" and "What should science look like?" are not merely questions referring to scientific methods or the validation of facts, but to normative aspects that are highly relevant for the status of science within the modern knowledge society. Ethical questions in current debates on science often revolve around the moral integrity of scientists, the ethical appropriateness of research projects and of their potentially harmful consequences.[31] But this way of looking at ethics in science offers a different, or more limited, understanding than what the idea of *ethos* actually entails.[32] According to Robert Merton's definition of "scientific ethos", it includes epistemological principles, methodological advice, values and virtues, that together guide science and scientists, behavioral norms and institutional legitimizations:

> The ethos of science is that affectively toned complex of values and norms which is held to be binding in the man of science. The norms are expressed in the form of prescriptions, proscriptions, preferences, and permissions. They are legitimatized in terms of institutional values. These imperatives, transmitted by precept and example and reenforced by sanctions are in varying degrees internalized by the scientist, thus fashioning his scientific conscience or, if one prefers the latter-day phrase, his superego.[33]

This concept of modern, collaborative science defines it not only as a systematically organized body of knowledge, but also as a distinct social process, influenced by communicative acts and collective sanctions, internal normative agreements and by a broader historical and social context. Consequently, social, political and ethical norms and values are considered essential constituents of science rather than external factors for evaluating the possible applications of science. In the end, science is neither value-neutral nor self-sufficient, but value-laden.[34] And as Merton emphasizes, such values, imperatives and principles constitute and legitimize norms of good scientific practice.[35] In this regard, science works just like any other institution, which extends self-organization by adding normative rules, institutionalized settings and an instrumental context of satisfying interests. Institutions thus represent normativity as a result of a "process by which social processes or structures come to take on a rule-like status in social thought and action".[36] Regarding scientific ethos, there are at least three levels. First, scientists collectively create binding norms for their profession. Second, these norms have retroactive effects on the self-identification of scientists. Third, norms are only accepted if they are institutionalized. Since scientists are also members of society, general social values influence scientific norms. Normativity and values are – occasional claims to the contrary notwithstanding – inherent in science. They define and structure what scientists should or should not do. By defining the settings of scientific (and social) practice and at the same time requiring compliance from scientists, evidence is part of the scientific ethos.

To sum up: Scientific ethos is influenced by the institutionalized norms and values which are and have to be accepted by the majority of members of the scientific community. Hence, by stressing the importance of values and norms in science, the institutionalization of a scientific ethos is one key factor in the realization of science and its constitutive standards. Moreover, the scientific ethos serves as a guideline for the identity and self-confidence of scientists by creating the normative directives of science and the regulations of formal operations – and also the guiding ideas for understanding science as a profession. In addition to rational norms for the institutional search for truth, the ethos also contains social norms, including imperatives, rules of justification and proceeding, activity orientations, role expectations, virtues, reward and criticism systems.[37] The legitimatization of values and norms succeeds insofar scientists are willing to comply with them, even if they seem to be restrictive for their practice. Hence, the scientific ethos is not only the foundation but also the result of a jointly organized community, which leads to regulations, codices and guidelines that are created by scientific communities, universities and external funding sources.[38]

What we have said on scientific criticism so far could be integrated into the concept of scientific ethos. Actually, one of the Mertonian norms is the principle of organized skepticism. This norm implies a reciprocal willingness on the part of the scientists to be open to criticism and possible adjustments. Such attitudes effectively serve as a stabilizing factor in science because they enable the critical participation of others in a collaborative setting. For Helen Longino, scientific criticism requires pluralistic views and approaches and can lead to consensual and secured knowledge. Criticism can lead to adjustments of a theory or other parts of scientific inquiry. It is part of a process-based and open science and may result in a *Gestaltwechsel* in the sense of a self-correcting science. Scientific criticism entails not only following rules, but also has the potential to influence and create epistemological and normative settings. In other words: Criticism defines and changes the knowledge and the environment in which science is embedded.

Now in what sense is evidence to be understood as part of the scientific ethos? We would once again highlight the close relationship between evidence and critique, but in a different way. For modern science, the creation and application of evidence are tied to specific production processes and practical uses. Pleas to evidence can be found during the process of surveying, evaluating and justifying scientific data. For example, the 18th century chemist Antoine Lavoisier ends his essay "Reflections on Phlogiston" by appealing to empirical evidence rather than speculative ideas.[39] During the corona pandemic, pleas are also increasingly encountered in the political sphere, particularly when it comes to orienting political decisions toward scientific evidence.[40] Scientists working in the natural sciences, humanities and social sciences describe how they understand the term evidence quite differently to one another.[41] Examples range from the interpretation of evidence as empirical facts, as part of methodological practice or as a narrative strategy in science communication. Notably, evidence practices are also considered normative instructions on how science should be carried out. In this regard, evidence serves as an epistemological authority with *legitimizing power*. Evidence is rhetorically and epistemically connected with the criteria that determine the quality of science. Debates about basic principles can be bypassed by referring to evidence. Evidence can be directly anchored in the values of a scientific ethos. Evidence is a key factor in every scientific practice.[42]

We have argued so far that despite its interdisciplinary plurality, conceptual ambiguity and various epistemic, methodological and normative functions, the concept of evidence is crucial for science and its distinctive interactions and methods.[43] One of these methods is scientific criticism, whereby the wide range of meanings of evidence from epistemic claims to truth to indicators of usefulness makes it possible to criticize numerous aspects of scientific inquiry and outcomes. Nevertheless, existing evidence and criticism can be in both productive and inhibiting tension with each other. Following philosopher of science Thomas S. Kuhn, when "nature [...] fail[s] to conform entirely to expectation", anomalies arise.[44] In a way, the rise of anomalies is par for the course in science. But under some circumstances, anomalies can develop into a fundamental crisis or extraordinary new science.[45] Both the tension caused by everyday doubts about specific evidence and the revolutionary move toward radical new scientific practice are ultimately inherent in the project of modern science.[46] One reason why science is not entirely called into question by such criticism or even crisis lies, among other things, in the normative power the scientific ethos yields, of which the notion of evidence and evidence contestation is an integral part. Evidence, due to its ambivalent properties, provides possibilities for dynamic changes in scientific practice in a special way, as long as the principles of the normative setting are not questioned in their entirety. Evidence is thus *quid facti* both the result and equally the starting point for new scientific practices. To answer the question *quid iuris*, that is, the question of the justified appeal to evidence by scientists, the perspective of agnosticism offers a starting point. The *pre-rational decision* for evidence is justified because the normative dimension of the scientific ethos generates, stabilizes, harmonizes and contests different registers and functions of evidence. At the same time, evidence is a powerful plea within science. Evidence is therefore a weighty reference in a scientific environment, which in turn is based precisely on such evidence. Evidence not only refers to the correctness of data or an appropriate procedure, but also to shared norms and thus to an institutionalized self-image of science.

On Contingencies, Crisis and the Sources of Stability in Science

Organized skepticism and other norms of the scientific ethos, such as disinterestedness and universalism, are important sources for the public image of science as reliant and trustworthy. Hereby public trust is manifested in two ways: As general confidence in the self-regulating and self-controlling norms and practices of the scientific community and as a more specific epistemic trust in the validity and reliability of scientific expertise.[47] The reverse is also true: Violations of the scientific ethos, as they have become known to a broader public through prominent cases of scientific fraud, undermine society's trust in the sciences dramatically.[48] This means that the interaction between evidence and criticism manifested in the scientific ethos plays an important role in establishing scientifically produced knowledge as worth knowing, valuable and reliable both internally, i.e. within the scientific community and externally, i.e. for society as a whole. The scientific ethos regulates scientific criticism and thus allows disapproval and dissent about evidence to become fruitful. But what about the possible contingencies that can arise from a pluralistic understanding of science, i.e. the existence and acceptance of diverse critical positions and interests?

Contingencies normally refer to the possibility of something being different or that future events may or may not happen. This means that phenomenon *A* occurs, but it would also have been possible that *A* did not occur. For scientific predictions, contingencies seem to be a problem, because one cannot be absolutely sure that the prediction is correct, and this might conflict with the whole task of scientific inquiry. By providing alternative explanations and, at the same time, challenging the necessities of scientific insights, criticism can lead to contingencies. For science, this is a fundamental feature rather than a problem. But criticism is not the only factor that leads to contingencies: The ambiguity and openness of evidence can have the same effect. Again, from a philosophical perspective, this is not an issue: The ambiguity of evidence, which might be considered to be problematic from an analytic perspective, provides scientific practice with the means to produce knowledge and to address pluralistic insights.

The fact that evidence is not limited to a narrow range of meanings makes it an essentially contested concept – primarily for philosophers. What is important for us here is that the reasons why evidence can be considered a contested concept in the first place are to be understood as a key feature and practical value of evidence-claims in scientific processes. It is conceptual ambiguity that enables scientific progress and constructive communication by providing the possibility to dynamically adjust scientific practice. Due to its potential influence on general aspects of scientific practice (either intended or unintended), it guarantees the openness of scientific procedure without being arbitrary. It is *ambiguous but not arbitrary* because evidence itself ties scientific practice back to the scientific ethos. As evidence has an influence on the scientific ethos and can help determine the application of sanctions, it is used to demand compliance with standards. Evidence thus defines the standards and settings of scientific practice and therefore has a normative dimension. This is what we call its determining character as a powerful plea.

At the same time, evidence does not refer to a single phenomenon. Or to put it differently: Obviously there are different scientific understandings of the concept of evidence. Therefore, the question as to why something can be considered evidence can and probably must be answered differently. And these answers can and should not only refer to epistemic aspects of scientific knowledge but also to the social, cultural, procedural, contextual, institutional conditions of science, which are integrated into the socio-epistemic normative structure of the scientific ethos. Science is, after all, a social phenomenon and a cultural practice.[49] Evidence is a constitutive part of a coherent social network of science in which references and legitimations among data, methods, reasoning, norms etc. take place. Consequently, the historians of science Moritz Epple and Claus Zittel note that "modern sciences, which have grown to social networks of substantial size, are not only shaped by the social and cultural condition under which they exist but also generate their own pattern of social and cultural behavior that tie in with the patterns of social life at large".[50] Karin Zachmann and Sarah Ehlers emphasize the dynamic negotiation process. This means for disputes about what counts as evidence, too, that they take place in a "web of competing, parallel, or constructive successive actions".[51]

As we have seen, in general, the plurality and ambiguity of evidence and the contingencies that they may cause are not a problem for the plea of scientific evidence. Criticism is one way to contest evidence and can lead to contingencies which are just as much a part of science. Criticism is one key factor, firmly anchored in science, establishing new and old scientific standards for quality, reliability, objectivity and trustworthiness by connecting epistemic arguments while taking into account social practices and norms. Even if these criteria are not completely detached from societal developments (in parts they are even determined by them), a certain independence of science develops, which is tolerated by other social subsystems such as constitutional law, which may legally guarantee the autonomy of science and academic freedom.[52] Criticism and contingencies can even stabilize science as long as scientists are willing to reflect on their own convictions and to acknowledge the possibility that things could be different. In our view, the ambivalent and plural functions and approaches of evidence offer precisely this scope for action.

Obviously, criticism, contingencies and crises can act in a more destructive way, so that they destabilize science and the public trust in scientific knowledge. As we have seen during the COVID-19 pandemic, not all participants in the public discourse acknowledge that evidence contestation is a key feature of science and that self-correction is not only necessary, but also progressive. Conspiracy theorists and general critics of scientific institutions, for example, treated corrections of research as an indication of the untrustworthiness and relativity of science, rather than understanding them as a mechanism of self-correcting science, which actually makes science more trustworthy, especially in times of crisis.[53] The idea of "alternative facts", to take a prominent example, is regressive with regard to the normative conditions of science criticism. They are completely opposed to the potential of being considered a fact at all. To be sure, this does not imply that there may be reasonable alternative opinions about the interpretation of given facts or evidence. General criticism of science typically ignores scientific standards and ethos-norms. This form of skepticism will normally not have productive effects for science – in most cases, it does not even intend to anyway. Therefore, it is hardly possible for such objections against science, in which evidence is highly valued, to be accepted as scientifically relevant criticism of science. Factually, the "argumentation" of skeptics or deniers of science is usually not oriented to facts or evidence at all, but to an underlying political agenda. Maybe the proper way to deal with such hidden political disputes about the impact and worth of scientific knowledge for politics should be political, too.

Given the leading subject of this volume, we conclude that such general skepticism is not open to contingencies, because it claims to represent an exclusive truth that does not allow for the possibility of things being different, including one's own background assumptions.[54] Thus, general skepticism about science does not fulfill the conditions of a principled openness to criticism as a crucial part of the scientific ethos.

Notes

1 The trust in science rose globally. On average, the number of people who said they trusted science a lot increased by ten percentage points. See for example the study from Wellcome Trust published in 2020: *Wellcome Global Monitor. How Covid-19 Affected People's Lives and Their Views About Science* (London, 2020).

2 Helen E. Longino, *Science as Social Knowledge. Values and Objectivity in Scientific Inquiry* (Princeton: Princeton Univ. Press, 1990), 71–72.

3 Longino, *Science as Social Knowledge*, 76–78. One can deny that Longino makes general statements about criticism with these criteria, since she refers to a transformative dimension of critical discourse. However, this transformation within conceptual criticism is e.g. an expression of a response to criticism through modification of fundamental background assumptions.

4 See for example René Descartes, *Discours de la Méthode Pour bien conduire sa raison, et chercher la vérité dans les sciences* (1637) or Jean Bodins, *Methodus ad facilem historiarum cognitionem* (1535). For the etymology and history of criticism/critique, see Douglas Harper, "Etymology of Criticism", Online Etymology Dictionary, accessed February 7, 2022, https://www.etymonline.com/word/criticism; Claus von Bormann, Giorgio Tonelli and Helmut Holzhey, "Kritik", in *Historisches Wörterbuch der Philosophie online*, eds. Joachim Ritter, Karlfried Gründer and Gottfried Gabriel (Basel: Schwabe Verlag, 2017), 1249–1261.

5 "C'est-là qu'il seroit à souhaiter qu'un philosophe aussi ferme qu'éclairé, osât appeller au tribunal de la verité, des jugemens que la flatterie et l'intérêt ont prononcés dans tout les siècles". Quoted from Bormann, "Kritik". See also Elif Özmen, "Wahrheit und Kritik. Über die Tugenden der Demokratie", *Studia philosophica* 74 (2015): 57–73.

6 Immanuel Kant, *Critique of Pure Reason*, translated by Paul Guyer (Cambridge: Cambridge Univ. Press, 1998), B 779.

7 Rüdiger Campe, "Epoche der Evidenz. Knoten in einem terminologischen Netzwerk zwischen Descartes und Kant", in *Intellektuelle Anschauung. Figurationen von Evidenz zwischen Kunst und Wissen*, eds. Sibylle Peters and Martin J. Schäfer (Bielefeld: transcript, 2006), 26–27.

8 Hole Rößler, *Die Kunst des Augenscheins. Praktiken der Evidenz im 17. Jahrhundert* (Wien: Lit, 2012), 1.

9 See for example Roger Hahn, *The Anatomy of a Scientific Institution: The Paris Academy of Sciences, 1666–1803* (Berkeley: Univ. of California Press, 1971); Frederic L. Holmes, *Eighteenth-Century Chemistry as an Investigative Enterprise: Five Lectures Delivered at the International Summer School in History of Science, Bologna* (Berkeley: Office for History of Science and Technology, 1989).

10 Wilda C. Anderson, "The Rhetoric of Scientific Language: An Example from Lavoisier", *Modern Language Notes* 96, no. 4 (1981): 747.

11 The "experimental exploration of material processes" leads to the standardization of instruments and to debates about the role of instruments in reproducing nature. Holmes, *Eighteenth-Century Chemistry*, 20. See also: Willem Hackmann, "Scientific Instruments: Models of Brass and Aids to Discovery", in *The Uses of Experiment: Studies in the Natural Sciences*, eds. David Gooding, Trevor Pinch and Simon Schaffer (Cambridge: Cambridge Univ. Press, 1999), 40–41; Marco Beretta, "Between the Workshop and the Laboratory: Lavoisier's Network of Instrument Makers", *Osiris* 29, no. 1 (2014): 197–214.

12 This is especially true due to the mathematical, instrumental and methodological ideal of the understanding of chemistry at the end of the 18th century. See Mi Gyung Kim, *Affinity, That Elusive Dream: A Genealogy of the Chemical Revolution* (Cambridge: MIT Press, 2003), 442–443 and Martin Carrier, "Antoine L. Lavoisier und die Chemische Revolution", in *Das bunte Gewand der Theorie. Vierzehn Begegnungen mit philosophierenden Forschern*, ed. Astrid Schwarz (Freiburg: Alber, 2009), 36–38.

The changes that took place there are still controversially discussed in the history of science today. However, we see chemical science in this period as a paradigmatic example of how evidential practices and claims changed and how a changing bourgeois ideology of the Enlightenment interacted with conceptions of science and thus had a lasting impact on scientists' self-perception. This case shows how competing scientific theories can refer to different ideas of evidence and how the claims and tasks of science are negotiated within society. First results see in Daniel Füger, "Die Umwälzung der wissenschaftlichen und zivilisierten Welt. Zum Verhältnis von Evidenz und Normativität in der frühen Chemiewissenschaft", in *Philosophie zwischen Sein und Sollen: Normative Theorie und empirische Forschung im Spannungsfeld*, eds. Alexander Bauer and Malte Meyerhuber (Berlin: De Gruyter, 2019), 159–178.

13 This also became clear at the international and interdisciplinary conference of the DFG Research Group 2448 "Practicing Evidence – Evidencing Practice" in February 2020. Daniel Füger, "Conference Report: Practicing Evidence – Evidencing Practice. How is (Scientific) Knowledge Validated, Valued and Contested? Munich 2020", *H-Soz-Kult*, accessed April 27, 2020, https://www.hsozkult.de/conferencereport/id/tagungsberichte-8741.

14 For a short introduction see Thomas Kelly. "Evidence", in *The Stanford Encyclopedia of Philosophy, Winter 2016*, ed. Edward N. Zalta (Stanford, CA: Metaphysics Research Lab, Stanford Univ., 2020), https://plato.stanford.edu/archives/win2016/entries/evidence/ and Victor DiFate, "Evidence", *Internet Encyclopedia of Philosophy* (2016), accessed January 18, 2022, https://iep.utm.edu/evidence/.

15 Hans J. Sandkühler, "Kritik der Evidenz", in *Wissen, was wirkt. Kritik evidenzbasierter Pädagogik*, eds. Johannes Bellmann and Thomas Müller (Wiesbaden: VS Verlag für Sozialwissenschaften, 2011), 34.

16 Perhaps the most prominent example would be Edmund Husserl. Thomas Mautner, "evidence", in *A Dictionary of Philosophy*, ed. Thomas Mautner (Cambridge: Blackwell Publishers, 1996), 139.

17 Peter Achinstein, *The Book of Evidence* (Oxford: Oxford Univ. Press, 2003); Gernot Kamecke, "Spiele mit den Worten, aber wissen was richtig ist! Zum Problem der Evidenz in der Sprachphilosophie", in *Sehnsucht nach Evidenz*, eds. Karin Harrasser, Helmut Lethen and Elisabeth Timm (Bielefeld: transcript, 2009), 11.

18 Jan-Dirk Müller, "Evidentia und Medialität. Zur Ausdifferenzierung von Evidenz in der frühen Neuzeit", in *Auf die Wirklichkeit zeigen. Zum Problem der Evidenz in den Kulturwissenschaften. Ein Reader*, eds. Helmut Lethen, Ludwig Jäger and Albrecht Koschorke (Frankfurt am Main: Campus Verlag, 2015), 266–271.

19 Based on these two main meanings, it is possible to sketch a semantic map of the technical term. In our research project "Evidence for Use – Evidence for Us. About the Legitimizing Function of Evidence for Science", we use a cartographic method to present the concepts, meanings and functions of evidence, and to relate them to each other. A map of this kind does not simply represent – it orders and models the different concepts of evidence. Our evidence map is thus neither complete, finished, nor "correct". It rather shows not only the branching, links and connections, but also the differences and contradictions between the core meanings of evidence. Throughout the project, we use 18th-century chemistry as a foil to exemplify the development of evidence and related practices during the Enlightenment. More information at: www.uni-giessen.de/fbz/fb04/institute/philosophie/praktphil/EvidenceForUse_DfG.

20 See for example Timothy Williamson, "Knowledge as Evidence", *Mind* 106 (1997): 717–742; Earl Conee and Richard Feldman, *Evidentialism* (Oxford: Oxford Univ. Press, 2004); Susan Haack, *Evidence and Inquiry. A Pragmatist Reconstruction of Epistemology* (Amherst: Prometheus Books, 2009); Thomas Kelly, "Evidence: Fundamental Concepts and the Phenomenal Conception", *Philosophy Compass* 3, no. 5 (2008): 933–955.

21 "All unser Argumentieren, Ableiten, Widerlegen, Überprüfen ist ein ununterbrochener Appell an Evidenzen". Stegmüller, *Metaphysik, 168* – all translations by the authors.

22 Jaegwon Kim, "What Is Naturalized Epistemology?", *Philosophical Perspectives*, no. 2 (1988): 390–391.

23 Wilhelm Halbfass and Klaus Held, "Evidenz", in *Historisches Wörterbuch der Philosophie online*, eds. Joachim Ritter, Karlfried Gründer and Gottfried Gabriel (Basel: Schwabe Verlag, 2017).

24 Heather E. Douglas, "Values in Science", in *The Oxford Handbook of Philosophy of Science*, ed. Paul Humphrey (New York: Oxford Univ. Press, 2019), 609–630. The fundamental debate about empirical science and its validity is often framed in terms of the inductive gap, underdetermination or supervenience of empirical practice to theory. However, this cannot be discussed further in this chapter.

25 Douglas, "Values in Science", 618.

26 This applies especially to the position of (logical) positivists, such as Moritz Schlick or Rudolf Carnap. But also adherents of critical realism, above all Nicolai Hartmann, sharply criticize certain uses of evidence.

27 Walter B. Gallie, "Essentially Contested Concepts", *Proceedings of the Aristotelian Society* 56 (1956): 169.

28 Max Weber, *Wissenschaft als Beruf. Max Weber-Studienausgabe* (Tübingen: Mohr Siebeck, 1994).

29 Wolfgang Stegmüller, *Metaphysik, Skepsis, Wissenschaft* (Berlin: Springer, 1969), 168: "[I]m wissenschaftlichen Verkehr [können wir] ohne die Annahme einer Evidenz oder Einsicht nicht auskommen […]. Daß wir nicht anders tun können als so, ist aber keine Rechtfertigung".

30 Stegmüller, *Metaphysik*, 2.

31 See for example Carl Mitcham, *Encyclopedia of Science, Technology, and Ethics* (Detroit: Macmillan Reference, 2005); Ana S. Iltis and Douglas MacKay, eds., *Research Ethics* (Oxford: Oxford Handbooks Online, 2020). There are projects concerning ethical aspects in a broader sense and these draw attention to how philosophers, politicians and scientists can benefit from a fundamental consideration of ethics. For example, Donna Haraway, "Situated Knowledges: The Science Question in Feminism and the Privilege of Partial Perspective", *Feminist Studies* 14, no. 3 (1988): 575–599; Elisabeth A. Lloyd, "Feminism As Method: What Scientists Get That Philosophers Don't", *Philosophical Topics* 23, no. 2 (1995): 189–220; Nancy Tuana, "Leading with Ethics, Aiming for Policy: New Opportunities for Philosophy of Science", *Synthese* 177, no. 3 (2010): 471–492.

32 Elif Özmen, "Wissenschaft. Freiheit. Verantwortung. Über Ethik und Ethos der freien Wissenschaft und Forschung", *Ordnung der Wissenschaft* 2 (2015): 65–72.

33 Robert K. Merton, "The Normative Structure of Science", in *The Sociology of Science: Theoretical and Empirical Investigations*, ed. Norman W. Storer (Chicago: Univ. of Chicago Press, 1973), 268–269.

34 See for example Heather E. Douglas, *Science, Policy, and the Value-Free Ideal* (Pittsburgh: Univ. of Pittsburgh Press, 2009); Elif Özmen, "Die

normativen Grundlagen der Wissenschaftsfreiheit", in *Freiheit der Wissenschaft. Beiträge zu ihrer Bedeutung, Normativität und Funktion*, ed. Friedemann Voigt (Berlin: De Gruyter, 2012), 111–132; Gerhard Schurz and Martin Carrier, eds., *Werte in den Wissenschaften. Neue Ansätze zum Werturteilsstreit* (Berlin: Suhrkamp, 2013).

35 Most prominently the CUDOS principles. These are four institutionalized imperatives that have both a technical and a moral dimension: Communism, universalism, disinterestedness, organized skepticism (Merton, "Normative Structure", 270–278).

36 Jeffrey W. Lucas and Michael J. Lovaglia, "Legitimation and Institutionalization as Trust-Building: Reducing Resistance to Power and Influence in Organizations", *Social Psychology of the Workplace* 23 (2006): 232.

37 Özmen, "Normative Grundlagen", 114.

38 See for example "Proposals for Safeguarding Good Scientific Practice", German Research Foundation (DFG, accessed September 14, 2021, http://www.dfg.de/download/pdf/dfg_im_profil/reden_stellungnahmen/download/empfehlung_wiss_praxis_1310.pdf; "Guidelines for Safeguarding Good Research Practice. Code of Conduct", German Research Foundation (DFG), accessed September 14, 2021, https://www.dfg.de/download/pdf/foerderung/rechtliche_rahmenbedingungen/gute_wissenschaftliche_praxis/kodex_gwp_en.pdf; "Guidelines and Rules of the Max Planck Society on a Responsible Approach to Freedom of Research and Research Risks", Max Planck Society (MPG), accessed September 14, 2021, https://www.mpg.de/197392/researchFreedomRisks.pdf; "The European Code of Conduct for Research Integrity", European Science Foundation/All European Academies (ESF/ALLEA), accessed September 14, 2021, https://www.allea.org/wp-content/uploads/2017/05/ALLEA-European-Code-of-Conduct-for-Research-Integrity-2017.pdf.

39 Antoine Lavoisier, "Lavoisier's 'Reflection on Phlogiston' II: On the Nature of Heat", translated by Nicholas W. Best *Foundations of Chemistry* 18, no. 1 (2016): 12.

40 Evidence was used both by advocates of stricter protective measures and by critics. See for example Philipp May, "Lockerung der Corona-Maßnahmen. Thorsten Lehr 'Blindflug ohne große Evidenz,'" *Deutschlandfunk* March 4, 2021, https://www.deutschlandfunk.de/lockerung-der-corona-massnahmen-thorsten-lehr-blindflug-100.html; "COVID-19: Wo ist die Evidenz?", Deutsches Netzwerk Evidenzbasierte Medizin e.V., accessed February 22, 2022, https://www.ebm-netzwerk.de/de/veroeffentlichungen/pdf/stn-20200903-covid19-update.pdf.

41 See for example the study of Eva-Maria Engelen, Christian Fleischhack, Giovanni C. Galizia and Katharina Landfester, eds., *Heureka – Evidenzkriterien in den Wissenschaften. Ein Kompendium für den interdisziplinären Gebrauch* (Heidelberg: Spektrum, 2010). Members of the *Junge Akademie* from different scientific disciplines were asked what they understood by the terms evidence, proof and thesis.

42 James Chandler, Arnold Davidson and Harry Harootunian, eds., *Questions of Evidence. Proof, Practice, and Persuasion Across the Disciplines* (Chicago: Univ. of Chicago Press Journals, 1994); Gary Smith and Matthias Kroß, eds., *Die ungewisse Evidenz. Für eine Kulturgeschichte des Beweises* (Berlin: Einstein Bücher, 1998); William Twining and Ian Hampsher-Monk, *Evidence and Inference in History and Law. Interdisciplinary Dialogues* (Evanston: Northwestern Univ. Press, 2003); Philip Dawid, William Twining and Mimi Vasilaki, eds., *Evidence, Inference and Enquiry* (Oxford, New York: British Academy, 2011).

43 Karin Zachmann and Sarah Ehlers, eds., *Wissen und Begründen. Evidenz als umkämpfte Ressource in der Wissensgesellschaft* (Baden-Baden: Nomos, 2019).
44 Thomas S. Kuhn, "Historical Structure of Scientific Discovery", *Science* 136 (1962): 762.
45 Thomas S. Kuhn, *The Structure of Scientific Revolution* (Chicago: Univ. of Chicago Press, 1970), 82.
46 At this point, we must mention that the term *scientific revolution* is controversial. For the example of the chemical revolution, see for example: William R. Newman, *Atoms and Alchemy: Chymistry and the Experimental Origins of the Scientific Revolution* (Chicago: Univ. of Chicago Press, 2006), 2.
47 Both dimensions of trust seem to be connected with an ideal of autonomy of science (and scientists) and academic freedom. See for example Torsten Wilholt, "Epistemic Trust in Science", *The British Journal for the Philosophy of Science* 64, no. 2 (2013), 233–253; Donald A. Downs and Chris W. Surprenant, eds., *The Value and Limits of Academic Speech. Philosophical, Political, and Legal Perspectives* (New York: Routledge, 2018).
48 Examples range from individual misconduct of scientists to studies that have become discredited due to questionable political and economic connections. See for example Stuart Ritchie, *Science Fictions. How Fraud, Bias, Negligence, and Hype Undermine the Search for Truth* (New York: Metropolitan Books, 2015); Stefan T. Siegel and Martin Daumiller, eds., *Wissenschaft und Wahrheit: Ursachen, Folgen und Prävention wissenschaftlichen Fehlverhaltens* (Leverkusen: Budrich, 2020). During the Corona pandemic, in particular, some studies came into disrepute, e.g., studies indicating that smoking protects against corona infections or the supposed potential for infection treatment by the parasite drug Ivermectin.
49 Torsten Wilholt, "Collaborative Research, Scientific Communities, and the Social Diffusion of Trustworthiness", in *The Epistemic Life of Groups: Essays in the Epistemology of Collectives*, eds. Michael S. Brady and Miranda Fricker (Oxford: Oxford Univ. Press, 2016), 218–233.
50 Moritz Epple and Claus Zittel, eds., *Science as Cultural Practice. Vol. 1: Cultures and Politics of Research from the Early Modern Period to the Age of Extremes* (Berlin: Akademie Verlag, 2010), 9.
51 "Geflecht konkurrierender, paralleler oder auch aufeinander aufbauender Handlungen". Zachmann and Ehlers, *Wissen und Begründen*, 23.
52 Elif Özmen, ed., *Wissenschaftsfreiheit im Konflikt. Grundlagen, Herausforderungen und Grenzen* (Berlin: Springer, 2021).
53 Rainer Bromme, Niels G. Mede, Eva Thomm, Bastian Kremer and Ricarda Ziegler, "An Anchor in Troubled Times: Trust in Science Before and within the COVID-19 Pandemic", PLoS One 17, no. 2 (2022).
54 Avishai Margalit, "Die Ethik von Hintergrundüberzeugungen", in *Die ungewisse Evidenz. Für eine Kulturgeschichte des Beweises*, eds. Matthias Kroß and Gary Smith (Berlin: Akademie Verlag, 1998), 183–184.

References

Anderson, Wilda C. "The Rhetoric of Scientific Language: An Example from Lavoisier". *Modern Language Notes* 96, no. 4 (April 1981): 746–770.
Beretta, Marco. "Between the Workshop and the Laboratory: Lavoisier's Network of Instrument Makers". *Osiris* 29, no. 1 (April 2014): 197–214.

Bormann, Claus von, Giorgio Tonelli, and Helmut Holzhey. "Kritik". In *Historisches Wörterbuch der Philosophie online*, edited by Joachim Ritter, Karlfried Gründer, and Gottfried Gabriel. Basel: Schwabe Verlag, 2017.

Bromme, Rainer, Niels G. Mede, Eva Thomm, Bastian Kremer, and Ricarda Ziegler. "An Anchor in Troubled Times: Trust in Science before and within the COVID-19 Pandemic". *PLoS One* 17, no. 2 (2022): 1–27.

Bunnin, Nicholas, and Jiyuan Yu. "Evidence". In *The Blackwell Dictionary of Western Philosophy*, edited by Nicholas Bunnin, and Jiyuan Yu, 233–234. Malden: Blackwell, 2004.

Campe, Rüdiger. "Epoche der Evidenz. Knoten in einem terminologischen Netzwerk zwischen Descartes und Kant". In *Intellektuelle Anschauung. Figurationen von Evidenz zwischen Kunst und Wissen*, edited by Sibylle Peters, and Martin Jörg Schäfer, 25–43. Bielefeld: transcript, 2006.

Carrier, Martin. "Antoine L. Lavoisier und die chemische Revolution". In *Das bunte Gewand der Theorie: Vierzehn Begegnungen mit philosophierenden Forschern*, edited by Astrid Schwarz, 12–42. Freiburg: Alber, 2009.

Chandler, James, Arnold Davidson, and Harry Harootunian, eds. *Questions of Evidence: Proof, Practice, and Persuasion Across the Disciplines*. Chicago: Univ. of Chicago Press Journals, 1994.

Conee, Earl, and Richard Feldman. *Evidentialism*. Oxford: Oxford Univ. Press, 2004.

Dawid, Philip, William Twining, and Mimi Vasilaki, eds. *Evidence, Inference and Enquiry*. Oxford, New York: British Academy, 2011.

Deutsches Netzwerk Evidenzbasierte Medizin e.V. "COVID-19: Wo ist die Evidenz?" Accessed February 22, 2022. https://www.ebm-netzwerk.de/de/veroeffentlichungen/pdf/stn-20200903-covid19-update.pdf.

DiFate, Victor. "Evidence". In *Internet Encyclopedia of Philosophy*. 2016. Accessed January 18, 2022. https://iep.utm.edu/evidence/.

Douglas, Heather E. *Science, Policy, and the Value-Free Ideal*. Pittsburgh: Univ. of Pittsburgh Press, 2009.

Douglas, Heather E. "Values in Science". In *The Oxford Handbook of Philosophy of Science*, edited by Paul Humphrey, 609–630. New York: Oxford Univ. Press, 2019.

Downs, Donald A., and Chris W. Surprenant, eds. *The Value and Limits of Academic Speech. Philosophical, Political, and Legal Perspectives*. New York: Routledge, 2018.

Engelen, Eva-Maria, Christian Fleischhack, Giovanni C. Galizia, and Katharina Landfester, eds. *Heureka – Evidenzkriterien in den Wissenschaften. Ein Kompendium für den interdisziplinären Gebrauch*. Heidelberg: Spektrum, 2010.

Epple, Moritz, and Claus Zittel, eds. *Science as Cultural Practice. Vol. 1: Cultures and Politics of Research from the Early Modern Period to the Age of Extremes*. Berlin: Akademie Verlag, 2010.

European Science Foundation/All European Academies (ESF/ALLEA). "The European Code of Conduct for Research Integrity". Accessed September 14, 2021. https://www.allea.org/wp-content/uploads/2017/05/ALLEA-European-Code-of-Conduct-for-Research-Integrity-2017.pdf.

Füger, Daniel. "Die Umwälzung der wissenschaftlichen und zivilisierten Welt. Zum Verhältnis von Evidenz und Normativität in der frühen Chemiewissenschaft". In *Philosophie zwischen Sein und Sollen: Normative Theorie und empirische*

Forschung im Spannungsfeld, edited by Alexander Bauer, and Malte Meyerhuber, 159–178. Berlin: De Gruyter, 2019.

Füger, Daniel. "Conference Report: Practicing Evidence – Evidencing Practice. How is (Scientific) Knowledge Validated, Valued and Contested? Munich 2020". *H-Soz-Kult*.

Gallie, Walter B. "Essentially Contested Concepts". *Proceedings of the Aristotelian Society* 56 (1956): 167–198.

German Research Foundation (DFG). "Proposals for Safeguarding Good Scientific Practice". Accessed September 14, 2021. http://www.dfg.de/download/pdf/dfg_im_profil/reden_stellungnahmen/download/empfehlung_wiss_praxis_1310.pdf.

German Research Foundation (DFG). "Guidelines for Safeguarding Good Research Practice. Code of Conduct". Accessed September 14, 2021. https://www.dfg.de/download/pdf/foerderung/rechtliche_rahmenbedingungen/gute_wissenschaftliche_praxis/kodex_gwp_en.pdf.

Haack, Susan. *Evidence and Inquiry: A Pragmatist Reconstruction of Epistemology*. Amherst: Prometheus Books, 2009.

Hackmann, Willem. "Scientific Instruments: Models of Brass and Aids to Discovery". In *The Uses of Experiment: Studies in the Natural Sciences*, edited by David Gooding, Trevor Pinch, and Simon Schaffer, 31–65. Cambridge: Cambridge Univ. Press, 1999.

Hahn, Roger. *The Anatomy of a Scientific Institution: The Paris Academy of Sciences, 1666 – 1803*. Berkeley: Univ. of California Press, 1971.

Haraway, Donna. "Situated Knowledges: The Science Question in Feminism and the Privilege of Partial Perspective". *Feminist Studies* 14, no. 3 (April 1988): 575–599.

Harper, Douglas. "Etymology of Criticism". *Online Etymology Dictionary*. Accessed February 7, 2022. https://www.etymonline.com/word/criticism.

Halbfass, Wilhelm, and Klaus Held. "Evidenz". In *Historisches Wörterbuch der Philosophie online*, edited by Joachim Ritter, Karlfried Gründer, and Gottfried Gabriel. Basel: Schwabe Verlag, 2017.

Holmes, Frederic L. *Eighteenth-Century Chemistry as an Investigative Enterprise: Five Lectures Delivered at the International Summer School in History of Science, Bologna*. Berkeley: Office for History of Science and Technology, 1989.

Iltis, Ana S., and Douglas MacKay, eds. *Research Ethics*. Oxford: Oxford Handbooks Online, 2020.

Kamecke, Gernot. "Spiele mit den Worten, aber wissen was richtig ist! Zum Problem der Evidenz in der Sprachphilosophie". In *Sehnsucht nach Evidenz*, edited by Karin Harrasser, Helmut Lethen, and Elisabeth Timm, 11–26. Bielefeld: transcript, 2009.

Kant, Immanuel. *Critique of Pure Reason*, translated by Paul Guyer. Cambridge: Cambridge Univ. Press, 1998.

Kelly, Thomas. "Evidence: Fundamental Concepts and the Phenomenal Conception". *Philosophy Compass* 3, no. 5 (April 2008): 933–955.

Kelly, Thomas. "Evidence". In *The Stanford Encyclopedia of Philosophy, Winter 2016*, edited by Edward N. Zalta. Stanford, CA: Metaphysics Research Lab, Stanford Univ., 2020. https://plato.stanford.edu/archives/win2016/entries/evidence/.

Kim, Jaegwon. "What Is Naturalized Epistemology?". *Philosophical Perspectives* 2 (1988): 381–405.

Kim, Mi Gyung. *Affinity, That Elusive Dream: A Genealogy of the Chemical Revolution*. Cambridge: MIT Press, 2003.

Kuhn, Thomas S. "Historical Structure of Scientific Discovery". *Science* 136 (1962): 760–764.

Kuhn, Thomas S. *The Structure of Scientific Revolution*. Chicago: Univ. of Chicago Press, 1970.

Lavoisier, Antoine. "Lavoisier's 'Reflection on Phlogiston' II: On the Nature of Heat", translated by Nicholas W. Best. *Foundations of Chemistry* 18, no. 1 (2016): 3–13.

Lucas, Jeffrey W., and Michael J. Lovaglia. "Legitimation and Institutionalization as Trust-Building: Reducing Resistance to Power and Influence in Organizations". *Social Psychology of the Workplace* 23 (2006): 229–252.

Lloyd, Elisabeth A. "Feminism As Method: What Scientists Get That Philosophers Don't". *Philosophical Topics* 23, no. 2 (April 1995): 189–220.

Longino, Helen E. *Science as Social Knowledge. Values and Objectivity in Scientific Inquiry*. Princeton: Princeton Univ. Press, 1990.

Margalit, Avishai. "Die Ethik von Hintergrundüberzeugungen". In *Die ungewisse Evidenz: Für eine Kulturgeschichte des Beweises*, edited by Matthias Kroß, and Gary Smith, 173–202. Berlin: Akademie Verlag, 1998.

Mautner, Thomas. "Evidence". In *A Dictionary of Philosophy*, edited by Thomas Mautner, 139. Cambridge: Blackwell Publishers, 1996.

Max Planck Society (MPG). "Guidelines and Rules of the Max Planck Society on a Responsible Approach to Freedom of Research and Research Risks". Accessed September 14, 2021. https://www.mpg.de/197392/researchFreedom Risks.pdf.

May, Philipp. "Lockerung der Corona-Maßnahmen. Thorsten Lehr 'Blindflug ohne große Evidenz'". *Deutschlandfunk*. Accessed March 4, 2021. https://www.deutschlandfunk.de/lockerung-der-corona-massnahmen-thorsten-lehr-blindflug-100.html.

Merton, Robert K. "The Normative Structure of Science". In *The Sociology of Science: Theoretical and Empirical Investigations*, edited by Norman W. Storer, 267–278. Chicago: Univ. of Chicago Press, 1973.

Mitcham, Carl. *Encyclopedia of Science, Technology, and Ethics*. Detroit: Macmillan Reference, 2005.

Müller, Jan-Dirk. "Evidentia und Medialität. Zur Ausdifferenzierung von Evidenz in der frühen Neuzeit". In *Auf die Wirklichkeit zeigen: Zum Problem der Evidenz in den Kulturwissenschaften. Ein Reader*, edited by Helmut Lethen, Ludwig Jäger, and Albrecht Koschorke, 261–289. Frankfurt am Main: Campus Verlag, 2015.

Newman, William R. *Atoms and Alchemy: Chymistry and the Experimental Origins of the Scientific Revolution*. Chicago: Univ. of Chicago Press, 2006.

Özmen, Elif. "Die normativen Grundlagen der Wissenschaftsfreiheit". In *Freiheit der Wissenschaft: Beiträge zu ihrer Bedeutung, Normativität und Funktion*, edited by Friedemann Voigt, 111–132. Berlin: De Gruyter, 2012.

Özmen, Elif. "Wissenschaft. Freiheit. Verantwortung. Über Ethik und Ethos der freien Wissenschaft und Forschung". *Ordnung der Wissenschaft* 2 (2015): 65–72.

Özmen, Elif. "Wahrheit und Kritik. Über die Tugenden der Demokratie". *Studia philosophica* 74 (2015): 57–73.

Özmen, Elif, ed. *Wissenschaftsfreiheit im Konflikt. Grundlagen, Herausforderungen und Grenzen.* Berlin: Springer, 2021.

Ritchie, Stuart. *Science Fictions. How Fraud, Bias, Negligence, and Hype Undermine the Search for Truth.* New York: Metropolitan Books, 2015.

Rößler, Hole. *Die Kunst des Augenscheins. Praktiken der Evidenz im 17. Jahrhundert.* Wien: Lit, 2012.

Sandkühler, Hans J. "Kritik der Evidenz". In *Wissen, was wirkt. Kritik evidenzbasierter Pädagogik*, edited by Johannes Bellmann, and Thomas Müller, 33–54. Wiesbaden: VS Verlag für Sozialwissenschaften, 2011.

Schurz, Gerhard, and Martin Carrier, eds. *Werte in den Wissenschaften. Neue Ansätze zum Werturteilsstreit.* Berlin: Suhrkamp, 2013.

Smith, Gary, and Matthias Kroß, eds. *Die ungewisse Evidenz. Für eine Kulturgeschichte des Beweises.* Berlin: Einstein Bücher, 1998.

Siegel, Stefan T., and Martin H. Daumiller, eds. *Wissenschaft und Wahrheit: Ursachen, Folgen und Prävention wissenschaftlichen Fehlverhaltens.* Leverkusen: Budrich, 2020.

Stachura, Mateusz. "Institution". In *Wörterbuch der Soziologie*, edited by Günter Endruweit, 200–201. Konstanz, Stuttgart: UVK/UTB, 2014.

Stegmüller, Wolfgang. *Metaphysik, Skepsis, Wissenschaft.* Berlin: Springer, 1969.

Tuana, Nancy. "Leading with Ethics, Aiming for Policy: New Opportunities for Philosophy of Science". *Synthese* 177, no. 3 (April 2010): 471–492.

Twining, William, and Ian Hampsher-Monk. *Evidence and Inference in History and Law. Interdisciplinary Dialogues.* Evanston: Northwestern Univ. Press, 2003.

Weber, Max. *Wissenschaft als Beruf. Max Weber-Studienausgabe.* Tübingen: Mohr Siebeck, 1994.

Wellcome Trust. "Wellcome Global Monitor. How Covid-19 Affected People's Lives and Their Views About Science". London, 2020.

Wilholt, Torsten. "Epistemic Trust in Science". *The British Journal for the Philosophy of Science* 64, no. 2 (April 2013): 233–253.

Wilholt, Torsten. "Collaborative Research, Scientific Communities, and the Social Diffusion of Trustworthiness". In *The Epistemic Life of Groups: Essays in the Epistemology of Collectives*, edited by Michael S. Brady, and Miranda Fricker, 218–233. Oxford: Oxford Univ. Press, 2016.

Williamson, Timothy. "Knowledge as Evidence". *Mind* 106 (1997): 717–742.

Zachmann, Karin, and Sarah Ehlers, eds. *Wissen und Begründen. Evidenz als umkämpfte Ressource in der Wissensgesellschaft.* Baden-Baden: Nomos, 2019.

2 Questioning Evidence
Three Modes of Contestation

Eva Barlösius and Eva Ruffing

In recent years, the validity of scientific knowledge and the adequacy of scientific expertise have been increasingly contested, which is why they are considered less trustworthy.[1] Thus, societal consent on the reference to scientific knowledge and expertise as best suited for adequate problem-solving – which is typical for knowledge societies – becomes more and more fragile. By increasing the fragility of that consensus, contestation spells trouble for the authority and legitimacy of scientific evidence within the political process. We argue that such an analysis is too crude: Scientific knowledge and scientific expertise are not always disputed; in the vast majority of cases, they are accepted without causing problems in making decisions. Furthermore, it is important that scientific knowledge and scientific expertise are challenged in a variety of ways, which is why the evidence attributed to them becomes fragile in various ways.

In this chapter, we present a heuristic for distinguishing between three modes of contestation, highlighting for each the sort of difficulties it can cause for scientific knowledge and scientific expertise. These modes of contestation are the questioning of (1) the validity and reliability of scientific knowledge, (2) the extent to which the specific scientific expertise is adequate for solving the political problem at hand and (3) the scientific field's exposed position in matters of decision- and policy-making and administrative processes. All three modes of contestation cause trouble for the recognition of evidence; in the first case, because the prerequisite for evidence no longer persists; in the second case, by judging the claimed evidence as irrelevant; and in the third case, by the fundamental rejection of evidence. Whereas earlier publications of ours focused on approaches to handling these different modes of contestation,[2] we now emphasize the consequences of the contestation in terms of evidence.

Scholars in science studies, political science, sociology and other disciplines have been exploring the challenges to scientific knowledge and scientific expertise intensively for more than 15 years.[3] Our review of the vast relevant research shows that the challenges have rarely been differentiated according to what exactly is in dispute or what the contestations are driving at. To fill this gap, we have developed a heuristic that

DOI: 10.4324/9781003273509-4

intertwines perspectives from the sociology of science, political science and administrative science.[4] Our heuristic rests on the observation that research on scientific knowledge and scientific expertise in the political and administrative process usually starts (depending on the scientific perspective) either from analyzing the genesis of scientific knowledge and its transformation into scientific expertise or from examining the role of scientific expertise in the political decision-making process. The heuristic we have developed, by contrast, makes clear that the various forms of questioning are not differentiated along these lines, but rather according to what is being questioned in each case, and what the aim of the questioning is.

Striving for Evidence

Evidence has become a booming topic in science studies in the past few years. Key research areas are the requirements for scientific evidence[5] and how evidence is produced and practiced.[6] Our research is inspired by Max Weber's perspective on evidence. His conception allows us to observe the different challenges to evidence in the scientific and political fields and over the process of transformation to a knowledge society. It is not by chance that Weber began his famous work *Economy and Society* with a paragraph in which he spelt out how interpretive sociology produces scientific *evidence*, a term arguably translatable as "insight and comprehension".[7] As he puts it there, "all interpretation of meaning, like all scientific observations, strives for clarity and verifiable accuracy of insight and comprehension".[8] Clarity and verifiable accuracy can be thought of as a means of generating evidence, which to Weber is "the basis for certainty in understanding".[9] Although this phrasing shows that Weber interpreted the striving for evidence in a broad sense, including emotional, empathic, comprehensive personal convictions, values, norms or artistic ideas, he primarily had scientific evidence in mind.

Weber regarded the scientific sphere as the social field with the "greatest and most principled" grade of rationality and with the highest degree of evidence.[10] To him, rationality and scientific evidence were nearly identical, and he was convinced that this kind of evidence would dominate in modern societies. Indeed, over the past 100 years, the value that societies attach to scientific evidence has increased enormously, and this importance has spread from the scientific field into almost all other social fields. It has permeated the decision- and policy-making community through scientific policy advice, especially in the form of scientific expertise. All other forms of evidence have waned in value, as ways of arriving at and justifying political decisions that are held to be problem-adequate and legitimate are judged by their basis in scientific evidence.

With the diffusion of scientific evidence into nearly all social fields, the way it is understood has undergone great change. Within the scientific field, Weber saw the quest for evidence as a *process*, and as the generation of reliable and valid scientific knowledge. In this process, the methodological criteria of science and the requirements for validity, reliability and transparency have to be followed to generate scientific knowledge for which scientific evidence is recognized. Thus, scientific evidence emerges from the process of generating scientific knowledge, which is why we call it *process evidence*. Scientific knowledge is always to be considered preliminary, which equally applies to scientific evidence. The political *use* of scientific knowledge casts it as a scientific outcome and conceives it as fact having the highest possible degree of veracity and authority. The consideration of scientific outcomes as fact legitimizes scientific knowledge becoming the basis for political decisions. This shift in understanding and use of scientific knowledge involves a switch from *process evidence* to *outcome evidence*, which draws its clarity and verifiable accuracy from the point that scientific outcomes are considered factual.

This switch has coincided with a decline in the political value and relevance accorded to other forms of knowledge for decision-making. Instead, the difference between scientific knowledge and scientific expertise has become very important. Scientific knowledge and scientific expertise are difficult to define. They will be more precisely described in the following sections. Here, however, we discuss how they are related to evidence. Scientific knowledge draws its reliability from the scientific process. Douglas Walton and Nanning Zhang define scientific knowledge "as something that is achieved through a process of marshalling in a scientific inquiry".[11] Scientific expertise is based on scientific knowledge but provides *Begründungswissen* (reasoning knowledge) from which political decisions are derived. Its evidence refers to its provision of adequate solutions for political problems. Scientific expertise ranks as the source of argumentation with both the highest degree of evidence and the highest degree of political authority and legitimacy. In the following, we describe the three modes of contestation of scientific knowledge and scientific expertise and analyze the consequences they have for the recognition of the evidence they produce.

A Critique of the Validity of Scientific Knowledge

The first mode of contesting scientific evidence involves doubting the validity and reliability of the scientific knowledge on which scientific evidence is based and hence its evidence. Indeed, there are mounting claims that scientific knowledge is simply wrong.[12] These challenges are convoluted; they essentially bark up the wrong tree by criticizing something to which science stakes no claim in the first place, at least no legitimate one. Scientific knowledge has no absolute validity; it is not 100 percent reliable, and there is certainly no assertion that it proclaims the truth. Rather, scientific knowledge differs from most other forms of knowledge precisely in that its validity and reliability derive explicitly from the state of research at a particular moment, and it is, therefore, always subject to review. At most, one can say that scientific knowledge is only correct in a preliminary sense. It is to be regarded as a particularly reliable form of knowledge precisely because its validity and reliability are (ideally) constantly under question. Scientific scrutiny is not only about the tentative correctness of any scientific knowledge, but also about the existing and alleged gaps in knowledge and shortcomings in research. Additional gaps, it goes without saying, surface each time new knowledge appears, confirming the adage that research constantly raises more questions than it answers.[13]

The question of justified and unjustified claims of validity and reliability of scientific knowledge is a perennial focus of science studies. The question of scientific objectivity, for instance, was a concern of Max Weber's.[14] A similar question troubled both Karl Mannheim[15] and Ludwik Fleck[16], who asked whether it was possible to adopt a scientific standpoint that allows objective scientific observations. These authors came to more or less the same conclusion as Weber did, stating that such a scientific standpoint is difficult to access. Other studies of science have shown that the processes of generating knowledge are influenced by social, cultural, habitual and many other contextual factors, such as social structures, thinking styles, ethical and religious values, emotions and power, professional insecurity, as well as economic and legal restrictions.[17]

Alongside the limited validity and reliability of scientific knowledge, its context-bound nature has also provoked debate.[18] None of these limitations stand for "incorrectness", "inaccuracy" or "arbitrariness". In order to count as scientific knowledge, a study has to conform to accepted standards of scholarly work in the relevant field of science at the time. These standards include the expectation that knowledge's limited validity, reliability and its context-bound nature will be pointed out, as well as the concomitant knowledge gaps and research deficits.

The limited validity and reliability for which scientific knowledge is reproached in the first mode of contestation (meaning that it is only correct in a preliminary sense) are part of the "epistemological core" of the sciences and show therefore no deficiency.[19] Scholarliness constitutes the framework of this type of inquiry. The manner in which scientific knowledge is questioned within that framework is a key characteristic of science, and is inscribed in the field's nomos, that constitutes the specific nature of science itself and guarantees its continuation. This approach to questioning *is* the scientific process.[20]

Such questioning is intended to dispute the *correctness of* scientific knowledge, but the standards of the scientific endeavor are not rejected in principle. Of course, these standards, too, must be examined and readjusted again and again. On that score, scientific evidence runs into no trouble at all in the first mode of contestation.[21] Following Weber, it can be said that it is a quest for evidence as a scientific process that includes perennial scientific contestation of the knowledge generated by the field. Consider an example to illustrate this nexus: If toxicological research finds higher residues of an environmental toxin, such as glyphosate, than previously, thanks to new detection methods, this result is a new scientific outcome, but it does not call into question the criteria for its scientific character. However, this shows that scientific knowledge, which serves as the basis for scientific expertise, has to meet particularly high requirements in terms of validity and reliability.

Researchers in science studies have proposed ways to improve the validity and reliability of identifying and characterizing scientific knowledge. Bruno Latour, for example, introduced the distinction between *science faite* (science made) and *science en train de se faire* (science in the making), the former being scientific knowledge that is solid, or cold, and the latter being new.[22] To Latour, "science in the making" represents new research results, which are too hot and risky to be taken as a basis for scientific expertise. In contrast, solid or cold scientific knowledge has been tested many times and in many different ways and has proven to be coherent and consistent with other bodies of knowledge, as in the context of meta-studies. An example for demonstrating the robustness of scientific knowledge by a meta-study is the famous research letter "Consensus on Consensus: A Synthesis of Consensus Estimates on Human-Caused Global Warming" written by 16 scientists who analyzed 2,412 papers on global warming.[23] They found a 97 percent consensus on global warming in the scientific articles, which represents a highly robust scientific outcome.

Contesting Evidence in the Context of the Decision- and Policy-Making Communities and Administrative Actors

Criticizing Problem Adequacy

The second mode of questioning evidence is to cast doubt on what is known as the problem adequacy of scientific expertise, that is, the ability and suitability of such expertise to address or solve a given problem in decision-making on political and administrative matters.[24] At stake is the way in which that expertise has been processed and used by decision makers.[25] The question here is not so much whether the expertise is based on knowledge that is right but whether the right expertise has been used, taking into account the fact that policymakers might disagree about roots and possible solutions of policy problems, and therefore also about the question of which expertise is most adequate to solve the problem at hand. To differentiate between scientific knowledge and scientific expertise, we adopt the view taken by philosopher of science Philippe Roqueplo, to whom the essential difference is that scientific expertise is underpinned by a rationale (*connaissance de cause*) and thus aims less at explaining and understanding than scientific knowledge does.[26] To Roqueplo, what transforms "the utterance of scientific knowledge into expertise is its inscription within the dynamics of decision making".[27] Above all, scientific experts are obliged to answer questions posed by politicians and administrators.[28] We add another serious difference: In making the transition to scientific expertise, scientific knowledge becomes politically framed as well.

By political framing, we mean that in practice problems are processed in policy domains, each of which is characterized by different constellations of actors and conflicts, political models and a stock of institutional regulations that influence politico-administrative action,[29] though different policy domains and problems are also intertwined.[30] Often, separate and distinct administrative structures – a domain's own ministry and authority – are constituting elements of a policy domain.[31] However, policy fields cannot necessarily be clearly delineated, which is why political problems can, in principle, be assigned to different ones.

The framing of the particular problem determines the policy domain to which it is assigned, the department that is given responsibility for it, and the research field from which scientific knowledge is sought as expertise. Glyphosate, for example, can be politically framed as a problem of health, nature conservation or food. Depending on which understanding of glyphosate prevails, separate departments become responsible for obtaining scientific expertise from medicine, biodiversity research or nutrition science.

However, these sciences look at glyphosate from very different research perspectives, develop correspondingly different test criteria and detection methods and, when asked for scientific expertise, come to different conclusions on the hazards and risks of this substance. None of the procedures or criteria is scientifically more correct than another. Rather, they result from the different scientific approach that each field takes to the problem. Accordingly, the reports by scientific experts concentrate on different impacts, propose different permissible levels and usually do not arrive at unanimous and unambiguous overall assessments, some of which are even contradictory. The inconsistency of the results is therefore not due to an insufficient quality of scientific evidence, nor to its limited validity and reliability; the lack of uniformity by no means results from insufficient scientific evidence. Rather, different political frameworks bring different kinds of scientific expertise to bear, each of which specializes in different consequences.

The second mode of contestation takes place within a political framework involving various kinds of scientific expertise. As already pointed out, this form of questioning takes aim at the problem-solving adequacy of that expertise and casts doubt on it. We distinguish three variants in challenging this adequacy and identify the transitions between the first and the second, in particular. In the first variant, proponents of counter-expertise accumulate knowledge other than that cited by the scientific expertise initially presented. In the second variant, the political framing is criticized as inappropriate. The point of the third variant is to contest the adequacy of scientific expertise in general to justify a specific political decision. These three possibilities for questioning scientific expertise are explained in more detail below. It is important to note, however, that none of them fundamentally questions the relevance of scientific expertise for decision makers. The second mode of contesting evidence is merely a dispute about whether the politically and administratively used scientific expertise is adequate for problem-solving – whether it is correctly framed politically. The third variant is a dispute of whether it is a political problem that should be "solved" on the basis of scientific expertise. This is where the insight and understanding offered by scientific expertise runs into trouble, not because it is said to be wrong, but rather because it is the wrong sort of expertise to solve the identified problem. The problem adequacy of the expertise is assessed in light of the policy problem's political framing. This juncture is where the aforementioned switch from process evidence to outcome evidence comes in, as the scientific results are assessed according to whether they provide an adequate basis for decision-making and administrative action.

Contestation through Counter-Expertise

A typical strategy by which to challenge scientific expertise is to establish scientific counter-knowledge and counter-expertise based on it. Opponents of one expertise advocate new research methods and topics, and often a completely different view of the problems. Transdisciplinary and socioecological research are examples of the creation of other modes of research not commonly found in the established sciences. Topics that have not been studied in the established sciences or that have been researched too little and, often, one-sidedly, include issues of gender, diversity and sustainability. Opponents of established research often help to draw attention to these gaps and shortcomings, by building grounded knowledge and expertise that runs counter to the mainstream.[32] For this purpose, they often establish their own scientific institutions, such as the Öko-Forschungsinstitute (Eco-Research Institutes) in Germany in the late 1970s.[33] It is not always the primary goal of these opponents to advance scientific research. Instead, the focus can also be to underpin their own political position scientifically.

In this variant of questioning, scientific expertise meets counter-expertise, but across these divides, there is widespread agreement that political and administrative decisions should rest on scientific knowledge, and that they derive legitimacy on that basis. The dispute is over what kind of scientific knowledge is adequate for the problem-solving that forms the basis of political decisions and administrative action. Scholarliness and scientific expertise are not fundamentally criticized or rejected as tools of policy-making and administrative action. Even the arguments rooted in counter-knowledge and counter-expertise acknowledge that scientific knowledge and scientific expertise are the basis for objectively adequate, problem-solving, political and administrative decisions. The disagreement is about the evidence of the scientific knowledge used for the expertise, as it is said to arrive at results (outcomes) that may be scientifically correct, but not suitable for solving the identified political problem. The scientific knowledge used by policy- and decision makers and administrative actors is contrasted with other scientific expertise for which scientific evidence is also claimed, but which is said to be more adequate for problem-solving, and which therefore possesses a higher level of evidence.

Sticking with the example of glyphosate, the following example illustrates how contestation is expressed via counter-expertise. The European Food Safety Authority (EFSA) was commissioned to apply its scientific expertise to the prolongation of the authorization of glyphosate. Based on its own risk assessment, the EFSA proposed the renewal of the license for the pesticide to the European Commission (EC).[34] The International Agency for Research on Cancer (IARC), which is part of the World Health Organization (WHO), also assessed glyphosate. Its report came to the conclusion that glyphosate is "probably carcinogenic to humans".[35] The EFSA was asked to review the IARC report but came to the conclusion that there is no causal link to cancer. The different assessments have a number of reasons, and a particularly important one is that the EFSA and the IARC were looking at the same phenomenon from different scientific perspectives: The EFSA viewed glyphosate as a food risk and analyzed pesticide residues, whereas the IARC considered the pesticide as a health risk and examined whether it could potentially cause cancer in an organism.

Contestation through Problem-Shifting

Another typical strategy for questioning the objective adequacy of scientific expertise is problem-shifting – the denial that the problem has been properly framed, assigned to the appropriate policy domain and handed over to a suitable department, regardless of how the problem is comprehended. As a result, it is argued that scientific expertise that is suitable to solve the problem has not been used. This criticism aims at a political shift of the problem to a different policy domain and/or different department, a change that then justifies the recourse to other scientific expertise and, if necessary, buttresses one's own political position.[36] Shifting the problem can mean basing scientific expertise on research knowledge from other sciences or other fields of research. These other scientific perspectives may arrive at considerations and assessments that diverge from the previous scientific presentation and evaluation. In the case of glyphosate, a very powerful contestation occurred by problem-shifting how harmful the pesticide is from agriculture and food policy to environmental policy, and by using biological expertise to prove how dangerous it is for biodiversity.

However, the questioning inherent in problem-shifting does not allege inadequacy of the knowledge being brought to bear, but rather of the problem's political framing. In other words, it criticizes the evidence by which the problem addressed was assigned to a particular policy domain, and by which scientific expertise was duly commissioned. In the political process, however, this remonstration is often argued as though it were about a lack of scientific evidence, though in fact this dispute is about inconsistencies between scientific expertise from different disciplines, which results from the diverse research perspectives. In this case, too, the reasoning is based on outcome evidence.

Contestation through Questioning the Ability to Solve a Specific Problem

A characteristic of the third variant of questioning evidence within the second mode is that a particular problem cannot be solved through scientific expertise alone. However, this feature in no way implies a fundamental rejection of the significance and importance of scientific expertise in the political process. As with the two other variants, it is acknowledged in principle that scientific expertise can contribute to adequate solutions to problems. The aim in this third variant of its critique is to keep scientific expertise from being used to decide politically contentious matters, or to cover them up scientifically. The main thrust of the argument is that conflicting interests and/or differing preferences give rise to the dissent.

In this variant of questioning, the criticism is that decision makers, policymakers and administrative actors should not use scientific evidence to bypass political dissent. In this case, too, scientific evidence runs into less trouble than do the decision-making, policy-making and administrative communities, which end up being confronted by divergent political views. The criticism is that political trouble is deliberately disguised or obfuscated with scientific evidence. For instance, the protest group called the Yellow Vests in France agreed with the scientific analyses that glyphosate is bad for biodiversity, and that the use of these pesticides should be reduced.[37] They protested against the scientific expertise exhibited by the French National Institute for Agriculture (INRA), which recommended that the farmers and winegrowers revert to using weed hooks.[38] The protestors pointed out the collision of conflicting interests: Elite Parisians were talking about the end of the world, while the farmers and winegrowers who were in economic difficulty were talking about the end of the month.

Criticism of the Social Position of Science

The third mode of contestation arises from the transformation to a knowledge society, and the accompanying societal upheavals, especially socio-structural ones. It is thus not enough to simply state that such inquiry fundamentally questions the relevance of scientific knowledge and scientific expertise for political and administrative decision-making. At the same time, this mode of contestation is driven by discontent about the position science has acquired in society, and the authority it has been granted. This third mode of contesting scientific expertise opposes the idea that the adequacy of political and administrative decisions can be measured by whether existing scientific knowledge has been taken into account. Thus, it rests on an objection to what in modern societies is the typical coupling of science and politics as established through scientific expertise. It is a linkage intended to guarantee that "all accessible knowledge about the relevant subject area is used and taken into account" in decision-making, in order to arrive at appropriate and expedient solutions.[39]

The relationship is designed to give scientific knowledge priority over other ways of knowing and other forms of epistemological content. However, it certainly does not mean that this priority is always, or even predominantly, conferred in political and administrative practice. The third mode of contestation questions the direct coupling of scientific knowledge and political decision-making. The argument against this coupling is that scientific knowledge is assigned with authority and legitimates power in decision- and policy-making and administrative action, allegedly enabling scientific evidence to rule in those spheres.

In recent years, various labels have taken root in common parlance to characterize the way scientific knowledge is typically handled in the third mode of contesting scientific expertise. These include "post-truth",[40] misinformation and conspiracy theories. These labels make clear that something incorrect, false and erroneous is being disseminated. They characteristically delegitimize an argument before it is even formulated, always implying in advance that "the view thus designated is wrong".[41] Their counterpart in some places is "Trotzpositivismus",[42] which may be translated as "defiant positivism", the message of which is "for alternativeless facts, for scientific evidence, for truth in politics".[43] These extremes clearly reveal a typical fault line running through the knowledge society, as the more scientific expertise is used to explain and justify political and administrative action, the more it becomes the subject of social and political disputes.

With the validity and reliability of scientific knowledge, as well as the problem-solving adequacy of scientific expertise, which is always under scrutiny, and with scientific knowledge having no inherent claim to truth, the third mode of contesting scientific expertise encounters a sort of "institutional" doubt, which it escalates into a matter of principle. However, the accusation that science spreads untruth should not be misconstrued as denial of the validity and reliability of scientific knowledge. The reproach is instead aimed at the power of authorization that science has in the knowledge society.[44]

Previous studies on "Post-Truth"[45] and misinformation identify three main reasons for the success of the third mode of contesting scientific expertise. One of them is the special position of science in the knowledge society, which is the basis for the cultural, political and social supremacy of those who work in the field of science and speak for it to the outside world as experts: "Public dismissal of science, or public distrust of experts should be seen in the context of public discontent with authorities and elites that exert power over citizens' lives".[46] These disputes are less about the validity and reliability of scientific knowledge than about criticism of politics and society.

A second reason for the enormous success of the third mode of contesting scientific expertise seems to be that political decisions are often presented "without alternative" with reference to scientific expertise. The aim of circulating opposing "alternative facts", denouncing scientific findings as untruths, and spreading deliberately false statements of fact is less to lend them recognition and practical effectiveness than to bring about new rules of the game for public discussions.[47] Axel Freimuth argues similarly: "scientific results are often used to identify political [and other] decisions as having no alternative".[48] This viewpoint figures in Jan Söffner's analysis that such representations often come across as though there were a discourse of truth, accessible only to experts, which is encroaching on the field of political decision-making.[49] He argues that the opinion-formation that used to take place is now often supplanted by factual analysis only. Such a purely technocratic use of scientific expertise contributes to the emaciation of political and administrative processes. In the political science debate, representatives of the so-called agonistic theory of democracy criticize this use of scientific expertise and call for repoliticization in the sense of a stronger emphasis on conflict resolution within the framework of democratic processes.[50]

Increased social fragmentation and exclusion are the third reason for the success of contesting scientific expertise by challenging scientific expertise's exclusive ability to solve identified problems. The social groups affected by social fragmentation and exclusion will seek to make themselves heard by means of such questioning.[51] They feel powerless, fear downward mobility and perceive themselves as disadvantaged. They also feel exposed to pressures that are highly normative socially and culturally. Consequently, both the public rejection of scientific knowledge and the publicly articulated mistrust of scientific experts must be understood in the context of real and perceived socio-structural discrimination.

If these three reasons correctly describe the core of the controversies, the third mode of contestation represents a dispute over the manner in which the relative positions of politics and science should be determined, and the extent to which scientific expertise should be a part of decision-making. Furthermore, this is apparently a socio-structural struggle against the way in which science and the decision-making, policy-making and administrative communities are entangled with social elites and the resulting power of authorization. It is a question of the relevance of scientific evidence to the decision-making, policy-making and administrative communities. The argument is that certain social groups seek to have the primacy of scientific evidence recognized, and that favoring this kind of evidence enables them to use their social positions to exercise power. Argumentation with scientific evidence and the coupling of science and politics is thus declared to be in the interest of certain socio-structural groups. It is not accepted as the best option for arriving at justified and effective political decisions. The causes underlying both the third mode of contesting scientific expertise and knowledge as a basis for guaranteeing appropriate consideration of political solutions extend far beyond the validity and reliability of scientific knowledge and scientific expertise.

Conclusion

Our chapter has presented a heuristic for systematically describing the different modes of questioning scientific knowledge and scientific expertise, by elaborating their causes, and discussing their consequences for the accepted reliability of scientific knowledge and expertise. As we have seen, the first mode starts from scientific knowledge and casts doubt on its certainty. However, this kind of doubt is inscribed in the understanding of scientific evidence as a methodologically conducted process oriented toward generating valid scientific knowledge in a transparent manner. According to its basic understanding, scientific results are always preliminary, and the validity and reliability of scientific knowledge are to be regarded as tentative.

In the second mode of contestation, politically and administratively produced and asserted evidence is the focus of criticism. The criticism derives primarily from the political and administrative framing of the problem, which is objected to as unsuitable for solving the policy problem at hand. This objection shows that the understanding of what is certain in the political and administrative process is not identical to what is taken to be certain or well-founded in the context of research. On the one hand, the argument is based mainly on outcome evidence, which means to take scientific knowledge as fact, and to make it the basis for scientific expertise. On the other hand, framed evidence is used in the political and administrative process to establish legitimacy for solutions to certain problems.

The third mode of contestation denies that scientific evidence guarantees a high degree of validity and reliability. It thus rejects the coupling of scientific evidence and politico-administrative decisions. Reasoning on the basis of scientific evidence is not seen as a procedure for arriving at the most acceptable solutions, but rather as a typical means by which social elites exercise power. This development marks the end of societal consent to basing political and administrative decisions in a rational and problem-adequate way on expertise. Drawing on Max Weber, one can say that this mode of contestation breaks with society's typical quest for scientific evidence and rationality, resulting in broken evidence.

One could assume that the three forms of contestation clearly indicate that actors no longer trust the validity of scientific knowledge and scientific expertise. However, as argued above, the evidence of scientific knowledge and scientific expertise are hardly contested in the majority of cases. In contrast, we observe two phenomena in parallel: Trust and mistrust in scientific evidence. The COVID-19 pandemic serves as an instructive example. On the one hand, "hot" scientific knowledge on the virus was eagerly awaited, as well as the newly developed vaccines and the policies that rested on that hot knowledge, which were consequently applied in daily life. On the other hand, mistrust in this type of science-led policies grew. The existence of COVID-19 was denied, vaccines were believed to carry higher risks than the virus and the policies for controlling the pandemic were seen as harbingers of an autocratic turn. These forms of denial and the resulting clashes are a manifestation of socio-structural and political struggles and belong to the third form of contestation in our heuristic. These struggles are not about questioning the evidence of scientific knowledge, but rather a dispute about how political solutions come about.

Notes

1 This chapter builds on two previous publications (Eva Barlösius and Eva Ruffing, "Für einen vorausschauenden Umgang mit der Infragestellung wissenschaftlicher Expertise". Impulspapier, *Institutionelles Repositorium der Leibniz Universität Hannover*, (2020); Eva Barlösius and Eva Ruffing, "Die Infragestellung von wissenschaftlichem Wissen und Expertise: Eine sozialwissenschaftliche Heuristik", *Sonderband Leviathan* 38 (2021): 113–134). Here we analyzed the consequences of the three modes of contestation of scientific knowledge and expertise for political and administrative decision-making. The heuristic of the three forms of contestation was developed during our research project on "Scientific Expertise as Basis for Political Decision-Making", https://www.lcss.uni-hannover.de/de/forschung/unsere-forschungsprojekte/mwk-projekt-wiss-expertise/.

2   Barlösius and Ruffing, "Für einen vorausschauenden Umgang", Barlösius and Ruffing, "Infragestellung".

3   Alexander Bogner, *Die Epistemisierung des Politischen. Wie die Macht des Wissens die Demokratie gefährdet* (Stuttgart: Reclam, 2021); Mark B. Brown, Justus Lentsch, and Peter Weingart, "Representation, Expertise, and the German Parliament: A Comparison of Three Advisory Institutions", in *Democratization of Expertise? Exploring Novel Forms of Scientific Advice in Political Decision-Making*, eds. Sabine Maasen and Peter Weingart (Dordrecht: Springer, 2005), 81–100.

4   Barlösius and Ruffing, "Infragestellung".

5   Susan Haack, "Clues to the Puzzle of Scientific Evidence", *Principia 5*, no. 1–2 (2001): 253–281; Alfred Moore and Jack Stilgoe, "Experts and Anecdotes: The Role of 'Anecdotal Evidence' in Public Scientific Controversies", *Science, Technology, & Human Values 43*, no. 5 (2009): 654–677; Kent Staley, "Robust Evidence and Secure Evidence Claims", *Philosophy of Science 71*, (2004): 467–488; Douglas Walton and Nanning Zhang, "The Epistemology of Scientific Evidence", *Artificial Intelligence and Law* 21, no. 2 (2012): 173–219.

6   Karin Zachmann and Sarah Ehlers, eds., *Wissen und Begründen. Evidenz als umkämpfte Ressource in der Wissensgesellschaft* (Baden-Baden: Nomos, 2019).

7   Max Weber, *Economy and Society: An Outline of Interpretive Sociology* (Berkeley: Univ. of California Press, 1978).

8   Weber, *Economy and Society*, 5.

9   Weber, *Economy and Society*, 5.

10 Max Weber, *Essays in Sociology* (Oxford: Oxford Univ. Press, 1946).

11 Walton and Zhang, "The Epistemology of Scientific Evidence", 215.

12 Ralf Hohlfeld, Michael Harnischmacher, Elfi Heinke, Lea Lehner, and Michael Sengl, eds., *Fake News und Desinformation. Herausforderungen für die vernetzte Gesellschaft und die empirische Forschung* (Baden-Baden: Nomos, 2020).

13 Stefan Böschen, "Risikogenese. Metamorphosen von Wissen und Nicht-Wissen", *Soziale Welt 53*, no. 1 (2002): 67–85.

14 Max Weber, *Collected Methodological Writings* (London: Routledge, 2012).

15 Karl Mannheim, *Die Strukturanalyse der Erkenntnistheorie* (Berlin: Reuther & Reichard, 1922).

16 Ludwik Fleck, *Entstehung und Entwicklung einer wissenschaftlichen Tatsache. Einführung in die Lehre vom Denkstil und Denkkollektiv* (Basel: Benno Schwabe, 1935).

17 Miranda Fricker, *Epistemic Injustice: Power and the Ethics of Knowing* (Oxford: Oxford Univ. Press, 2009); Donna Haraway, *Primate Visions: Gender, Race and Nature in the World of Modern Society* (New York: Routledge, 1989); Sheila Jasanoff, *Science and Public Reason* (London: Routledge, 2012); Helen E. Longino, *Science as Social Knowledge: Values and Objectivity in Scientific Inquiry* (Princeton: Univ. Press, 1990).

18 Erin Leahey, "Overseeing Research Practice: The Case of Data Editing", *Science, Technology & Human Values 33*, no. 5 (2008): 605–630.

19 Helga Nowotny, Peter Scott, and Michael Gibbons, *Re-Thinking Science: Knowledge and the Public in an Age of Uncertainty* (Cambridge: Polity Press, 2001).

20 Pierre Bourdieu, *Science de la Science et Refléxivité* (Paris: Raison d'Agir, 2001).

21 See also the contribution of Daniel Füger and Elif Özmen in this volume.

22 Bruno Latour, *Le métier de chercheur: Regard d'un anthropologue* (Versailles: Editions Quae, 2001), 14.

23 John Cook, William R.L. Anderegg, Naomi Oreskes, J. Stuart Carlton, Peter T. Doran, Stephan Lewandowsky, Sarah A. Green et al., "Consensus on Consensus: A Synthesis of Consensus Estimates on Human-Caused Global Warming", *Environmental Research Letter* 11 (2016): 048002.

24 Brown, "Expertise"; Harry Collins, Robert Evans, and Martin Weinel, "STS as Science or Politics", *Social Studies of Science* 47, no. 4 (2017): 580–586; Max Krott and Michael Suda, eds., *Macht Wissenschaft Politik? Erfahrungen wissenschaftlicher Beratung im Politikfeld Wald und Umwelt* (Wiesbaden: VS Verlag für Sozialwissenschaften, 2007).

25 Renate Mayntz, "Speaking Truth to Power: Leitlinien für die Regelung wissenschaftlicher Politikberatung", *dms-der moderne Staat* 2, no. 1 (2009): 5–16; Frank Nullmeier, "Knowledge and Decision-Making", in Democratization of Expertise? Exploring Novel Forms of Scientific Advice in Political Decision-Making, eds. Sabine Maasen and Peter Weingart (Wiesbaden: Springer, 2005), 123–134; Coral Weiss, "The Many Meanings of Research Utilization", *Public Administration Review* 39, no. 5 (1979): 426–431.

26 Philippe Roqueplo, *Entre savoir et décision, l'expertise scientifique* (Paris: inra éditions, 1997).

27 Philippe Roqueplo, "Scientific Expertise Among Political Powers, Administrations and Public Opinion", *Science and Policy Studies* 22, no. 3 (1995): 175.

28 Roqueplo, *Entre savoir et décision*, 20.

29 Marian Döhler, *Die politische Steuerung der Verwaltung* (Baden-Baden: Nomos, 2007).

30 Brian Head and John Alford, "Wicked Problems: Implications for Public Policy and Management", *Administration & Society* 47, no. 6 (2015): 711–739.

31 Döhler, *Die politische Steuerung*.

32 Les Levidow, Kean Birch, and Theo Papaioannou, "Divergent Paradigms of European Agro-Food Innovation: The Knowledge-Based Bio-Economy (KBBE) as an R&D Agenda", *Science, Technology, & Human Values* 38, no. 1 (2012): 94–125.

33 See the chapter by Stefan Esselborn and Karin Zachmann in this book.

34 Irina Lock, "Debating Glyphosate: A Macro Perspective on the Role of Strategic Communication in Forming and Monitoring a Global Issue Arena Using Inductive Topic Modelling", *International Journal of Strategic Communication* 14, no. 4 (2020): 223–245.

35 IARC, "Some Organophosphate Insecticides and Herbicides", *Monographs on the Evaluation of Carcinogenic Risks to Humans Volume* 112 (2017), 30.

36 Michael Böcher, "Wissenschaftliche Politikberatung im politischen Prozess", in *Macht Wissenschaft Politik? Erfahrungen wissenschaftlicher Beratung im Politikfeld Wald und Umwelt*, eds. Michael Krott and Max Suda (Wiesbaden: VS Verlag für Sozialwissenschaften, 2007), 14–42.

37 "Glyphosate, comment s'en sortir?", France 2, December 13, 2018, https://www.france.tv/actualites-et-societe/magazines-d-actu/860147-glyphosate-comment-s-en-sortir.html.

38 "Usages et alternatives au glyphosate dans l'agriculture française", INRA, accessed February 7, 2022, https://www.actu-environnement.com/media/pdf/news-30183-rapport-inra-glyphosate.pdf.

39 Andreas Voßkuhle, "Sachverständige Beratung des Staates", in *Handbuch des Staatsrechts* (3rd ed.), eds. Joseph Isensee and Paul Kirchhof (Heidelberg: C. F. Müller, 2005), 425–476, own translation.

40 Vittorio Bufacchi, "Truth, Lies and Tweets: A Consensus Theory of Post-Truth", *Philosophy and Social Criticism* 47, no. 3 (2021): 347–361.

41 Michael Butter, *"Nichts ist, wie es scheint". Über Verschwörungstheorien* (Frankfurt am Main: Suhrkamp, 2020).

42 Alexander Bogner, "Glaubwürdigkeit und Dissens. Zur Mikropolitik ethischer Expertise", Lecture delivered at *Wissenschaft in der Verlässlichkeitsfalle? Praktiken der Konstruktion von Relevanz und Neutralität* Conference of the Leopoldina-Zentrum für Wissenschaftsforschung (April 11–12, 2019).

43 Peter Strohschneider, Über Wissenschaft in Zeiten des Populismus. www.dfg.de/download/pdf/dfg_im_profil/reden_stellungnahmen/2017/170704_rede_strohschneider_festveranstaltung.pdf, 2017, p. iv.

44 Gili S. Drori, John W. Meyer, and Hokyu Hwang, eds., *Globalization and Organization, World Society and Change* (Oxford: Oxford Univ. Press, 2006).

45 Lee McIntyre, *Post-Truth* (Cambridge: MIT Press, 2018).

46 Bart Penders, "Why Public Dismissal of Nutrition Science Makes Sense. Post-Truth, Public Accountability and Dietary Credibility", *British Food Journal* 120 (2018): 1956.

47 Silke van Dyk, "Krise der Faktizität? Über Wahrheit und Lüge in der Politik und die Aufgabe der Kritik", *PROKLA* 47, no. 188 (2017): 355.

48 Axel Freimuth, "Vertrauenswürdige Wissenschaft: Alternativlosigkeit, Impact und andere Verführungen", in *Vom Umgang mit Fakten. Antworten aus Natur-, Sozial- und Geisteswissenschaften*, eds. Günter von Blamberger, Axel Freimuth, and Peter Strohschneider (Paderborn: Wilhelm Fink, 2018), 104, own translation.

49 Jan Söffner, "Truth Politics: Warum wissenschaftliche Expertise und parlamentarische Demokratie gleichzeitig in die Krise geraten sein könnten", in *Vom Umgang mit Fakten. Antworten aus Natur-, Sozial- und Geisteswissenschaften*, eds. Günter von Blamberger, Axel Freimuth, and Peter Strohschneider (Paderborn: Wilhelm Fink, 2018), 138.

50 Chantal Mouffe, *Agonistics: Thinking the World Politically* (London: Verso, 2013).

51 Gordon Gauchat, "Politicization of Science in the Public Sphere: A Study of Public Trust in the United States, 1974 to 2010", *American Sociological Review* 77, no. 2 (2012): 167–187; Gordon Gauchat, "Public Views of Science-based Policy and Funding. The Political Context of Science in the United States: Public Acceptance of Evidence-based Policy and Science Funding", 94, no. 2 (2015): 723–746.

References

Barlösius, Eva, and Eva Ruffing. "Für einen vorausschauenden Umgang mit der Infragestellung wissenschaftlicher Expertise". Impulspapier, *Institutionelles Repositorium der Leibniz Universität Hannover*, 2020.

Barlösius, Eva, and Eva Ruffing, "Die Infragestellung von wissenschaftlichem Wissen und Expertise: Eine sozialwissenschaftliche Heuristik". *Sonderband Leviathan. Umstrittene Expertise* 38 (2021): 113–134.

Böcher, Michael. "Wissenschaftliche Politikberatung im politischen Prozess", In *Macht Wissenschaft Politik? Erfahrungen wissenschaftlicher Beratung im*

Politikfeld Wald und Umwelt, edited by Michael Krott, and Max Suda, 14–42. Wiesbaden: VS Verlag für Sozialwissenschaften, 2007.

Bogner, Alexander. "Glaubwürdigkeit und Dissens. Zur Mikropolitik ethischer Expertise", Lecture delivered at *Wissenschaft in der Verlässlichkeitsfalle? Praktiken der Konstruktion von Relevanz und Neutralität. Conference of the Leopoldina-Zentrum für Wissenschaftsforschung*, Halle, April 11–12, 2019.

Bogner, Alexander. *Die Epistemisierung des Politischen. Wie die Macht des Wissens die Demokratie gefährdet*. Stuttgart: Reclam, 2021.

Böschen, Stefan, "Risikogenese. Metamorphosen von Wissen und Nicht-Wissen". *Soziale Welt* 53, no. 1 (April 2002): 67–85.

Bourdieu, Pierre. *Science de la science et refléxivité*. Paris: Raison d'Agir, 2001.

Brown, Mark B. Justus Lentsch, and Peter Weingart. "Representation, Expertise, and the German Parliament: A Comparison of Three Advisory Institutions", In *Democratization of Expertise? Exploring Novel Forms of Scientific Advice in Political Decision-Making*, edited by Sabine Maasen, and Peter Weingart, 81–100. Dortrecht: Springer, 2005.

Brown, Mark B. "Expertise", In *Science and the Politics of Openness. Here be Monsters*, edited by Brigitte Nerlich, Sarah Hartley, Sujatha Raman, and Alexander Smith, 169–175. Manchester: University Press, 2018.

Bufacchi, Vittorio. "Truth, Lies and Tweets: A Consensus Theory of Post-Truth". *Philosophy and Social Criticism* 47, no. 3 (April 2021): 347–361.

Butter, Michael. *"Nichts ist, wie es scheint." Über Verschwörungstheorien*. Frankfurt am Main: Suhrkamp, 2020.

Collins, Harry, Robert Evans, and Martin Weinel. "STS as Science or Politics", *Social Studies of Science* 47, no. 4 (April 2017): 580–586.

Cook, John, William R.L. Anderegg, Naomi Oreskes, J. Stuart Carlton, Peter T. Doran, Stephan Lewandowsky, Sarah A. Green, et al., "Consensus on Consensus: A Synthesis of Consensus Estimates on Human-Caused Global Warming". *Environmental Research Letter* 11 (2016): 048002.

Döhler, Marian. *Die politische Steuerung der Verwaltung*. Baden-Baden: Nomos, 2007.

Drori, Gili S., John W. Meyer, and Hokyu Hwang, eds. *Globalization and Organization, World Society and Change*. Oxford: Oxford Univ. Press, 2006.

Ehlers, Sarah, and Karin Zachmann. "Wissen und Begründen. Evidenz als umkämpfte Ressource in der Wissensgesellschaft". In *Wissen und Begründen. Evidenz als umkämpfte Ressource in der Wissensgesellschaft*, edited by Karin Zachmann, and Sarah Ehlers, 9–29. Baden-Baden: Nomos, 2019.

Fleck, Ludwik. *Entstehung und Entwicklung einer wissenschaftlichen Tatsache. Einführung in die Lehre vom Denkstil und Denkkollektiv*. Basel: Benno Schwabe, 1935.

France 2, "*Envoyé spécial, soirée spéciale 'Glyphosate, comment s'en sortir?'*", December 13, 2018. Paris: 118 minutes. https://www.france.tv/actualites-et-societe/magazines-d-actu/860147-glyphosate-comment-s-en-sortir.html.

Freimuth, Axel. "Vertrauenswürdige Wissenschaft: Alternativlosigkeit, Impact und andere Verführungen", In *Vom Umgang mit Fakten. Antworten aus Natur-, Sozial- und Geisteswissenschaften*, edited by Günter von Blamberger, Axel Freimuth, and Peter Strohschneider, 99–110. Paderborn: Wilhelm Fink, 2018.

Fricker, Miranda. *Epistemic Injustice: Power and the Ethics of Knowing*. Oxford: Oxford Univ. Press, 2009.

Gauchat, Gordon. "Politicization of Science in the Public Sphere: A Study of Public Trust in the United States, 1974 to 2010". *American Sociological Review* 77, no. 2 (April 2012): 167–187.

Gauchat, Gordon. "Public Views of Science-Based Policy and Funding. The Political Context of Science in the United States: Public Acceptance of Evidence-Based Policy and Science Funding". *Social Forces* 94, no. 2 (April 2015): 723–746.

Haack, Susan, "Clues to the Puzzle of Scientific Evidence". *Principia* 5, no. 1–2 (April 2001): 253–281.

Haraway, Donna. *Primate Visions: Gender, Race and Nature in the World of Modern Society*. London: Routledge and Kegan Paul, 1989.

Head, Brian, and John Alford, "Wicked Problems: Implications for Public Policy and Management". *Administration & Society* 47, no. 6 (April 2015): 711–739.

Hohlfeld, Ralf, Michael Harnischmacher, Elfi Heinke, Lea Lehner, and Michael Sengl, eds. *Fake News und Desinformation. Herausforderungen für die vernetzte Gesellschaft und die empirische Forschung*. Baden-Baden: Nomos, 2020.

IARC, "Some Organophosphate Insecticides and Herbicides", Monographs on the Evaluation of Carcinogenic Risks to Humans *Volume* 112, 2017. https://monographs.iarc.who.int/wp-content/uploads/2018/07/mono112.pdf.

INRA. "Usages et alternatives au glyphosate dans l'agriculture française". Rapport 2017. Paris, 2017.

Jasanoff, Sheila. *Science and Public Reason*. London: Routledge, 2012.

Krott, Max, and Michael Suda, eds. *Macht Wissenschaft Politik? Erfahrungen wissenschaftlicher Beratung im Politikfeld Wald und Umwelt*. Wiesbaden: VS Verlag für Sozialwissenschaften, 2007.

Latour, Bruno. *Le métier de chercheur: Regard d'un anthropologue*. Versailles: Editions Quae, 2001.

Leahey, Erin. "Overseeing Research Practice: The Case of Data Editing". *Science, Technology & Human Values* 33, no. 5 (April 2008): 605–630.

Levidow, Les, Kean Birch, and Theo Papaioannou. "Divergent Paradigms of European Agro-Food Innovation: The Knowledge-Based Bio-Economy (KBBE) as an R&D Agenda". *Science, Technology, & Human Values* 38, no. 1 (April 2012): 94–125.

Longino, Helen E. *Science as Social Knowledge: Values and Objectivity in Scientific Inquiry*. Princeton, New Jersey: Princeton Univ. Press, 1990.

Lock, Irina. "Debating Glyphosate: A Macro Perspective on the Role of Strategic Communication in Forming and Monitoring. A Global Issue Arena Using Inductive Topic Modelling". *International Journal of Strategic Communication* 14, no. 4 (April 2020): 223–245.

Mannheim, Karl. *Die Strukturanalyse der Erkenntnistheorie*. Berlin: Reuther & Reichard, 1922.

Mayntz, Renate. "Speaking Truth to Power: Leitlinien für die Regelung wissenschaftlicher Politikberatung". *dms—der moderne Staat* 2, no. 1 (April 2009): 5–16.

McIntyre, Lee. *Post-Truth*. Cambridge: MIT Press, 2018.

Moore, Alfred, and Jack Stilgoe. "Experts and Anecdotes: The Role of 'Anecdotal Evidence' in Public Scientific Controversies". *Science, Technology, & Human Values* 43, no. 5 (April 2009): 654–677.

Mouffe, Chantal. *Agonistics: Thinking the World Politically*. London: Verso, 2013.

Nowotny, Helga, Peter Scott, and Michael Gibbons. *Re-Thinking Science: Knowledge and the Public in an Age of Uncertainty*. Cambridge: Polity Press, 2001.

Nullmeier, Frank. "Knowledge and Decision-Making", In *Democratization of Expertise? Exploring Novel Forms of Scientific Advice in Political Decision-Making*, edited by Sabine Maasen and Peter Weingart, 123–134. Wiesbaden: Springer, 2005.

Penders, Bart. "Why Public Dismissal of Nutrition Science Makes Sense. Post-Truth, Public Accountability and Dietary Credibility". *British Food Journal* 120 (2018): 1953–1964.

Roqueplo, Philippe. *Entre savoir et décision, l'expertise scientifique*. Paris: inra éditions, 1997.

Roqueplo, Philippe. "Scientific Expertise Among Political Powers, Administrations and Public Opinion". *Science and Policy Studies* 22, no. 3 (April 1995): 175–182.

Söffner, Jan. "Truth Politics: Warum wissenschaftliche Expertise und parlamentarische Demokratie gleichzeitig in die Krise geraten sein könnten", In *Vom Umgang mit Fakten. Antworten aus Natur-, Sozial- und Geisteswissenschaften*, edited by Günter von Blamberger, Axel Freimuth, and Peter Strohschneider, 131–142. Paderborn: Wilhelm Fink, 2018.

Staley, Kent. "Robust Evidence and Secure Evidence Claims". *Philosophy of Science* 71 (2004): 467–488.

Van Dyk, Silke. "Krise der Faktizität? Über Wahrheit und Lüge in der Politik und die Aufgabe der Kritik". *PROKLA* 47, no. 188 (April 2017): 347–367.

Voßkuhle, Andreas. "Sachverständige Beratung des Staates", In *Handbuch des Staatsrechts* (3rd ed.), edited by Joseph Isensee and Paul Kirchhof, 425–476. Heidelberg: C.F. Müller, 2005.

Walton, Douglas, and Nanning Zhang. "The Epistemology of Scientific Evidence". *Artificial Intelligence and Law* 21, no. 2 (April 2012): 173–219.

Weber, Max. *Collected Methodological Writings*, edited by Hans Henrik Bruun and Sam Whimster. London: Routledge, 2012 (Original work published 1904).

Weber, Max. *Essays in Sociology*, translated, edited and with an introduction by H. H. Gerth, and C. Wright Mills. Oxford: Univ. Press, 1946.

Weber, Max. *Economy and Society: An Outline of Interpretive Sociology*. Berkeley: University of California Press, 1978.

Weiss, Carol. "The Many Meanings of Research Utilization". *Public Administration Review* 39, no. 5 (April 1979): 426–431.

Zachmann, Karin, and Sarah Ehlers, eds. *Wissen und Begründen. Evidenz als umkämpfte Ressource in der Wissensgesellschaft*. Baden-Baden: Nomos, 2019.

Part II

Striving for Diverse Evidence

Ecological and Biological Evidence Critique

3 How Many Plots Can the Data Hold? Reconciling Stories and Evidence in Evolutionary Biology

Martha Kenney

The universe is not only queerer than we suppose,
but queerer than we can suppose.

- J.B.S. Haldane

This is a chapter about the sometimes uneasy, often generative relationship between storytelling and scientific evidence. As a feminist science studies scholar, I argue that we not only need to critique politically charged narratives lurking in scientific knowledge claims, but, in order to tell better stories about the world, we must embrace storytelling as an important component of scientific practice.[1] In an era characterized by the unprecedented circulation of fake news, disinformation, science denialism and conspiracy theories, this might feel like a dangerous proposition. Writing from the United States in late 2021, where 30 percent of adults are still not vaccinated for COVID-19, it seems particularly urgent to separate scientific fact from fiction, opinions or "just stories". However, this stark opposition between meaning, esthetics and subjectivity, on the one hand, and facts and objectivity, on the other, is fundamentally an artificial distinction. Although it may be epistemologically comforting, it is impossible to banish socially and culturally situated forms of meaning-making from our knowledge-making practices. As feminist philosopher Helen Longino argues, background beliefs, which include the narrative substrate of scientific inquiry, are, in fact, "an enabling condition of the reasoning process" and therefore intrinsic to scientific inquiry.[2]

In this chapter, I argue that stories and storytelling are not the enemy of empiricism and are often at the heart of generating both evidence and evidence criticism. In particular, attention to stories can help us contest established orthodoxies in the natural sciences and stimulate new avenues of empirical investigation. To build this argument, I draw on the work of philosopher Isabelle Stengers, who argues for the importance of calling dominant scientific narratives into question in order to create what Donna Haraway has characterized as "a more adequate, richer, better account of a world".[3] Stengers writes: "I am convinced we need other

DOI: 10.4324/9781003273509-6

narratives, narratives that populate our worlds and imaginations in a different way".[4] Importantly, for Stengers, this call is not in opposition to empirical inquiry but in the service of a more capacious empiricism that actively resists "explaining away what would complicate our judgement"[5] and is capable of contesting received categories and explanations. She calls her philosophical proposition, "a materialism without eliminativism".

In what follows, I briefly outline Stengers's materialism without eliminativism and then turn to the work of evolutionary biologist Joan Roughgarden to illustrate this proposition, paying particular attention to the relationship between story and evidence in Roughgarden's books *Evolution's Rainbow* (2004) and *The Genial Gene* (2009) as well as her many articles. This central case study is based on a literature review of Roughgarden's scientific and popular work critiquing sexual selection, as well as reviews of Roughgarden's books by evolutionary biologists and the evolutionary biology literature that has since cited Roughgarden. I outline the controversy her critique of sexual selection garnered and how, despite her even-handed and empirically grounded defense of her work, Roughgarden's proposal has largely been dismissed by her colleagues in evolutionary biology. Honoring the intimate relationship between narrative and evidence in Roughgarden's work, I conclude by offering the speculative question, "How many plots can the data hold?" as a provocation to embrace storytelling as an active part of scientific practice, and to open up new pathways of narration, relation and investigation in evolutionary biology and beyond. Finally, I provide a short epilogue to demonstrate how the arts and humanities, in this case poetry, can help generate promising alternatives to dominant biological narratives.

Materialism without Eliminativism

In "Diderot's Egg: Divorcing Materialism from Eliminativism", Isabelle Stengers argues for a demanding materialism, a materialism that can explain "how, with matter, we get sensitivity, life, memory, consciousness, and thought".[6] She tells the story of French philosopher Denis Diderot, who produces an egg and declares, "Do you see this egg? With this you can overthrow all the schools of theology, all the churches of the earth".[7] Stengers argues that Diderot's egg not only offers a materialist challenge to theology, but also a challenge for materialists to embrace a concept of matter that can account for complex phenomena: "The challenge of the egg points to what is required from the egg in order for the development of the chicken not to be a miracle".[8] Stengers imagines a materialism capable of causing wonder, surprise and alarm – a materialism that affirms that there are more things in heaven and earth than are dreamt of in our philosophy, to paraphrase Hamlet.

Stengers's proposition is not only philosophical and empirical, but also political. She argues that we must divorce materialism from eliminativism in order to connect materialism with worldly, political struggle. A materialism without eliminativism would not use the decree "science says" to shut down controversy but would embrace an empiricism that can actively contest "what matters and what is excluded from mattering".[9] Stengers gives the example of debates over genetically modified organisms (GMOs), where "the certainties of lab biologists silenc[e] those colleagues who work outside of the lab and ask different and perplexing questions".[10] She characterizes Vandana Shiva's anti-GMO campaigns for food sovereignty in the Global South, in part, as a demand for "a relevant science, a science that would actively take into account the knowledge associated with those agricultural practices that are in the process of being destroyed in the name of progress".[11] Shiva's claims are often subject to scientific "debunking" on behalf of powerful actors such as Monsanto; Stengers yearns for a science capable of supporting activists like Shiva, rather than powerful, settled interests such as corporate global biotech. A materialism without eliminativism would be capable of unsettling the status quo and challenging a "closed definition of a rational science".[12]

In the face of this definition of matter, a matter that always exceeds our ability to capture it, we need a more capacious understanding of empiricism, an empiricism where we can open up the question of what counts as evidence and ask what other stories about the natural world are possible.

Joan Roughgarden's Alternative to Sexual Selection: Social Selection

To illustrate the intimate relationship between story and evidence in a materialism without elimativism, I turn to evolutionary biologist Joan Roughgarden, Professor Emerita at Stanford University in California. Her work is particularly exemplary of a methodology that is simultaneously narrative and empirical. In her 2009 book *The Genial Gene*, Roughgarden boldly critiques Darwin's theory of sexual selection as "inadequate to address the diversity of bodies, behaviors, and life histories that actually exist [in nature]".[13] The theory of sexual selection, which Darwin presents in *The Descent of Man and Selection in Relation to Sex* (1871), seeks to explain ornaments and armaments such as a stag's antlers or the bright feathers of a peacock's tail that do not appear to enhance survival. Instead, Darwin argues that these traits bestow a reproductive advantage. As males compete for access to females, females choose to mate with males who have the best example of that trait and, therefore, over time, it is selected for in the population at large, despite causing, for example, an increased risk of predation due to heavy antlers or colorful plumage. As Darwin argues, the peahen's "continued preference for the most beautiful males, rendered the peacock the most splendid of living birds".[14] Today, sexual selection remains the central theory for explaining ornaments and armaments and other traits arising from competition for mates.[15] In many Neo-Darwinian accounts, male ornaments are considered indicators of the genetic quality of males; the elaborateness of the peacock's tail can be seen as advertising his "good genes" to potential mates.[16]

While there are many species that do not conform to this template, this preponderance of evidence has not been seen as a challenge to sexual selection. For example, not all species of birds have bright males and cryptic females; there are many sexually monomorphic species such as the emperor penguin, in which biologists cannot identify the difference between males and females without examining their sex organs. Similarly, males do not always compete for access to females; there are species, where females solicit males for sex, such as the alpine accentors. However, these species are cast as outliers that do not challenge sexual selection itself but can be accounted for with some small amendments; the alpine accentor, is considered a "sex-role reversed" species, an exception that seemingly proves the rule. However, as Roughgarden points out, these species are not seen as challenging the central narrative of sexual selection: "Sexual selection advocates resist a critique that emphasizes what they regard as exceptional species, because such a critique ignores the supposedly many cases in which they claim the sexual selection narrative is correct".[17]

For this reason, Roughgarden's critique of sexual selection not only enumerates the many outliers, but also returns to the species that seem to conform most with the Darwinian template. Roughgarden goes back into the most potent fables of sexual selection and retells them, looking for alternative hypotheses that explain the trait or behavior in question. In the first chapter of *The Genial Gene*, she revisits the peacock, the textbook example of sexual selection. Here, Roughgarden offers an alternative account of peacock evolution proposed by a team of Japanese researchers who found, in a six-year study of 105 free-ranging Indian peafowl, that peahens do not, in fact, prefer peacocks with more elaborate tails. Takahashi et al. provide an alternative evolutionary explanation for their results that directly contradicts the Darwinian account in which both peacocks and peahens were originally drab and the peacock became bright through sexual selection.[18] In this new story, it is the reverse: Both males and females were originally brightly colored. However, since females were more vulnerable to predation while incubating eggs, their coloring became more cryptic over time due to natural selection, while the males remained bright. This new explanation is supported by evidence from their study, which found that despite their cryptic coloring females were twice as vulnerable to predation than males. They also cite other studies that show that bright plumage color in galliform birds like the peacock is inhibited by estrogen, suggesting that the cryptic coloring of the female is the more recently derived trait. When sexual selection could not account for the empirical evidence collected in this peacock study, there was a narrative opening: What can account for this phenomenon if not sexual selection? Is there evidence to support a different story?

The findings from Takahashi et al. raise a larger evolutionary question: Why do peacocks continue to have brightly colored plumage, when brightly colored plumage is more likely to lead to predation? Roughgarden suggests an alternative hypothesis. Rather than communicating information about a male's "good genes" to potential mates, Roughgarden suggests that brightly colored tails act instead as "admission tickets to power-holding cliques that control the resources for successful rearing of offspring".[19] In Roughgarden's account, bright tails are part of a wider set of "social inclusionary traits" that not only consist of physiological traits that serve social functions, but social behaviors like grooming that promote teamwork among individuals. This new story is about maintaining the social structure to successfully rear young; here behaviors such as teamwork emerge as evolutionarily significant, in opposition to the Darwinian story that always centers reproductive success.

Roughgarden's peacock story is one illustration of "social selection", the theory that she has devised to replace sexual selection. Social selection is defined as "selection for, and in the context of, the social infrastructure of a species within which offspring are produced and reared"[20]; here evolutionary success is defined not as individual reproductive success but is based on the number of offspring that are reared into the next generation. When we pay attention to the social structure in which offspring are produced and reared, mating is no longer a privileged behavior; rather it constitutes one "reproductive social behavior" among many:

> Social selection views reproductive social behavior as comprising an 'offspring-rearing system'. Within this system, natural selection arises from differences in the number of offspring successfully reared, and particular behaviors are understood by how they contribute to building, or maintaining, the social infrastructure within which offspring are reared.[21]

Within this framework, animal behaviors that are poorly accounted for by sexual selection, such as same-sex sexual behavior, become intelligible as reproductive social behavior. Sexual selection explains away same-sex sexual behavior by claiming it is "an inadvertent mistake, a deception, or a deleterious trait";[22] in social selection, same-sex sexuality is a reproductive social behavior that promotes intimacy and teamwork,[23] therefore strengthening the social infrastructure within which offspring are reared.

While Roughgarden believes social selection offers a better account for the morphology and behaviors of animals than sexual selection, she ultimately argues that empirical inquiry should be privileged over her narrative instincts. Throughout *The Genial Gene*, Roughgarden emphasizes the need to work against confirmation bias in evolutionary biology and to test multiple hypotheses. She writes: "I can't overemphasize enough that evolutionary theories, like theories in any other area of biology, such as molecular biology, genetics, and physiology are destined for testing".[24] Roughgarden presents social selection not as a complete or confirmed theory, but a set of hypotheses to be proven. At the end of *The Genial Gene*, she writes: "Time will tell whether social selection is indeed correct or whether some substantial modification or third approach is needed".[25] *The Genial Gene* is an invitation for more hypotheses about evolutionary biology that can account for the sheer diversity of social behaviors and physical traits in the natural world. "The welcoming door is open", she concludes, "come on in".[26]

Described this way, Roughgarden's argument does not, at first, appear controversial. Indeed, it is firmly rooted in a conventional understanding of scientific rigor: Review evidence in the existing literature, test additional hypotheses, avoid confirmation bias, publish both positive and negative results, consider alternative explanations when the data do not support the dominant theory. However, when she published her critique of sexual selection in *Science* in 2006,[27] over 40 prominent figures in evolutionary biology submitted 14 letters to the editors of *Science* that viciously criticized her claims and methods,[28] writing that her work represents an "attack on Darwin",[29] that she fails to understand the scientific method, and threatens to set evolutionary biology back by some 30 years.[30] After this initial attack, Roughgarden's views have been actively ignored and marginalized in the peer-reviewed literature, often being called "idiosyncratic",[31] "unusual",[32] "unorthodox",[33] representing only a "fringe viewpoint"[34] and "marred by bias".[35] Given her reasonable, welcoming and scientifically sound rhetoric, why have all of these evolutionary biologists responded to her work as if to a threat?

One obvious answer is that Roughgarden's 2004 book *Evolution's Rainbow* makes an explicitly political intervention into evolutionary biology. She investigates the many species that differ from the male and female templates of sexual selection, the many species that engage in same-sex sexual behavior and the many species that raise young in social structures that have little in common with heteronormative monogamy or the nuclear family. In this earlier book, Roughgarden speaks openly about her own identity as a transgender woman and about the need for evolutionary biologists to pay attention to the wealth of primary data on animal diversity for both political and scientific reasons. Although *The Genial Gene* forwards a very different project that speaks specifically to Darwin's theory of sexual selection and Roughgarden's proposed alternative, social selection, her colleagues' distaste for her arguments is clearly colored by their reaction to *Evolution's Rainbow*. This view is exemplified by the remarks of evolutionary biologist Troy Day who claimed that many people believe Roughgarden's work to be "shoddy science and poor scholarship, all motivated by a personal agenda".[36]

Here we see a familiar story, where scientific objectivity is only accessible to some subject positions, and everyone else's claims are considered to be "just politics". "Personal agenda" is a barely disguised euphemism for Roughgarden's openly LGBTQ+ politics, which she dared to include in a book about evolutionary biology. Her personal and political motivations are wielded against her as the incontrovertible evidence for her lack of objectivity and her "shoddy science", which can be explained away as an effect of politics.[37] As Donna Haraway writes in her *Primate Visions* chapter on feminist paleoanthropologist Adrienne Zilhman: "woman the scientist becomes the trope figuring bias".[38] And here, for "trans woman the scientist" this is doubly so – as evidenced by the blatant transphobia in some biologists' responses to *Evolution's Rainbow*.[39] According to these "critics", scientific *impropriety* is her cardinal sin. First, she immodestly goes after Darwin's master narrative of sexual selection from a marginalized subject position; Roughgarden explains that as an LGBTQ+ scientist, "it would [have been] okay to add a little fluff to sexual selection to account for gay and gender-bending animals, so long as I [did] not touch the central narrative".[40] Secondly, by insisting on speaking politically in a scientific context, she is stubbornly "not in accordance with good manners, modesty, or decorum; unbecoming, unseemly; indecorous, indecent".[41] Put more simply: *How dare you?*

Roughgarden's Empirical Defense: Test More Hypotheses

Despite Roughgarden's explicit political stakes, she is careful to defend her work in strictly empirical terms, stressing the importance of testing hypotheses and finding the truth. In interviews, talks and popular articles, Roughgarden has responded to criticism, not by entrenching further into her own position but by inviting more evolutionary inquiry. This strategy can be seen, for example, in an online video interview with science journalist Robert Wright.[42] Wright, who does not hide his skepticism of Roughgarden's claims, presses her about her critiques of evolutionary psychology studies that are based in sexual selection theory. Wright cites a well-known study by Martin Daly and Margo Wilson, which claims that in humans, men are more jealous about the sexual infidelity of their partners, whereas women are more jealous of emotional infidelity.[43] Their theory is based on a sexual conflict model that posits that males and females have opposing goals in reproduction and child rearing. The difference in the kind of jealousy experienced by men and women, they argue, is an evolutionary adaptation that helps males ensure that they raise their own offspring and helps females ensure that the males will stick around to provide for their children. They write:

> Sociobiologists would expect male jealousy to be more specifically focused upon the sexual act than female jealousy. This is because the reproductive threat in a wife's infidelity lies in the risk of alien insemination, whereas the reproductive threat in a husband's infidelity lies more in the risk of lost resources.[44]

Wright is convinced that this theory makes excellent intuitive sense and is also borne out by the "raw data". What interests me here is not only Wright's certainty about the "raw data" from a decades old study, but Roughgarden's response to Wright. She does not point out the gender stereotypes and heteronormative assumptions embedded in Daly and Wilson's jealousy story. Instead she asks Wright "whether there was an alternative hypothesis that was on the table, which they could have confirmed, instead of the one they claimed to have confirmed". This strikes me as a tactful and potentially disarming response to Wright's line of questioning. Roughgarden suggests that Daly and Wilson's study, or at least the way Wright remembers it, might be "bad science". Although she makes it clear that she is a proponent of evolutionary psychology as a field, she is concerned about the prevalence of studies where the "data are mined to effect an appearance of the confirmation of [a single] hypothesis". In her response to Wright, Roughgarden demands a more rigorous relationship between narrative and evidence. She insists that evolutionary psychologists test more hypotheses.

This raises the question of why scientific stories that make intuitive sense are perceived not only to be true, but supported by the empirical evidence. Historian of science Daryn Lehoux argues that it is common for us to credit empirical investigation when we are actually relying on commonsense beliefs.[45] His example comes from the ancient world. Greek and Roman authors like Plutarch, Ptolemy and Pliny the Elder believed – according to the laws of sympathy and antipathy – that if you rub garlic on a magnet, you can disable its attractive force. From today's perspective this is obviously false, so why did these illustrious classical thinkers believe it? Were the ancients simply dimwitted? Lehoux argues that, no, the ancients were not dimwitted and, in fact, our belief that garlic cannot disable a magnet is no more empirically sound than the ancients' belief that it can. Have you ever rubbed garlic on a magnet? Probably not. Lehoux suggests, "our argument against the garlic-magnet antipathy is no stronger, and, more importantly, no more or less empirical, than Plutarch's argument for it".[46] Both of us are using categorical thinking to ground our beliefs: "Magnets used to be the *kind of thing* that was sympathetic, as was garlic. Now magnets are the kind of thing that are magnetic, and garlic in our experience is not".[47]

In both cases, Lehoux argues, "classification was doing the epistemolog-ical work, but experience was getting the credit".[48] When Roughgarden suggests that evolutionary biologists test more hypotheses, she encour-ages empirical investigation in a field where commonsense beliefs about the world are doing some of the heavy lifting. She has proposed a field called "sexual selection studies" specifically for the purpose of testing new hypotheses that diverge from the Neo-Darwinian orthodoxy.[49]

However, although Roughgarden's proposal offers new empirical directions for evolutionary biologists, her call to test more hypothe-ses does not fully capture the role of personal experience and political commitments in Roughgarden's work. Although her personal and polit-ical motivations are clear in *Evolution's Rainbow*, she often dismisses their influence when discussing the importance of empirical inquiry. For example, writing about the dominance of sexual conflict narratives in evolutionary biology that make males and females into enemies and reproduction into warfare, Roughgarden writes:

> Yet again, the issue before us is not whether one finds these thinly disguised rape narratives appealing or repugnant. The issue is whether a kind of rape actually does underlie all male and female relationships throughout nature. Sexual-conflict advocates do not acknowledge even the possibility of alternative hypotheses springing from a different point of view. Nonetheless, the scientific method requires alternative hypotheses.[50]

It is clear from her use of the description "thinly disguised rape narra-tives" that Roughgarden does indeed find these stories repugnant and this repugnance motivates her desire to search for evidence to furnish other narratives that can "populate our worlds and imaginations in a different way".[51] As she distances herself from her own emotions in this passage ("the issue is not whether we find these narratives appealing or repugnant"), I cannot help but wonder if it is possible to affirm these personal, political and empirical desires at the same time so that we can avoid "splitting our life, our bodies, our language, our breath into several worlds".[52]

Reconciling Story and Evidence

In this chapter, I move away from Roughgarden's exclusively empirical defense of her work to argue that her political instincts are one of her essential epistemological strengths as an evolutionary biologist. If, as Isabelle Stengers argues, a materialism without eliminativism allows us to connect with the struggle over what matters and what is excluded from mattering, this is surely a materialism that allows us to imagine story-telling as a significant part of empirical inquiry. As a trans woman with a commitment to LGBTQ+ politics, Roughgarden is acutely attuned to the hetero and cis normativity lurking within the sexual selection literature that automatically pathologizes any departure from the universal male and female templates of sexual selection. Here she identifies the moralizing language used to describe different forms of sexual behavior and morphology:

> In the primary peer-reviewed literature, males are described as being "cuckolded", females as "faithful" or "promiscuous", offspring as "legitimate" or "illegitimate", males who do not hold territory as "floaters" or "sneakers" (code for "sneaky fuckers") all of whom are "sexual parasites", small males as "gigolos", feminine males as "female mimics" or even as "transvestite serpents" or "she-males" (a pornographic reference) and so forth.[53]

Examples of this kind of moralizing language in evolutionary biology and science media are not hard to find. For example, there is a short segment from the popular BBC documentary series *Life* (2009) that illustrates the ubiquitous pathologization of feminine males. This segment features the giant cuttlefish – a species with two distinct phenotypes of males: A large-territory holding male and a small feminine male. We see a female with a large male cuttlefish, being approached by a small male. David Attenborough's familiar and sonorous voice reads:

> Cuttlefish are good communicators, but there is a flip side; they can also be *masters of deception*. This male is too small to fight for a mate. But he has another plan and it's *sneaky*. He approaches the couple cautiously, holding his tentacles tucked up at the front, *mimicking a female* that wants to mate. To complete his *disguise*, he changes color to appear even more like a female. The guarding male seems *convinced*. Maybe he thinks his luck is in; another female to add to his conquests. *The sly crossdressing male* edges closer and closer to the female, holding his nerve. As long as he avoids being grabbed in a mating embrace, *the sneak* is safe. At what point the female guesses his true identity is unclear, but she isn't choosey and *surreptitiously* mates with him right under the larger male's tentacles. It's time for the female to lay her eggs. Using the sperm from both males, she fertilizes her eggs one by one…with luck, some may become *masterful males* and others, *little sneaks* – she'll have all the bases covered.
>
> (emphasis mine)

The italicized language used to speak about the small male cuttlefish relies on an unspoken and arbitrary parallel between this male phenotype of cuttlefish and human trans women. This narrative, furthermore, employs a particularly ugly stereotype, namely that trans women are illegitimate women, who mimic "real" women in order to fool hetero-sexual men and sneak into women-only spaces – an ideology increas-ingly contested, but nonetheless omnipresent in American culture from *The Crying Game*, to the Michigan Womyn's Music Festival, to the recent proliferation of so-called Bathroom Bills.[54] When this rhetoric is isolated and subjected to even the most rudimentary critical thought, it becomes clear that evolutionary biologists have grafted a repugnant human stereotype onto the natural world and left it there, hidden in plain sight. How could this narrative be anything but woefully inade-quate to describe the behavior of these socially complex cephalopods? As Emily Martin argues in her classic article "The Egg and the Sperm", it is not only politically important to identify harmful stereotypes in scientific literature, but that this activity also benefits empirical inquiry: "Waking up such metaphors, by becoming aware of when we are pro-jecting cultural imagery onto what we study, will improve our ability to investigate and understand nature".[55]

Having identified the transphobia lurking in the literature on species with multiple male phenotypes, Roughgarden is able to subject the claim to empirical scrutiny. Is the large male actually duped by the smaller male? Roughgarden does not think so: "The territory-holding male is often a visual predator with well-honed skills at sizing up and identi-fying prey from a distance; he is not likely to be fooled by a feminine male who only imperfectly resembles a female".[56] Here Roughgarden's personal and political distaste for language that polices gender, sexual-ity and kinship is the motivation for narrative speculation and further empirical inquiry: *Does the evidence support this narrative? Is there another story? Can the evidence support another story?* Is there a way to affirm, rather than deny the importance of narrative – narrative affin-ity, narrative repulsion – in Roughgarden's work without also losing the scientific rigor or the strength of her knowledge claims? To embrace storytelling as scientific practice without fear of relativism or perma-nently de-stabilizing evidence in evolutionary biology? To grapple with the political questions of biological storytelling rather than pretending that science is free of politics? These are the questions that remain for me at the end of *The Genial Gene*. I am left wanting a different way of imagining the relation between story and evidence, where narrative is not ultimately subordinate to data.

How Many Plots Can the Data Hold?

The question that incites me is not Roughgarden's question: "Is this story true?", but something more speculative, a question that opens up rather than forecloses narrative possibilities in evolutionary biology. Working from Roughgarden's commitment to testing more hypotheses, the question I want to ask instead is: "How many plots can the data hold?"[57] Here I use Sabina Leonelli's definition of "data" as "any product of research activities...which is collected, stored and disseminated in order to be used as evidence for knowledge claims".[58] With this question, with this refrain,[59] I'm playing with the multiple meanings of the word "plot". Plotting a graph is the practice of drawing relations between data points. Data points are not meaningful on their own, but must be brought into relation with one another to become evidence. There are different ways of connecting the dots, as it were; some better than others. What is included, what is excluded, what becomes signal, what is noise is decided through a process of interpretation. As Leonelli suggests, "the same set of data can act as evidence for a variety of phenomena, depending on how they are interpreted – a feature...central to understanding the epistemic power of data as research components".[60]

"How many plots can the data hold?" uses graphing as a visual metaphor for thinking about the relationship between narrative and evidence. Plotting is, of course, also the practice of storytelling. Of identifying meaningful actors and actions, of ordering them in time, of discerning cause and effect. "How many plots can the data hold?" can open up new pathways of narration, relation and investigation. To plot a graph and to plot a story are both meaning-making activities. "How many plots can the data hold?" asks us to multiply our meaning-making strategies. To try on different stories, categories and metaphors to make sense of our world. The answer to this question might be more than one.[61] But it is also not a relativist question; the answer is not infinite.

This refrain is my attempt to pose a question at the intersection of story and evidence. A question that is simultaneously speculative and pragmatic. A question that feels actionable. I developed this formulation to find another way to approach scientific work like *The Genial Gene*, where we would not have to separate the empirical/scientific from the narrative/political. What is exciting to me in *The Genial Gene* is that it takes us inside the most familiar biological stories and insists another story is possible. The peacock's tail might not be the ultimate example of a world in which males compete for access to females, who choose their mates based on bright colors or good genes. When Takahashi et al. found that their data would not hold the sexual selection narrative, they looked for evidence that would furnish a different story. They investigated the role of estrogen in plumage color; Roughgarden investigated the social structure in which peafowl rear their offspring. Going back into the peer-reviewed literature, into studies that have already been completed, and asking "How many plots can the data hold?" offers different empirical and narrative possibilities for the practice of doing evolutionary biology.

Asking "How many plots can the data hold?" invites the kind of empirical, narrative, speculative and political inquiry, where evolutionary biology could be called upon to contest rather than confirm an established order. Not by dropping science and picking up politics, but by working with story and data simultaneously, with equal skill and seriousness. This is what Isabelle Stengers means by a materialism without eliminativism. To approach this question with skill and seriousness, I believe, we cannot rely on familiar knowledge-making practices. We would need to devise new connections between the sciences and the arts and humanities, new ways of working together and new kinds of accuracy.

Accuracy is a term that we tend to associate with the methods of natural sciences, with statistical practices such as excluding outliers from datasets and calculating margins of error. However, in the arts and humanities, we also experience our activities as a process of working with and toward accuracy. Take, for example, this excerpt from an interview with Canadian poet Anne Carson on the topic of writing:

> we're talking about the struggle to drag a thought over from the mush of the unconscious into some kind of grammar, syntax, human sense; every attempt means starting over with language. starting over with *accuracy*. i mean, every thought starts over, so every expression of a thought has to do the same. every accuracy has to be invented.[62]

Carson's accuracy is not a scientific accuracy that is about bringing a measurement as close as possible to the correct value, but a linguistic accuracy. It pertains to the difficulty of capturing, or maybe *rendering*, something with language. The poetic injunction to avoid hackneyed signifiers and to begin again every time with language as if from scratch.

This is why the empirical and narrative question "How many plots can the data hold?" is fundamentally an interdisciplinary question. It requires people trained in different kinds of disciplinary accuracy, consummate storytellers and consummate observers, experimenters, statisticians, working together on a practice where neither the data nor the story is foundational, but where they must be handled together. "How many plots can the data hold?" is an invitation not just to mix preexisting disciplinary practices, but to *invent new forms of accuracy* that might be unfamiliar or awkward but could be epistemologically, narratively and politically generative.

This proposal requires a fundamentally different kind of interdisciplinary, one that is open-ended and nonhierarchical. As I have argued elsewhere,[63] the pressure on scientists to publish disincentivizes lengthy and uncertain collaboration. Making space for scientists and humanities scholars to learn from one another, to pursue mutual curiosity and devise new ways of working together is necessary for true interdisciplinarity, as is allowing time for frustrations, dead-ends and failures. With its focus on productivity and quantifiable outcomes, the contemporary neoliberal university is ill-equipped to support these kinds of uncertain and exploratory collaborations. However, if we want to resist the status quo and forge new relations between stories and evidence, we must value the slow work of coming together around problems that we thought had already been settled and ask questions like, "How many plots can the data hold?"

Of course this is not the only way to imagine the relation between story and evidence. Data does not neatly precede story; data is always already theory-laden.[64] Stories guide how we pay attention to the world and what seems significant in the first place. And sometimes the work to craft just one good story or even a partial story from the chaos of the data is the real epistemological and political challenge. I offer "How many plots can the data hold?" as a seed, a meme, a spark. A call for fewer "Just So" stories and more "What if?" stories.[65] To use the speculative to crack open the ordinary. To embrace rather than disavow the storytelling practices of scientists like Joan Roughgarden. To fashion a possible attachment site for cross-disciplinary collaboration. To invent new forms of accuracy. To practice a materialism without eliminativism.

Epilogue: The Oulipo Cuttlefish

I conclude my chapter, not with a further critique of the dominant accounts of sexual selection, but with an alternative feminist narrative about the cuttlefish from BBC *Life*. To write this story, I used the Oulipo technique of constraint-based writing. The Oulipo are a group of mostly French poets and mathematicians, who write using self-imposed constraints or protocols, the most famous of which is Georges Perec's novel *La Disparition* (1969), written without the use of the letter "e", the most common letter in the French language. However, the purpose of Oulipo writing is not only creating works that demonstrate linguistic playfulness or formalist mastery but, as their name suggests,[66] devising procedures for thinking otherwise:

> For the Oulipo, constrained writing does not necessarily aspire to the creation of a literary work; rather, it participates in a general research program invested in interdisciplinary invention, collaborative innovation, ludic approaches to writing and reading, and the elaboration of new economies of expression, complex literary forms that become the springboard for a speculative lit[erature].[67]

For this task, I decided to describe what I was seeing on the screen, while using only the vowels A and I (no E, O, or U). I chose this lipogram exercise for generating my counter-narrative, because of the way constraints help us to avoid the most naturalized and clichéd language. As I was composing my script, I was most struck by how the frustration of not being allowed to use a desired word quickly gave way to new opportunities, how unanticipated possibilities emerged when I found my narrative pathway blocked. Occasionally an appropriate synonym was available – but even then, my intended meaning was already shifting, leading in different directions than I had initially planned; I found "many subtle channels"[68] opening up before me.

In this short narrative, which begins with the small male approaching the female and the large male, I tried to imagine cuttlefish mating without recourse to pathological language or the concept of deception:

> Radiant fish, with his skin alight, all arms and charisma. Swimming in amid this captivating pair, his arms in tight, shifting his skin, flashing dark and bright. His timing is right. This striking fish with rippling fins, inviting his mating arm in, is brilliant and satisfying against his skin. As is this vigilant watchman draping his arms, incasing this activity with his vast calm. It is as if it was always like this. *S. apama* bliss.

In this narrative, the cuttlefish, who were previously possessive and sneaky, become serene and alluring; each individual participates actively and knowingly in the intimacy of the scene. Without the words "male" or "female", without the full complement of English pronouns, it becomes necessary to find other ways to name the animals and describe their behaviors. In the absence of pronouns, a multitude of adjectives rush in, beckoning the listener into the encounter: Radiant, striking, rippling, brilliant, satisfying, vigilant and vast. Unlike Attenborough's narration, this story asks us to consider the subjectivity of animals without recourse to a mechanistic nature that is red in tooth and claw. While it may feel anthropomorphic to ascribe particular moods and thoughts to the cuttlefish, is it any more anthropomorphic than Attenborough's story? Why does a deceptive cuttlefish seem more sober and realistic than a blissed-out cuttlefish? Why is it so risky for evolutionary biology to "admit pleasure, play, or improvisation within or among species"?[69]

How can a story like this sensitize us differently to the lives of animals? What do we notice that we did not see before? What captures our curiosity when we do not feel we already know the story before it has been told? This is, of course, not the only alternative to the transphobic sexual selection narrative. What story would you tell me about the cuttlefish if I gave you an E and an O?

Oulipo writing is just one possibility for exploring resistances and potentials of language, as we seek to tell better stories about the natural world. In answering the question "How many plots can the data hold?", we can experiment with form, not in pursuit of the avant-garde, but as a way of activating different bio-poetic and therefore bio-political possibilities. Although these stories might ultimately not account for complex natural phenomena, such as a species with multiple sex phenotypes, these poetic hypotheses can reorient our attention and reroute our curiosities. In this way, evolutionary stories can act as fables of attention, helping us ask animals better questions,[70] which in turn can generate new evolutionary hypotheses. When it is obvious that the dominant story is not enough, as is the case with the cuttlefish, we need to open up the relationship between story and evidence, to generate new questions and see where these questions can take us. A materialism without eliminativism affirms this kind of "interpretive adventure",[71] over and against forces that would settle scientific facts once and for all. In this way, Stengers encourages us to disavow scientism and affirm a science capable of connecting with the diverse and ongoing struggles for more livable technoscientific worlds.

Acknowledgments

Thank you to the organizers and participants of "Story as Evidence: Communicating Science" at Duke University, "Experiments in E/valuation Workshop" at Leiden University and "Critiquing Evidence Criticisms" at the Technical University of Munich for engaging with earlier versions of this paper, as well as Donna Haraway, Karen Barad and Jenny Reardon for feedback on my dissertation work that provided the basis for this chapter.

Notes

1 Martha Kenney, "Fables of Response-ability: Feminist Science Studies as Didactic Literature", *Catalyst: Feminism, Theory, Technoscience 5*, no. 1 (April 2019).
2 Helen Longino, *Science as Social Knowledge: Values and Objectivity in Scientific Inquiry* (Princeton: Princeton Univ. Press, 1990). See also: Donna Haraway, *The Haraway Reader* (New York: Routledge, 2004), 201.
3 Donna Haraway, "Situated Knowledges: The Science Question in Feminism and the Privilege of Partial Perspectives", *Feminist Studies* 14, no. 3 (1988): 575–599.
4 Isabelle Stengers, "Diderot's Egg: Divorcing Materialism from Eliminativism", *Radical Philosophy* 144 (July/August 2007): 9.
5 Stengers, "Diderot's Egg", 11.
6 Stengers, "Diderot's Egg", 10.
7 Stengers, "Diderot's Egg", 9.
8 Stengers, "Diderot's Egg", 10.
9 Karen Barad, *Meeting the Universe Halfway: Quantum Physics and the Entanglement of Matter and Meaning* (Durham: Duke Univ. Press, 2007), 220.
10 Stengers, "Diderot's Egg", 8.
11 Stengers, "Diderot's Egg", 8.
12 Stengers, "Diderot's Egg", 10.

13 Joan Roughgarden, *The Genial Gene: Deconstructing Darwinian Selfishness* (Berkeley: Univ. of California Press, 2009), 169.

14 Charles Darwin, *The Descent of Man and Selection in Relation to Sex* (Samford, Australia: Emereo Publishing, 2012 [1871]), 178.

15 Joan Roughgarden, "Sexual Selection: Is there Anything Left?" in *Current Perspectives in Sexual Selection: What's Left After Darwin*, ed. Thierry Hoquet (Dordrecht: Springer, 2015), 85–102.

16 In the most gene-centrist version of female choice in sexual selection, females choose males only for the quality of their genes, so that they can pass those genes on to their offspring. Two popular versions of this theory are called the "Good Genes Hypothesis" and the "Sexy Son Hypothesis". William D. Hamilton and Marlene Zuk, "Heritable True Fitness and Bright Birds: A Role for Parasites?" *Science* 218, no. 4570 (1982): 384–387; Patrick J. Weatherhead and Raleigh J. Robertson, "Offspring Quality and the Polygyny Threshold: 'The Sexy Son Hypothesis'", *American Naturalist* 113, no. 2 (1979): 201–208.

17 Roughgarden, *The Genial Gene*, 20.

18 Takahashi, Mariko, Hiroyuki Arita, Mariko Hiraiwa-Hasegawa, and Toshikazu Hasegawa, "Peahens do not Prefer Peacocks with More Elaborate Trains", *Animal Behavior* 75, no. 4 (2008): 1209–1219.

19 Joan Roughgarden, "Challenging Darwin's Theory of Sexual Selection", *Daedalus* 136, no. 2 (Spring 2007): 29.

20 Roughgarden, "Challenging Darwin's Theory", 24.

21 Roughgarden, "Challenging Darwin's Theory", 26.

22 Roughgarden, *The Genial Gene*, 244.

23 This is also true for most heterosexual sex for humans and other animals. As Roughgarden points out: "Mating initiation and frequency is much more extensive than needed for offspring production, suggesting...that mating serves social purposes". Reproductive necessity remains the alibi for heterosexual sex, but is usually/often not its purpose. Joan Roughgarden, "Sexual Selection", 90; Lee Edelman, *No Future: Queer Theory and the Death Drive* (Durham: Duke Univ. Press, 2004), 13.

24 Roughgarden, *The Genial Gene*, 69.

25 Roughgarden, *The Genial Gene*, 248.

26 Roughgarden, *The Genial Gene*, 248.

27 Joan Roughgarden, Meeko Oishi and Erol Akçay, "Reproductive Social Behavior: Cooperative Games to Replace Sexual Selection", *Science* 311, no. 5763 (2006): 965–969.

28 While the majority of these letters were published in the May 5, 2006 issue of *Science*, the most hyperbolic letters were published online only. These are no longer available on the *Science* website, but can be accessed through the Internet Archive: https://web.archive.org/web/20111104225638/http://www.sciencemag.org/content/311/5763/965.abstract/reply.

29 Sasha R.X. Dall, John M. McNamara, Nina Wedell, and David J. Hosken, "Sexual Selection Cannot be Replaced by Cooperative Game Theory (and It Doesn't Need Replacing)", *Science*, April 6, 2006.

30 Pizzari, Tommaso, Tim R. Birkhead, Mark W. Blows, Rob Brooks, Katherine L. Buchanan, Tim H. Clutton-Brock, Paul H. Harvey et al., "Reproductive Behavior: Sexual Selection Remains the Best Explanation", *Science*, April 6, 2006. Beyond this initial 2006 skirmish in *Science*, the controversy continued in reviews of *The Genial Gene* and a meeting on sexual selection at the National Evolutionary Synthesis Center (NESCent) in Durham, NC in 2013, convened by philosopher Thierry Hoquet. Many

96 *Martha Kenney*

of the contributions were compiled in *Sexual Selection: What's Left After Darwin?* (2015) edited by Hoquet. Other than Hoquet's balanced introduction, none of the other articles in the edited volume show support for Roughgarden's critique of sexual selection or her alternative, social selection. See Joan Roughgarden, Elizabeth Adkins-Regan, Erol Akçay, Jeremy Chase Crawford, Raghavendra Gadagkar, Simon C. Griffith, Camilla A. Hinde et al., "Sexual Selection Studies: A NESCent Catalyst Meeting" (2015).

31 Randolph M. Nesse, "Runaway Social Selection for Displays of Partner Value and Altruism", *Biological Theory* 2, no. 2 (2007): 144; Samir Oshaka, Ken Binmore, Jonathan Grose, and Cedric Patternot, "Cooperation, Conflict, Sex, and Bargaining", *Biology & Philosophy* 25 (2010): 258.

32 Tim Clutton-Brock, "Sexual Selection in Males and Females", *Science* 318, no. 5858 (2007): 1882.

33 Oshaka, "Cooperation", 260.

34 Dustin R. Rubenstein, "Sexual and Social Competition: Broadening Perspectives by Defining Female Roles", *Philosophical Transactions of the Royal Society B: Biological Sciences* 367 (2012): 2249.

35 Geoff A. Parker and Tommaso Pizzari, "Sexual Selection: The Logical Imperative", in *Current Perspectives on Sexual Selection: What's Left After Darwin*, ed. Thierry Hoquet (New York: Springer, 2015), 138; unfortunately, these quotes are representative of the response in evolutionary biology to Roughgarden's work since *Evolution's Rainbow*. Her concept of social selection has not been taken up by biologists beyond those in her inner circle. My interpretation of this outcome is that Roughgarden's rejection of Darwin's sexual selection is seen as beyond the pale. From my perspective, Roughgarden and her critics agree on many points; however, her critics argue that cases that don't conform with Darwin's sexual selection can be incorporated into sexual selection theory, whereas Roughgarden argues that sexual selection is fundamentally bankrupt.

36 Nick Atkinson, "Sexual Selection Alternative Slammed", *The Scientist*, May 5, 2006.

37 Stengers, "Diderot's Egg", 11.

38 Donna Haraway, *Primate Visions. Gender, Race, and Nature in the World of Modern Science* (New York: Routledge, 1989), 346; "Zilhman's science cannot be allowed to cohabit with her feminism, which has turned an already marked gender into politics, which is quintessentially marked 'other' to unmarked science. Either feminism or science must be evicted". Donna Haraway, *Primate Visions*, 345.

39 See Roughgarden, "Challenging Darwin's Theory", 33–35 for examples.

40 Roughgarden, "Challenging Darwin's Theory", 36.

41 Oxford English Dictionary.

42 Joan Roughgarden and Robert Wright, "Percontations: Evolving Explanations for Human Nature", uploaded May 4, 2018, http://bloggingheads.tv/videos/2063.

43 Martin Daly, Margo Wilson and Suzanne J. Weghorst, "Male Sexual Jealousy", *Ethology and Sociobiology* 3 (1982): 11–27.

44 Daly, Wilson, and Weghorst, "Male Sexual Jealousy", 17.

45 Daryn Lehoux, "Tropes, Fact, and Empiricism", *Perspectives on Science* 11, no. 3 (2003): 326–345.

46 Daryn Lehoux, "Tropes", 338.

47 Daryn Lehoux, "Tropes", 340. Italics original.

48 Daryn Lehoux, "Tropes", 340.
49 Roughgarden, "Sexual Selection", 2015.
50 Roughgarden, *The Genial Gene*, 2005.
51 Stengers, "Diderot's Egg", 9.
52 Luce Irigaray, *An Ethics of Sexual Difference*, Translated by Carolyn Burke (Ithaca: Cornell Univ. Press, 1991), 72.
53 Roughgarden, *The Genial Gene*, 30.
54 I use these cultural touchstones as stand-ins for common cultural narratives about trans women. *The Crying Game* (1992) is a British film with an infamous scene where a trans woman reveals that she is trans during a sexual encounter and her partner, a cis man, acts with disgust and violence. The Michigan Womyn's Music Festival (1976–2015), was a women-only music festival that ended in 2015 amid controversy for its longstanding policy only to admit "womyn-born-womyn" and exclude trans women. "Bathroom Bills" refers to a spate of state-level US legislation (2016–present) intended to exclude trans people from bathrooms and locker rooms that correspond with their gender identity. Often the myth of trans women or men falsely claiming to be trans women accessing women's bathrooms to perpetrate sexual assault is used to incite moral panic.
55 Emily Martin "The Egg and the Sperm: How Science has Constructed a Romance Based on Stereotypical Male-Female Roles", *Signs* 16, no. 3 (1991): 501.
56 Roughgarden, "Challenging Darwin's Theory", 2007.
57 I developed this formulation remembering a quote from *Gravity's Rainbow* where one of the protagonists, Tyrone Slothrop, moves from a narcissistic or paranoid orientation to recognize that there are other stories simultaneously happening within immediately post World War II Germany (a territory Pynchon calls "the Zone"). Pynchon writes: "[Slothrop discovered that] the Zone can sustain many other plots besides those polarized upon himself". This meta-textual moment is an excellent distillation of the idea that a given territory can hold multiple true stories at once. Thomas Pynchon, *Gravity's Rainbow* (New York: Penguin Books, 2000 [1973]), 614.
58 Sabina Leonelli, "What Counts as Scientific Data: A Relational Framework", *Philosophy of Science* 82, no. 5 (2015): 817.
59 Isabelle Stengers, "Experimenting with Refrains: Subjectivity and the Challenge of Escaping Modern Dualism", *Subjectivity* 22, no. 1 (May 2008): 38–59.
60 Leonelli, "What Counts as Scientific Data", 818.
61 Marilyn Strathern, *Partial Connections* (New York: Altamira Press, 2005 [1991]); Annemarie Mol, *The Body Multiple: Ontology in Medical Practice* (Durham: Duke Univ. Press).
62 Sam Anderson, "The Inscrutable Brilliance of Anne Carson", *The New York Times Magazine*, March 17, 2013, 20.
63 Martha Kenney and Ruth Müller, "Of Rats and Women: Narratives of Motherhood in Environmental Epigenetics", *BioSocieties* 121 (2017): 23–46.
64 Leonelli, "What Counts as Scientific Data".
65 Stephen Jay Gould, "Sociobiology: The Art of Storytelling", *New Scientist*, November 16, 1978, 530–532.
66 OUvroir de LIttérature POtentielle (workshop of potential literature).
67 Jean-Jacques Poucel, "Oulipo", in *The Princeton Encyclopedia of Poetry and Poetics*, ed. Roland Greene (Princeton, NJ: Princeton Univ. Press, 2012), 988.
68 Daniel Levin Becker, *Many Subtle Channels: In Praise of Potential Literature* (Cambridge: Harvard Univ. Press, 2012).

69 Carla Hustak and Natasha Myers, "Involutionary Momentum: Affective Ecologies and the Science of Plant/Insect Encounters", *differences* 23, no. 3 (2012): 77.
70 Vinciane Despret, *What Would Animals Say if We Asked the Right Questions?* Translated by Brett Buchanan (Minneapolis: Univ. of Minnesota Press, 2016).
71 Stengers, "Diderot's Egg", 10.

References

Atkinson, Nick. "Sexual Selection Alternative Slammed", *The Scientist*, May 5, 2006.

Anderson, Sam. "The Inscrutable Brilliance of Anne Carson", *The New York Times Magazine*, March 17, 2013, 20.

Barad, Karen. *Meeting the Universe Halfway: Quantum Physics and the Entanglement of Matter and Meaning*. Durham: Duke Univ. Press, 2007.

Becker, Daniel Levin. *Many Subtle Channels: In Praise of Potential Literature*. Cambridge: Harvard Univ. Press, 2012.

Clutton-Brock, Tim. "Sexual Selection in Males and Females". *Science* 318, no. 5858 (2007): 1882–1885.

Dall, Sasha R.X., John M. McNamara, Nina Wedell, and David J. Hosken. "Sexual Selection Cannot be Replaced by Cooperative Game Theory (and It Doesn't Need Replacing)", *Science*, April 6, 2006.

Daly, Martin, Margo Wilson, and Suzanne J. Weghorst. "Male Sexual Jealousy". *Ethology and Sociobiology* 3 (1982): 11–27.

Darwin, Charles. *The Descent of Man and Selection in Relation to Sex*. Samford, Australia: Emereo Publishing, 2012 [1871].

Despret, Vinciane. *What Would Animals Say If We Asked the Right Questions?* Translated by Brett Buchanan. Minneapolis: Univ. of Minnesota Press, 2016.

Edelman, Lee. *No Future: Queer Theory and the Death Drive*. Durham: Duke Univ. Press, 2004.

Gould, Stephen Jay. "Sociobiology: The Art of Storytelling", *New Scientist*, November 16, 1978, 530–532.

Hamilton, William D., and Marlene Zuk. "Heritable True Fitness and Bright Birds: A Role for Parasites?" *Science* 218, no. 4570 (1982): 384–387.

Haraway, Donna. "Situated Knowledges: The Science Question in Feminism and the Privilege of Partial Perspectives". *Feminist Studies* 14, no. 3 (1988): 575–599.

Haraway, Donna. *Primate Visions. Gender, Race, and Nature in the World of Modern Science*. New York: Routledge, 1989.

Haraway, Donna. *The Haraway Reader*. New York: Routledge, 2004.

Hustak, Carla, and Natasha Myers. "Involutionary Momentum: Affective Ecologies and the Science of Plant/Insect Encounters". *differences* 23, no. 3 (2012): 74–118.

Irigaray, Luce. *An Ethics of Sexual Difference*. Translated by Carolyn Burke. Ithaca: Cornell Univ. Press, 1991.

Kenney, Martha. "Fables of Response-Ability: Feminist Science Studies as Didactic Literature". *Catalyst: Feminism, Theory, Technoscience* 5, no. 1 (April 2019).

Kenney, Martha, and Ruth Müller. "Of Rats and Women: Narratives of Motherhood in Environmental Epigenetics". *BioSocieties* 121 (2017): 23–46.

Lehoux, Daryn. "Tropes, Fact, and Empiricism". *Perspectives on Science* 11, no. 3 (2003): 326–345.

Leonelli, Sabina. "What Counts as Scientific Data: A Relational Framework". *Philosophy of Science* 82, no. 5 (2015): 810–821.

Longino, Helen. *Science as Social Knowledge: Values and Objectivity in Scientific Inquiry*. Princeton: Princeton Univ. Press, 1990.

Martin, Emily. "The Egg and the Sperm: How Science Has Constructed a Romance Based on Stereotypical Male-Female Roles". *Signs* 16, no. 3 (1991): 485–501.

Mol, Annemarie. *The Body Multiple: Ontology in Medical Practice*. Durham: Duke Univ. Press.

Nesse, Randolph M. "Runaway Social Selection for Displays of Partner Value and Altruism". *Biological Theory* 2, no. 2 (2007): 143–155.

Oshaka, Samir, Ken Binmore, Jonathan Grose, and Cedric Patternot. "Cooperation, Conflict, Sex, and Bargaining". *Biology & Philosophy* 25 (2010): 257–267.

Parker, Geoff A., and Tommaso Pizzari. "Sexual Selection: The Logical Imperative", In *Current Perspectives on Sexual Selection: What's Left After Darwin*, edited by Thierry Hoquet, 119–163. New York: Springer, 2015.

Pizzari, Tommaso, Tim R. Birkhead, Mark W. Blows, Rob Brooks, Katherine L. Buchanan, Tim H. Clutton-Brock, Paul H. Harvey et al. "Reproductive Behavior: Sexual Selection Remains the Best Explanation", *Science*, April 6, 2006.

Poucel, Jean-Jacques. "Oulipo", In *The Princeton Encyclopedia of Poetry and Poetics*, edited by Roland Greene, 987–988. Princeton: Princeton Univ. Press, 2012.

Pynchon, Thomas. *Gravity's Rainbow*. New York: Penguin Books, 2000 [1973].

Roughgarden, Joan. *Evolution's Rainbow: Diversity, Gender, and Sexuality in Nature and People*. Berkeley: Univ. of California Press, 2004.

Roughgarden, Joan. "Challenging Darwin's Theory of Sexual Selection". *Daedalus* 136, no. 2 (Spring 2007): 23–36.

Roughgarden, Joan. *The Genial Gene: Deconstructing Darwinian Selfishness*. Berkeley: Univ. of California Press, 2009.

Roughgarden, Joan. "Sexual Selection: Is There Anything Left?", In *Current Perspectives in Sexual Selection: What's Left After Darwin*, edited by Thierry Hoquet, 85–102. Dordrecht: Springer, 2015.

Roughgarden, Joan, and Robert Wright. "Percontations: Evolving Explanations for Human Nature". Uploaded May 4, 2018. http://bloggingheads.tv/videos/2063.

Roughgarden, Joan, Meeko Oishi, and Erol Akçay, "Reproductive Social Behavior: Cooperative Games to Replace Sexual Selection". *Science* 311, no. 5763 (2006): 965–969.

Roughgarden, Joan, Elizabeth Adkins-Regan, Erol Akçay, Jeremy Chase Crawford, Raghavendra Gadagkar, Simon C. Griffith, Camilla A. Hinde et al. "Sexual Selection Studies: A NESCent Catalyst Meeting" (2015).

Rubenstein, Dustin R. "Sexual and Social Competition: Broadening Perspectives by Defining Female Roles". *Philosophical Transactions of the Royal Society B: Biological Sciences* 367 (2012): 2248–2252.

Stengers, Isabelle. "Diderot's Egg: Divorcing Materialism from Eliminativism". *Radical Philosophy* 144 (July/August 2007): 7–15.

Stengers, Isabelle. "Experimenting with Refrains: Subjectivity and the Challenge of Escaping Modern Dualism". *Subjectivity* 22, no. 1 (May 2008): 38–59.

Strathern, Marilyn. *Partial Connections*. New York: Altamira Press, 2005 [1991].

Takahashi, Mariko, Hiroyuki Arita, Mariko Hiraiwa-Hasegawa, and Toshikazu Hasegawa. "Peahens Do Not Prefer Peacocks with More Elaborate Trains". *Animal Behavior* 75, no. 4 (2008): 1209–1219.

Weatherhead, Patrick J., and Raleigh J. Robertson. "Offspring Quality and the Polygyny Threshold: 'The Sexy Son Hypothesis'". *American Naturalist* 113, no. 2 (1979): 201–208.

4 Rethinking Evidence Practices for Environmental Decision-Making in the Anthropocene

What Can We Learn from Invasive Species Research and Policy?

Christoph Kueffer

We live in a time of multiple, interacting and accelerating crises, including climate change, overexploitation of natural resources, pollution, growing inequalities and injustice, political and social instabilities, war, migration and weakened democratic and truthful deliberations.[1] Many of these crises are directly or indirectly linked to the degradation of ecosystems and biodiversity loss.[2] Since 1993, the nations of the world have committed themselves to the protection and restoration of the Earth's diversity of life through the *UN's Convention on Biological Diversity* (CBD).[3] The *UN Decade on Ecosystem Restoration* from 2021 to 2030 further emphasizes the urgent need to reverse the degradation of ecosystems worldwide within the coming years.[4]

Evidently, the current economic system is a major driver of ecological degradation,[5] and technological solutions will not suffice to avert catastrophic climate change and biodiversity loss.[6] The UN's "17 Sustainable Development Goals" of the *2030 Agenda for Sustainable Development* instead recognize the need for integrative sociocultural and ecological solutions.[7] This will require that the voices and ecological competencies of diverse cultural groups, many of them marginalized or oppressed, must be strengthened with the help of expertise from the humanities, social sciences and arts.[8] Following an era dominated by economics and engineering, the 21st century must become a century of cultural diversity and ecological sensibilities. Indeed, ecology is reaching the status of a guiding natural science of our time. Since 2012, the *Intergovernmental Science-Policy Platform on Biodiversity and Ecosystem Services (IPBES)* has been coordinating experts around the world to assess ecological knowledge for policy-making.[9] While for most of the 20th century physics was seen as the paradigmatic model of scientific inquiry, in the 21st century, we must better appreciate the ontological, epistemological, methodological and pragmatic implications of an ecological view of nature and human-nature relationships.[10] Rich and thorough ecological expertise is essential for an urgently needed societal transformation toward a sustainable future.[11]

DOI: 10.4324/9781003273509-7

Such an ecological turn will have important implications for how we see the role of scientific evidence in resolving conflicts and legitimizing decisions. Ecological expertise is confronted with particularly difficult challenges. Expertise about the open and non-equilibrium environmental systems of the Anthropocene is inevitably highly uncertain. Open environmental systems are characterized by features such as nonlinearity, emergent properties, non-equilibrium and causal chains that span vast spatial scales that make robust prediction and reliable advice on effective system manipulation difficult. This makes it also hard for experts to demonstrate that their evidence is reliable.[12] Moreover, the experimental testing of hypotheses and refinement of solutions through learning-by-doing in a controlled setting such as a laboratory is often not possible.[13]

To circumvent these problems, the modern natural sciences have often used the strategy of turning open-system problems into closed-system problems. Accordingly, innovations have been developed in laboratories and their risks assessed based on highly simplified model systems, while the potential consequences on the environment have often been neglected.[14] Intensive agriculture and plantations, for instance, have been designed so that they can easily be controlled and manipulated, while cities and technical artifacts are considered separate from nature. This has often led to unintended consequences stemming from new technologies and other innovations on the natural world. It has hitherto been possible to neglect these negative externalities because the planet that provided us with free ecosystem services and goods quietly absorbed our pollutants and waste and allowed us to conduct our dangerous experiments and destructive activities in remote areas where those affected, whether human or non-human, were powerless.[15]

Meanwhile, and especially since World War II, we have lost most of the refugia of nature,[16] and we now live in a full world,[17] with no cheap nature left.[18] A key characteristic of the Anthropocene is that even the rich and powerful among us can no longer escape the causal interconnections between the environment and human systems. Whether in a laboratory, in relation to technical infrastructures, in cities, in intensive agriculture or in the way we imagine our social and culture life, nature is talking back. We have to relearn how to listen to nature, while accepting that our knowledge about nature is inevitably incomplete and ignorance widespread.

A second challenge is that our thinking about nature and human-nature relationships is undergoing a paradigm shift. Fundamental ontological, epistemological, methodological and ethical assumptions underlying ecological research and our understandings of nature and human-nature relationships are open for debate in our pluralistic and globalized society. When such a phase of cognitive indeterminacy occurs in a field of expertise so closely intertwined with deliberations in society, the situation further complexifies. Silvio O. Funtowicz and Jerome R. Ravetz[19] have called this type of science-policy nexus *post-normal* in reference to Thomas Kuhn's description of scientific revolutions.[20] Fundamental assumptions about what counts as relevant and relia-ble expertise and evidence as well as about the ontology of the study subjects and the ethics and goals of interventions are being questioned from multiple and conflicting perspectives from within the sciences and society at large.[21] Because these various assumptions are mutually inter-twined, it is difficult to separate political, cultural, ethical, epistemolog-ical and ontological aspects of a controversy.[22] And because conflicting perspectives are often incommensurable, there is no arbiter available to clarify debates.

This situation is further aggravated by the fact that our knowledge about nature is rarely based on direct observation accessible to a non-expert anymore; rather, nature increasingly speaks to us only indirectly through various specialized scientists and their diverse tools. Thus, sci-entific evidence about nature is increasingly more open to alternative and often conflicting interpretations.[23] When nature still speaks directly to us, many of us have lost the competencies to listen – we depend on interpreters to explain the ecological realities around us.

There is no easy way out of this bind. In particular, there is a growing recognition that reducing ambiguities by turning pluralistic and open-system problems into disciplinary and closed-system problems only worsens the situation.[24] Instead, we need to develop a new culture of evidence practices that embraces pluralism, ambiguity and ignorance. In some cases, previous strategies of evidence-based decision-making, such as the use of projections, risk assessments or cost-benefit anal-yses, still work.[25] In other cases, it might be more effective to design evidence-based decision-making processes and policy institutions that are more inclusive and transparent and facilitate continuous social learn-ing.[26] In ecology, for instance, there is a long tradition of adaptive man-agement processes that attempt to continuously improve interventions in nature through social learning-by-doing in the real-world settings of particular environmental problems.[27]

Often, however, it is not even clear what constitutes a scientific and societal problem, how it should be approached and who the relevant experts are. In such a situation, the formulation of the framing of the societal and scientific problem becomes in itself a critical step in the pro-duction of reliable and socially robust evidence.[28] Transdisciplinary and participatory research aims at clarifying contested problem structurings in pluralistic decision-making contexts.[29] Arguably, the situation is even more ambiguous in the case of ecological expertise in the Anthropocene because the epistemology and ontology of a whole research field – ecology – and even whole epistemes are exposed to heightened dis-agreement. It might therefore be necessary to embrace pluralism and disagreement as an opportunity for renegotiating the very fundament of our thinking.[30] The role of experts might become one of nurturing critical thinking, virtues and cultures of responsibility and empowerment and agency rather than of attempting to achieve a definite clarification of problem diagnosis, targets and solutions.[31]

In what follows, I present the example of invasive species research and policy as a model case of a scientific and societal issue that is characteristic of evidence-based deliberations in ecology and environmental decision-making in the 20th century. Biological invasions are the result of global environmental changes and globalization and are considered one of the main drivers of the biodiversity crisis. While there is a well-established expert community that addresses the issue through research rooted in a mainstream scientific discipline – ecology – the interpretation of the scientific evidence and the conclusions drawn for management action are increasingly contested from numerous angles; thus, biological invasions represent a case of post-normal science.[32]

Biological invasions were formally recognized as a specific scientific and societal issue after World War II. The framing of the problem is thus rooted in post-war ecological science and environmental decision-making. It was an era when ecological problems were framed as socially and epistemologically well-bounded issues amenable to clarification by academic and disciplinary ecologists alone and solved through policy-making that closely follows scientific assessments such as cost-benefit analyses or scenario analysis (mode 1 knowledge production *sensu* Helga Nowotny and colleagues).[33]

Ecology and the Science-Policy Nexus after World War II

The core of academic ecology after World War II contrasted strongly with early modern ecology in the 19th and early 20th centuries; thus, the expert culture of the post-war years was socially constructed in a specific way, and this shaped ecological thinking and decision-making related to the different environmental crises of the 20th century, including biodiversity loss and climate change.

Applied ecology was often institutionally isolated from basic ecology. It was widely distributed across diverse research institutions and departments of applied sciences such as natural resources management, fisheries, forestry or agriculture,[34] and scientists with an ecological expertise and research focus also worked at departments ranging from geography and anthropology to the environmental sciences. In contrast, basic ecology was increasingly separated from applied ecology and non-biological sciences, including geography and the social sciences, which in the 19th and early 20th centuries shared interests and regularly collaborated with ecology. A reflection of this separation was that humans were excluded from basic ecological theory as an agent integral to ecological systems. An interest in the ecology of human-made ecosystems such as cities, for instance, only re-emerged much later.[35] A gap opened between ecology and the social sciences and humanities. Partly this was a consequence of the episteme of modernity that assumed that nature and culture were separate realities, but it was also a result of more specific misunderstandings between natural and social scientists among others resulting from the heated sociobiology debates of the 1970s.[36]

Reductionist ontological frameworks increasingly shaped ecological theories.[37] The study of animal behavior came under the influence of behaviorism. Animals were interpreted as beings without consciousness and their behavior as purely mechanistic – following René Descartes' characterization of animals as machines. Animal behavior was studied in animals in captivity and often by harming them. The emergent properties of species communities were interpreted as the result of the interplay of autonomous individuals that compete for limited resources[38] – an ontological understanding of species coexistence that is interpreted by some historians of science as being rooted in an ideology of liberalism.[39] Ecosystem ecology that explained the overall workings of ecosystems as characterized by fluxes of energy, matter and information in analogy to physics solidified through major funding from the *US Atomic Energy Commission* with the goal of understanding the fate of radioactive isotopes in the environment.[40] It was further developed by building on the toolbox of systems science and cybernetics and with the help of computer simulation modeling.[41] Such systems ecology can be seen as technocratic,[42] and as an approach that characterizes ecosystems as a kind of a machine.[43]

After World War II, the ambition of basic ecologists was to advise on global-scale policies on biodiversity with context-independent and globally applicable knowledge intended to represent the consensus of a global scientific expert community, comparable to the advisory work of climate modelers in climate policy. However, it was not possible to make quantitative predictions as a basis for policy-making using computer simulations like those of climate scientists.[44] Instead, the hope was to deduce policy advice directly from ecological theory. For instance, mathematical models from population biology were used to determine minimal viable populations of threatened species;[45] species coexistence theory was used to show why local species diversity matters for ecosystem functioning;[46] biogeographic research on the correlation between the size of oceanic islands and the number of species present on these islands to advise on the design of nature protection areas;[47] and a combination of theoretical assumptions from biogeography, population ecology and evolutionary biology to argue that nonnative species – i.e. those introduced to a new geographic area by humans – pose high ecological risks (see below).

Deduction of expert advice and implicitly normative judgments from general scientific laws and thus underlying ontological and epistemological assumptions can be problematic, especially when implicit assumptions and how they shape sociopolitical discourses are not made transparent.[48] In particular, the assumption that non-anthropogenic, pristine ecological systems are characterized by a particularly high degree of biological organization implicitly influenced research and policy. In research, interpretations and generalizations of observational data were built on the assumption that ecological patterns represent well-designed adaptations. According to such a view, species traits represent optimized designs that help species to survive under particular environmental conditions, interactions of species are fine-tuned through coevolution, and the composition of species community is the result of ecological sorting so that coexisting species with complementary specializations (i.e. niches) fit together like pieces of a puzzle. This adaptionist interpretation has been criticized as empirically unjustified teleological thinking[49] and as based on an empirically unjustified assumption that there is some kind of harmonic balance in nature.[50] In policy, the view that pristine nature is particularly well-functioning thanks to long-term coevolution and ecological sorting led to the presumption that humans are by default a problematic disturbance factor in nature; and thus that protected natural areas are ecologically preferable to managed land and that species introduced by humans to an ecosystem – alien species – pose ecological risks. In the Anthropocene, these assumptions confront ecology with epistemological and pragmatic problems given non-equilibrium and anthropogenic ecological realities.[51] A second important implicit assumption was that ecological issues were framed as global rather than local policy issues. Ecology tried to fit into the emerging framework of climate change and global change science and policy.[52] The concept of biodiversity was meant to condense the overwhelming diversity of life across the multitudes of local places on Earth into one concise and highly generalized entity that could be used to talk to international decision makers. According to E.O. Wilson, the term was meant to become "the talisman of conservation, embracing every living creature".[53] This framing of ecological thinking and decision-making contributed to unequal und hegemonic globalized discourses about nature.[54]

In this context, biological invasions were conceived as a scientific problem and a societal issue. This background is important in understanding how invasion biologists were at first successful in framing a complex socioecological problem in such a way that a small and homogeneous group of scientists was accepted as the only legitimate experts and their expertise was largely undisputed, and how thereafter it developed into a highly contested post-normal issue characterized by incommensurable disagreements among experts and widespread contestation of evidence and policies by diverse stakeholders.[55]

Case Example: Invasive Species Research and Policy

As long as humans have migrated across the planet, they have carried other species to new places.[56] For instance, the successful expansion of indigenous people across the Pacific and the colonization of remote islands thousands of years ago depended on plants that they transported with them,[57] in Greek and Roman times, alien species were part of religious ceremonies,[58] and the redistribution of diseases, animals and plants played an important role in colonial expansion.[59] The transportation of species to new places was often deliberate because of their known usefulness, and thus they played an important subsistence role and were often perceived positively and integrated into daily life.[60] Alien species had manifold cultural and symbolic meanings, including as part of religious practices and as ornamentals.[61] These meanings often differed for different social groups.[62] Thus, throughout human history, alien species have been an integral part of livelihoods, and the perceptions of them have been pluralistic.

The Initial Framing of Invasive Species Research and Policy

When naturalists in the 18th century started to systematically document the diversity of the natural world, human-associated species were recognized as such but not seen as something fundamentally different from naturally occurring species.[63] Early 20th-century plant ecology further developed a differentiated conceptualization of different types of human-associated plant species, and they studied how humans shape local floras among others in urban areas.[64] Thus, these early naturalists addressed human-associated species as part of their broad interdisciplinary interests in the interplay of geographic, ecological and human factors in shaping the landscapes and biomes of the planet. It was only in the 1950s that introduced species began to be portrayed as a distinct scientific and societal problem. A book entitled *Ecology of Invasions by Animals and Plants* published in 1958 by the leading animal population ecologist at the time, Charles Elton, is generally seen as the birth of formalized research on invasive species.[65] The book initially triggered little interest in invasions as an environmental problem but was rather read as a contribution to basic ecology.[66] It was an international research program within the *Scientific Committee of Problems of the Environment (SCOPE)* framework[67] focused on biological invasions that triggered the rapid growth of a new research field specifically focused on biological invasions in the 1980s.[68] Elton's 1950s book and the subsequent international SCOPE research program in the 1980s are here treated together as the phase leading to the initial problem framing of formalized invasive species science and policy.

This initial framing of biological invasions as a scientific and societal problem has some interesting characteristics.[69] A diverse array of complex socioecological phenomena was subsumed under one unifying framework rooted in ecological theory. According to the broad scope of the postulated problem structuring, biological invasions encompass all alien organisms, ranging from animal and plant diseases to plants and mammals in all biomes of the world – from the Arctic to the tropics, both terrestrial and marine – that spread in all sorts of wild and man-made landscapes and are associated with diverse human activities. One achievement of Elton and subsequent invasion science was that insights from biogeography, population, community, ecosystem, landscape and evolutionary ecology were integrated to look at very diverse ecological phenomena through a single unifying lens, thereby contributing to theoretical synthesis in ecology. In contrast, there was little reciprocal conversation with applied research fields such as weed science, plant health, biological control or epidemiology that already had a long tradition of working on some of the issues that were now considered biological invasions. The new scientific framing thus redefined different applied ecological research questions as examples of the same kind of phenomenon, the essential workings of which should be clarified by basic ecology.

The underlying assumption that allowed for such a broad-brushed generalization of diverse real-world phenomena was that all natural ecosystems were considered to be uniformly characterized by the same ordering principles, and the modulating effects of the particular socioecological contexts were considered to be negligible in comparison to these universal ecological principles. In particular, humans were seen as an external disturbance to natural systems. Because alien species are by definition a human-induced change of the pre-human species composition of ecosystems, they were therefore by default considered a risk to the well-functioning of ecosystems. The (perceived) unusual population dynamics of invasive alien species, i.e. their rapid spread and tendency to reach high abundance, was attributed to their human-associated origin. Their nonnative origin was thus seen as the keystone of the causal interpretation of the dynamics of biological invasions – a view derived from particular ontological and epistemological assumptions about how nature works and should be studied. At the science-policy interface, these presumptions legitimated the normative claim that alien species are by default problematic and therefore should be prevented from entering new areas and where present should be controlled and if possible eradicated.

This normative prejudice gained special weight in decision-making because it was argued that, in line with the precautionary principle, the risk of biological invasions should be prevented proactively, i.e. newly arriving alien species should be controlled and if possible eradicated before an invasion could happen and therefore before any empirical data that demonstrated their negative impacts in a particular location became available. A normative principle of environmental policy – the precautionary principle – thus legitimized policy advice from ontological presumptions about nature without the need to refer to case-specific empirical data. The question of how to legitimize precautionary action has since been constantly renegotiated in invasive species research and management.

At first, however, invasion biologists were not forced to engage in deliberations about their implicit assumptions and were very successful in getting their perception of a new environmental risk integrated into policies at national and international levels. National and international legislation, science and policy networks and institutions and tools such as data inventories were quickly and widely established.[70] In 1993, the UN's *Convention on Biological Diversity* (CBD) came into force and included article 8(h) on invasive alien species, requiring parties to "prevent the introduction of, control or eradicate those alien species which threaten ecosystems, habitats or species", and the *World Conservation Union (IUCN)* established an invasive species specialist group.[71]

A number of hypotheses can be formulated to explain why the need to proactively address the problem of biological invasions according to a framing proposed by a rather small group of invasion biologists was initially not contested and explains why these scientists were effective in influencing policies. First, the recognition of a new ecological risk resonated well with the emerging environmental awareness and the growing interest in problems attributable to global change. Invasions were seen as a paradigmatic example of the ecological consequences of environmental degradation and globalization. The SCOPE research program, which was a key driver for the formation of institutionalized invasive species research, was along with, for instance, the *International Biological Program* (IBP), aimed at addressing global environmental problems through coordinated international efforts. It further helped that invasions proved to be an interesting global natural experiment that could be studied particularly well through internationally coordinated multisite research, which increased its attractiveness for basic ecologists. Essentially, during colonial expansion, the same set of species was introduced to North America, South Africa, Australia, islands in the Pacific etc., and after 50–200 years, their fate in different biogeographic regions and habitats could be compared and analyzed based on observational data in, as it were, a long-term, outdoors multisite experiment (i.e. a natural experiment).[72]

Second, in contrast to other applied ecological research, invasion biology had close institutional affinities to basic ecology and could profit from the social status of leading ecologists. Charles Elton, for instance, was the leading animal population ecologist of his time and his book built on three lectures he gave on BBC radio to a large audience.[73] Equally, the SCOPE program involved some of the leading ecologists of the 1980s, and invasions were seen as a model system to test and further develop ecological theory, thereby raising the status of invasion-related research within basic ecology.[74] Third, with the emergence of international biodiversity policy institutions and legal frameworks such as the *Convention on Biological Diversity* in the 1990s, the scientific results of the SCOPE program about biological invasions came at the right moment to be integrated into international and then national legislation, and invasion biologists were well networked with decision makers in international biodiversity policy. Fourth, the proposed management actions fitted with established institutional frameworks and interests of stakeholders. Legislation and institutional mechanisms from plant health and animal and human epidemiology for the precautionary regulation of the transportation of problem species between nations already existed. Authorities at borders were prepared to control transboundary movement of listed species, and invasion biologists, for their part, were in a position to develop risk assessment tools that identified problem species as a basis for preventative screening. Also, the control and if possible eradication of problematic species – pests and weeds – was a well-established strategy that nourished a large and profitable industry and profited from broad social acceptance. And, lastly, the inherent narrative of invasion biology brought together cultural stereotypes from across the political spectrum: To prevent the unregulated "invasion" by "nonnative" species from outside a nation, to weed out and kill problem species and to protect pristine nature from negative human influence.

Thus, in summary, the case of invasive species research and policy rooted in the 1950s and developed throughout the 1980s and 1990s turned a broad range of complex socioecological phenomena into a socially and epistemologically well-bounded one, thereby containing contestation of evidence and its interpretation within the sciences as well as in society. A small and homogeneous group of scientists – trained in ecology and working at natural sciences departments – was privileged as the relevant experts. They advised on policy by using generalized rules about the workings of ecology instead of digging into the muddy details of real-world management cases. Of course, in the same period, many real-world invasions were managed locally, but this case-specific management and associated expertise were treated as applied science of lower status and therefore the institutional and epistemological core of the discipline of invasion biology was not affected by how applicable its theory was to local, real-world cases.

The example of invasive species science and policy illustrates how discipline-based policy advice – mode 1 knowledge production *sensu* Nowotny and colleagues[75] – is maintained through the social construction of a whole regime of codependent cultural, social, institutional and epistemological elements. To what extent this example of the social construction of proactive action in response to an environmental risk should be seen as a successful or problematic model for reducing scientific and social complexities to enable effective action against an emerging risk requires a differentiated assessment. Some of its accomplishments and weaknesses became evident when it started to break apart in the late 1990s. This is the next phase of the story of invasive species science and management.

Post-Normal Disturbances of the Expert Consensus

Toward the end of the 1990s, the science-policy regime of invasive species research and policy increasingly ran into problems and dissent was voiced more loudly within the sciences and in society.[76] The definitions of an alien and an invasive species were questioned by different experts and stakeholders,[77] and the whole problem framing as well as the science-policy regime were being challenged. What human-assisted extra-range dispersal meant was no longer quite so clear. From how far must a species come so that its dispersal counts as extra-range? For instance, does the planting of a species outside of its ecological habitat, but within the same geographic area – for instance plantations of conifers that naturally occur in mountainous areas but are often planted in lowlands – also count as a case of extra-range and thus nonnative occurrence? Furthermore, is there a time duration after which a long-established nonnative species is considered a native species? And, when should a dispersal event be considered a human-assisted one? For instance, do species that migrate due to anthropogenic climate change, but without being transported by humans also count as nonnative species? After all, why is human assistance even a relevant dimension of a definition of an ecologically novel species? Especially in the Anthropocene, does a definition that considers humans separate from nature still make sense (Figure 4.1), or how can ecological novelty be better defined in a time of massive anthropogenic ecological changes?[78] Such critique of the problem framing came from within invasion science and ecology – including new subfields such as global change ecology that had started to compete with invasion science for expert status on the same issues – as well as from diverse other disciplines, including geography, social and cultural sciences, and from practitioners, stakeholders and decision makers.[79] Thus, in line with Thomas Kuhn's model of scientific revolutions, conceptual questions that had been treated as a taboo by the prevailing paradigm suddenly became the focus of scientific debate. These questions had occasionally been discussed before in the scientific literature but did not receive much attention, while now they led to energetic scientific correspondence among the leaders in the field. In the case of a real-world and policy-oriented

Figure 4.1 The Bosco Verticale building in Milan (Italy) – a high-rise building planted with trees in an urbanized area. In the Anthropocene,
human agency and man-made landscapes shape novel ecologies.
Photograph by Christoph Kueffer.

science, however, the post-normal phase was not confined to discussions among a small circle of specialized experts, but triggered more
wide-ranging debates about evidence-based decision-making on biological invasions.[80]

Indeed, the breakup of the paradigm opened space for more diverse
expert perspectives.[81] In particular, critical voices called for case-specific
evaluations of actual invasions and their management instead of assuming that all alien species should be treated equally as a problem independent of context.[82] As a result, a greater diversity of alternative policy
options and expert advice became available, which made it more difficult
to reach a consensus on management actions.[83] Furthermore, it was no
longer possible to neglect the contingencies and context-dependencies of
particular real-world invasions. Whether a particular mechanism is relevant in explaining a specific invasion depends on the ecological and
anthropogenic context of the invasion.[84] Invasion science theory was not
particularly well prepared to explain how confounding factors shape
real-world invasion dynamics.[85] Thus, the scientific robustness of the
available expert knowledge weakened. While broad expert consensus
supported general theory about invasions, reliable predictions of the outcomes of specific invasions in particular contexts were more difficult to
make. It also became more challenging to evaluate the impacts of particular invasions and the cost-benefits of their management.

One observation about the consequences of this post-normal disturbance of the biological invasion science-policy regime is that strong

and incommensurable disagreements within an established scientific discipline led to division among experts from the same discipline and research field, who otherwise agreed on the validity of the underlying scientific theory and evidence.[86] In 2011, for instance, Mark Davis and other ecologists published a fundamental critique of invasion science in the scientific journal *Nature*,[87] which triggered strong responses from the community of invasion biologists.[88] In the same time period, Davis published a textbook about invasion ecology that represented the mainstream thinking in the field;[89] and his coauthors were equally well rooted in mainstream ecology. Thus, although they taught the same science to their students, their interpretations of evidence became incommensurable with mainstream thinking. Davis et al. stated that "nativeness is not a sign of evolutionary fitness or of a species having positive effects", thereby challenging the most fundamental pillar of the paradigm of invasion science.[90] They also argued that "the conclusion made [...] that invaders are the second-greatest threat to the survival of threatened or endangered species after habitat destruction" was based on no empirical data. This mounted a fundamental challenge to the claims of invasion biologists in their role as policy advisors. Daniel Simberloff and Montserrat Vilà, in their response entitled "141 scientists object", emphasized that they represented the expert consensus and responded to the critique that empirical data was lacking by emphasizing the need for proactive action in line with the precautionary principle: "severe impact of non-native species [...] may not manifest for decades" and "some species may have only a subtle immediate impact but affect entire ecosystems, for example through their effect on soils".[91] Thus, while these different experts agreed on the nuts and bolts of the underlying science, they were forced into separate camps at the level of overarching perspectives on the science and policy of invasions.

Indeed, perceptions of alien species and their management – whether by experts or those affected by an invasion – can be influenced by a wide range of factors, including the involved actors, the attitude toward the affected biodiversity and targeted invasive species, the social and cultural context of the invasion or terminologies.[92] For instance, stateled actions against an alien species on private land might be opposed due to personal stances about the role of the state in solving problems. Depending on the framing of the problem, fault lines between supporters and dissenting voices can shift radically. For instance, while there are many biological similarities between the ecological risks of invasive alien species and novel species engineered through biotechnology,[93] these two types of ecological risks are evaluated very differently by different experts. Some experts see a high risk stemming from alien species but not from genetically modified organisms, and vice versa. Overall, examples of dissent related to biological invasions show that the reasons for disagreement are not necessarily linked to any inherent aspect of biological invasions. Rather, who agrees or disagrees depends largely on how the problem is framed: Who is considered a relevant expert or actor, what the envisioned solutions are, who has a voice in the process of developing the problem understanding and solutions and how the problem is communicated and by whom.

This leads to another important observation. The great flexibility of forming alliances in support of an environmental cause is both an opportunity for and a threat to scientific experts. It highlights that transparent, inclusive and careful deliberations about the social, political, cultural, ethical and emotional dimensions of an environmental issue can be effectively employed to foster consensus. But it also leaves open the possibility that public support and perceptions will shift. Indeed, the perception of a particular alien species – for instance the tree genus Tamarix over the course of the 20th century in the United States – can change fundamentally.[94] Such shifts in problem understandings can trigger a need for the rearrangement of the whole science-policy regime that interlinks scientific expertise, policy responses and public perceptions. New legislation might have to be formulated, new institutional arrangements financed, the public engaged through different communication strategies and practitioners might have to learn new management approaches. Such knock-on effects might cascade through science-policy regimes with time delays leading to asynchrony between expert thinking and implemented solutions. In many places, policy-makers at local and national levels are currently implementing essentially the framing of invasive species management formulated in the 1980s and 1990s,[95] while some scientists have moved on and are now questioning whether these solutions are still effective. Furthermore, once one group of scientists loses its unquestioned status as the only relevant expert group on a particular issue, alternative science-policy regimes, all with their own temporal dynamics, can coexist with regard to the same policy issue. In the case of urban tree planting policies, for instance, there are two positions: The first calls for a native-species-preference policy (in line with invasive species science and policy),[96] while the second calls for an alien-species-preference policy (in line with horticulture and urban design and with the goal of adapting to climate change).[97] Which position is taken up by a particular city seems to be at least partly coincidental, although intermediate perspectives that bridge between the two positions have been formulated.[98]

A further observation is that sometimes the reframing of a problem understanding can open space for more stable and less contested and therefore more effective science and policy approaches. This is indeed what happened in the case of biological invasions in the 1990s. In 1996, an inter- and transdisciplinary multi-stakeholder program focused on biological invasions – the *Global Invasive Species Programme (GISP)*[99] – was initiated.[100] In the wake of the GISP, a new problem framing of biological invasions developed in complement to the existing one[101] – so-called pathway or vector science.[102] While the traditional framing of biological invasions aimed at understanding and managing the risks posed by particular alien species individually, pathway science aimed at understanding how different socioeconomic pathways led to the transportation of alien species across landscapes and continents. Thus, the focus of research shifted from understanding the biology of alien species to understanding the socioeconomics and practicalities of trade relationships. This new focus enabled the development of targeted concepts and tools for mitigating ecological risks associated with different transport pathways, for instance, the transportation of aquatic organisms in ballast water in international shipping,[103] or of plants in horticultural trade,[104] leading to different scientific questions and policy options depending on pathway (Figure 4.2). In the case of ballast water – i.e. marine water that is transported in ships that are not fully packed with cargo to stabilize them – an effective risk mitigation strategy is to sterilize the ballast water before releasing it back into the ocean at a port,[105] while in the case of horticulture, the responsible use of alien species in garden design can be fine-tuned through close collaboration with actors in the green industry.[106] This might mean that garden centers inform their clients about invasion risks, alien species are not planted in the vicinity of a nature reserve, or alien trees with known benefits for native pollinators are preferred in urban plantings over alien trees without biodiversity benefits. Developing such fine-tuned solutions with experts and stakeholders from practice increases their acceptance and effectiveness.

Figure 4.2 *Lupinus polyphyllus* is an ornamental plant that can form monospecific stands in cold environments as an alien species. Photograph by Christoph Kueffer.

Indeed, more generally, differentiating one overarching problem framing into multiple context-specific ones can help to lead to more pragmatic and less ideological solutions, which are more effective and can be better integrated into existing institutional frameworks. Thus, pathway science is not a replacement for species-focused risk assessments but a complement. Preventing some particularly problematic alien species through border control and species-specific strategies may still be necessary alongside diverse additional measures implemented for different pathways.

In summary, this phase of heightened evidence contestation in invasive species research and policy illustrates a fundamental dilemma of evidence-based environmental decision-making: When is generalized knowledge and expert consensus sufficient to legitimize action and when is it necessary to invest the time needed to collect case-specific evidence and allow for societal deliberation? Especially in cases when preventative and coordinated actions are needed, it is often not possible to gain sufficient case-specific evidence by the time a decision is required, and inclusive and open-ended deliberations might not lead to coordinated action across large geographic spaces (e.g. at international levels) and among diverse stakeholders. However, the alternative – defending the consensus of a narrow group of experts as the sole basis for legitimizing decisions – is also problematic. To maintain such narrowly focused consensus among experts, there is a strong incentive to accommodate critique of the existing paradigm and keep the conceptual core of the research field as stable as possible.[107] For instance, although it is increasingly evident that invasions are inherently driven by humans and can only be effectively addressed through approaches that integrate an ecological understanding with expertise on social and cultural dimensions,[108] the social sciences are still of only marginal importance in the published literature on biological invasions.[109] There is thus a risk that expertise is not adaptive enough to respond flexibly to dynamic, complex and ambiguous challenges. Secondly, the defense of narrowly framed expertise in a context of messy real-world realities and pluralism risks becoming ideological. Indeed, invasion biologists were increasingly confronted with such critiques. It was said that the concept of invasion "appeals to political and social values but has no scientific meaning",[110] invasion biology was denounced as a pseudoscience,[111] it was suggested that scientists were demonizing certain alien species,[112] and invasion biologists were criticized for promoting their views in rhetoric redolent of xenophobic nationalism.[113]

Invasion biologists have in recent years attempted to walk the line between defending their established problem framing and giving space to a greater diversity of voices. Franz Essl et al.[114] argue that "many conflicts in the valuation of the impacts of alien species are attributable to differences in the framing of the issue and implicit assumptions" and they propose principles to make valuation of alien species impacts more socially robust. They refer to Roger Pielke's model of the honest broker,[115] thereby accepting the need for participatory deliberation but maintaining that ultimately "science must play a central role in providing information and advice to policymakers". This reflects a more general development in the environmental sciences and policy toward more inclusive and reflective decision-making frameworks,[116] processes[117] and policy and academic institutions,[118] inter- and transdisciplinary research processes,[119] adaptive management and social learning processes,[120] and training environmental scientists in the skills necessary for participatory and integrative approaches.[121]

Rethinking Ecology and Environmental Decision-Making for the Anthropocene

The case study could end here. But the story has continued. In recent years, the awareness has grown that the reshuffling of species communities through anthropogenic interference leads to fundamentally novel ecologies of the Anthropocene, so-called ecological novelty.[122] Not only biological invasions and alien species contribute to it, but all sorts of other global change drivers: Land use changes, urbanization, climate change, extinctions, rapid evolutionary responses to an anthropogenic world and biotechnology. Positions among experts range widely from rigid preservationists' views that hope to reverse the trend toward ecological novelty[123] to pragmatic ones that call for a balanced approach[124] and optimistic ones that see a new biodiversity of the Anthropocene emerging.[125] Fault lines in the debates shift. Some conservationists don't judge alien species by default as problematic anymore but rather try to find ways to weigh their positive and negative sides depending on context. So-called novel ecosystems characterized by alien species are considered an integral part of and sometimes an opportunity for nature conservation.[126] Alien species are considered to play important roles in wild to anthropogenic ecosystems, including by supporting the ecoevolutionary adaptation of species communities to novel ecologies.[127] Conservationists promote the deliberate transportation of alien species to new biogeographic regions to replace the ecological functions of extinct species (re-wilding)[128] or to help species track climate change in space (assisted migration).[129] Collaborations between conservationists and biotechnologists look into possibilities to resurrect extinct species, adapt threatened species through gene-editing to a changing environment, control invasive species through biotechnology, or release synthetic organisms to clean up pollution.[130] There is almost a feeling of anything goes.

In this highly ambiguous and dynamic situation, the current strategy of environmental science and policy to enable consensus building through inclusive deliberation processes has limits. We might have to fundamentally rethink what robust evidence about complex and socially contested environmental problems entails, and what role it should play in legitimizing environmentally responsible coordinated action. Heterogeneous and context-specific ecological knowledge, which cannot easily be generalized, should become a more central pillar of evidence-based decision-making. Invasion scientists have for instance started to adopt strategies such as the identification of syndromes to generalize knowledge in a more context-sensitive way.[131] However, these strategies might not suffice to effectively use locally rooted evidence at national and international scales and in decision-making contexts where vested interests play a dominant role, i.e. in situations where evidence is exposed to the manufacturing of truth and communication campaigns of interest groups or more generally alternative facts and fake news. Rather than adhering to an unrealistic ideal of irrefutable facts that can be defended against vested interests as a necessary condition for environmental valuation and actions, the task of clarifying the evidence basis of environmental issues should emphasize a continuous process of nurturing critical thinking based on society-wide ecological competencies and rooted in a shared ecological ethic.[132] Such a reappraisal of situated ecological knowledge challenges the established hierarchy of knowledge within ecology that attributes higher status to universal than case-specific ecological knowledge. Supported by work in social studies of science and epistemology, we must move toward an expert culture of real-world ecological expertise that cherishes the full diversity of ecological knowledge, competence and sensibilities: Of field ecologists as much as of experimental ecologists and system modelers, and of practitioners, amateur naturalists and holders of traditional and indigenous knowledge as much as of academic ecologists. It has been shown that case-specific integration of diverse evidence can lead to more robust invasive species policies and management.[133]

Furthermore, given that in the Anthropocene ecological processes are interwoven with human activities, ecological expertise must become inherently inter- and transdisciplinary and especially build on close collaborations with the social and cultural sciences. Biological invasions are by definition human-associated ecological phenomena and they play out in man-made ecosystems and landscapes, and therefore biological invasions can only be understood robustly and addressed effectively based on interdisciplinary perspectives that integrate biology with landscape sciences and the social sciences and humanities.[134] One reason why ecological novelty seems so difficult to grasp is that current research does not address it as an inherently socioecological phenomenon.

Ultimately, improving only the evidence base for understanding a messy world will not suffice to deal with the novel ecological realities of the Anthropocene. Foremost we are faced with a lack of shared values

Figure 4.3 Caretaking for nature. In the Terra Nostra gardens in the Azores (Portugal), plants from around the world are combined to design novel ecosystems, while on oceanic islands, also many remaining fragments of "wild" habitat depend on continuous weeding and re-planting. Photograph by Christoph Kueffer.

and visions: What are good human-nature relationships and what are realistic goals for ecological regeneration in the Anthropocene? We have to address a deficiency in our culture to engage in rich social, cultural, emotional and cognitive ways with our ecological environment and to express our deep dependence on nature (Figure 4.3). Engaging with ecological novelty might thus require us to rethink how we can responsibly care for the degraded ecosystems of the Anthropocene and their manifold living beings, with the aim to regenerate their functioning, instead of using invasive species as scapegoats for the inevitable consequences of environmental destruction.

Conclusions

Through the prism of invasive species research and policy, a rich picture of ecological research practices and associated environmental decision-making in the 20th century emerges. It is evident that the disciplinary and mode 1 science-policy approach employed by ecologists after World War II has limits in a pluralistic world and on a planet characterized by a perfect storm of environmental crises. A disciplinary problem framing does not do justice to the socioecological phenomena of the Anthropocene, and the monopolization of expert power by an exclusive circle of academics trained at and employed by natural science departments of universities in the Global North lacks legitimization in a globalized, post-colonial world. Building policy advice on generalized ecological knowledge risks ineffective solutions and imposing normative positions and ontological assumptions held by a small social group on holders of diverse values, worldviews, ontologies and interests.

However, there are no easy solutions. General ecological knowledge and expert judgments about the well-functioning of ecosystems must play an important role in societal decision-making to enable proactive and coordinated environmental action. We cannot found decisions on case-specific empirical data and deliberations in every single management case. It is evident that some ecosystems have higher ecological qualities than others and that modern forms of land use destroy ecological qualities. It is also evident that certain academic and non-academic experts have a more in-depth understanding of ecology than the rest of society – especially in our era, when many citizens live a life isolated from nature. We must find new ways to interweave ecological knowledge with cultural and social practices, narratives, norms and our personal lives. In pre-modern times, ecological knowledge was embedded in mythologies and everyday life, and in the early days of the Enlightenment period, the boundaries between storytelling, the arts and the social and cultural sciences on the one hand and ecology on the other were still permeable. Thereafter, fears of biological determinism and naturalistic fallacies and of anthropomorphism and a loss of scientific objectivity were easy ways out of sometimes difficult inter- and transdisciplinary conversations between the natural and human sciences. We cannot afford to avoid such a dialogue anymore.

Notes

1 Future Earth, *Our Future on Earth* (Future Earth, 2020); Ernst U. von Weizsäcker and Anders Wijkman, *Come On! Capitalism, Short-termism, Population and the Destruction of the Planet* (Heidelberg: Springer, 2018); Rob Nixon, *Slow Violence and the Environmentalism of the Poor* (Cambridge, MA: Harvard Univ. Press, 2011).
2 Malgorzata Blicharska, Richard J. Smithers, Grzegorz Mikusiński, Patrik Rönnbäck, Paula A. Harrison, Måns Nilsson, and William J. Sutherland, "Biodiversity's Contributions to Sustainable Development", *Nature Sustainability* 2, no. 12 (2019): 1083–1093.
3 *Convention on Biological Diversity*, United Nations, accessed March 14, 2022, https://www.cbd.int/.
4 *United Nations Decade on Ecosystem Restoration 2021–2030*, United Nations, accessed February 5, 2022, https://www.decadeonrestoration.org/.
5 Partha Dasgupta, *The Economics of Biodiversity: The Dasgupta Review* (London: HM Treasury, 2021); OECD, *Beyond Growth: Towards a New*

Economic Approach. New Approaches to Economic Challenges, (Paris: OECD Publishing, 2020); Kate Raworth, *Doughnut Economics, Seven Ways to Think Like a 21st-Century Economist* (London: Random House Business, 2017); Joseph E. Stiglitz, Amartya Sen, and Jean-Paul Fitoussi, *Mismeasuring Our Lives: Why GDP Doesn't Add Up* (New Press, 2010).

6   Cengiz Akandil, Sascha A. Ismail, and Christoph Kueffer, "No Green Deal Without a Nature-Based Economy", *GAIA* 4, no. 4 (2021): 281–283; Timothée Parrique, Jonathan Barth, François Briens, Christian Kerschner, Alejo Kraus-Polk, Anna Kuokkanen, and Joachim H. Spangenberg, *Decoupling Debunked: Evidence and Arguments against Green Growth as a Sole Strategy for Sustainability* (Brussels: European Environmental Bureau, 2019).

7   "The 17 Goals", United Nations, accessed February 5, 2022, https://sdgs.un.org/goals.

8   Kekuhi Kealiikanakaoleohaililani, and Christian P. Giardina, "Embracing the Sacred: An Indigenous Framework for Tomorrow's Sustainability Science", *Sustainability Science* 11, no. 1 (2016): 57–67; Zoe Todd, "Indigenizing the Anthropocene", in *Art in the Anthropocene: Encounters Among Aesthetics, Politics, Environments and Epistemologies*, eds. Heather Davis and Etienne Turpin (London: Open Humanities Press, 2015), 241–254; Esther Turnhout, Bob Bloomfield, Mike Hulme, Johannes Vogel, and Brian Wynne, "Listen to the Voices of Experience", *Nature* 488 (2012): 454–455; Nixon, *Slow Violence*; Lesley J.F. Green, "'Indigenous Knowledge' and 'Science': Reframing the Debate on Knowledge Diversity", *Archaeologies* 4, no. 1 (2008): 144–163.

9   "Intergovernmental Science-Policy Platform on Biodiversity and Ecosystem Services", IPBES, accessed February 5, 2022, https://ipbes.net/,.

10   Steward T.A. Pickett, Jurek Kolasa, and Clive G. Jones, *Ecological Understanding: The Nature of Theory and the Theory of Nature* (Amsterdam: Elsevier, 2010); Yrjö Haila and Peter Taylor, "The Philosophical Dullness of Classical Ecology, and a Levinsian Alternative", *Biology and Philosophy* 16 (2001): 93–102.

11   Akandil, Ismail and Christoph Kueffer, "No Green Deal Without a Nature-Based Economy", 281–283; Christoph Kueffer, Manuela di Giulio, Kathrin Hauser, and Caroline Wiedmer, "Time for a Biodiversity Turn in Sustainability Science", *GAIA* 29, no. 4 (2020): 272–274; UN. *Global Sustainable Development Report 2019: The Future is Now – Science for Achieving Sustainable Development* (New York: United Nations, 2019); WBGU. *World in Transition. A Social Contract for Sustainability* (Berlin: WBGU, 2011).

12   Steve Rayner and Daniel Sarewitz, "Policy Making in the Post-Truth World. On the Limits of Science and the Rise of Inappropriate Expertise", *The Breakthrough Institute Blog*, March 1, 2021; Horst W.J. Rittel and Melvin M. Webber, "Dilemmas in a General Theory of Planning", *Policy Sciences* 4, no. 2 (1973): 155–169.

13   Rayner and Sarewitz, "Policy Making in the Post-Truth World".

14   Paul Harremoes, David Gee, Malcom MacGarvin, Andy Stirling, Jane Keys, Brian Wynne, and Sofia Guedes Vaz, *The Precautionary Principle in the 20th Century: Late Lessons from Early Warning* (London: Routledge, 2013).

15   Nixon, *Slow Violence*.

16   Anna Tsing, "A Threat to Holocene Resurgence Is a Threat to Livability", in *The Anthropology of Sustainability. Beyond Development and Progress*, eds. Marc Brightman and Jerome Lewis (London: Palgrave Macmillan, 2017), 51–65.

17   Weizsäcker and Wijkman, *Come On!*

18 Jason W. Moore, "The End of Cheap Nature, Or: How I Learned to Stop Worrying About 'the' Environment and Love the Crisis of Capitalism", in *Structures of the World Political Economy and the Future of Global Conflict and Cooperation*, eds. Christian Suter and Christopher Chase-Dunn (Berlin: LIT, 2014), 285–314.

19 Silvio O. Funtowicz, and Jerome R. Ravetz, "Science for the Post-Normal Age", *Futures* 25, no. 7 (1993): 739–755.

20 Thomas S. Kuhn, *The Structure of Scientific Revolutions* (Chicago: Univ. of Chicago Press, 1962).

21 Sanford D. Eigenbrode, Michael O'Rourke, J.D. Wulfhorst, David M. Althoff, Caren S. Goldberg, Kaylani Merrill, Wayde Morse et al., "Employing Philosophical Dialogue in Collaborative Science", *BioScience* 57, no. 1 (2007): 55–64.

22 Rayner and Sarewitz, "Policy Making in the Post-Truth World"; Andy Stirling, "Keep it Complex", *Nature* 468 (2010): 1029–1031; Funtowicz and Ravetz, "Science for the Post-Normal Age"; Brian Wynne, "Uncertainty and Environmental Learning: Reconceiving Science and Policy in the Preventive Paradigm", *Global Environmental Change* 2, no. 2 (1992): 111–127.

23 Rayner and Sarewitz, "Policy Making in the Post-Truth World".

24 Stirling, "Keep it Complex"; Wynne, "Uncertainty and Environmental Learning".

25 Stirling, "Keep it Complex".

26 Jennifer Pontius and Alan McIntosh, *Critical Skills for Environmental Professionals. Putting Knowledge into Practice* (Berlin: Springer, 2020); Esther Turnhout, Willemijn Tuinstra, and Willem Halffman, *Environmental Expertise. Connecting Science, Policy, and Society* (Cambridge, UK: Cambridge Univ. Press, 2019); Sven Ove Hansson, and Gertrude Hirsch Hadorn, *The Argumentative Turn in Policy Analysis. Reasoning about Uncertainty* (Heidelberg: Springer, 2016); Turnhout et al., "Listen to the Voices of Experience"; Christoph Kueffer, Evelyn Underwood, Gertrude Hirsch Hadorn, Rolf Holderegger, Michael Lehning, Christian Pohl, Mario Schirmer et al., "Enabling Effective Problem-Oriented Research for Sustainable Development", *Ecology and Society* 17, no. 4 (2012): 8; Stirling, "Keep it Complex"; Gertrude Hirsch Hadorn, Holger Hoffmann-Riem, Susette Biber-Klemm, Walter Grossenbacher-Mansuy, Dominique Joye, Christian Pohl, Urs Wiesmann, and Elisabeth Zemp, *Handbook of Transdisciplinary Research* (Heidelberg: Springer, 2008); Roger A. Pielke Jr., *The Honest Broker: Making Sense of Science in Policy and Politics* (Cambridge, UK: Cambridge Univ. Press, 2007); Donald Ludwig, Marc Mangel, and Brent Haddad, "Ecology, Conservation, and Public Policy", *Annual Review of Ecology and Systematics* 32 (2001): 481–517; Wynne, "Uncertainty and Environmental Learning"; Peter Checkland and Jim Scholes, *Soft Systems Methodology in Action* (Chichester: Wiley, 1990).

27 Matthias Gross, *Ignorance and Surprise. Science, Society, and Ecological Design* (Cambridge, MA: MIT Press, 2010); Carl Folke, Thomas Hahn, Per Olsson, and Jon Norberg, "Adaptive Governance of Social-Ecological Systems", *Annual Review of Environment and Resources* 30 (2005): 441–473; Ludwig et al., "Ecology, Conservation, and Public Policy"; Carl J. Walters, and Crawford Stanley Holling, "Large-Scale Management Experiments and Learning by Doing", *Ecology* 71 (1990): 2060–2068.

28 For example, Gertrude Hirsch, "Beziehungen zwischen Umweltforschung und disziplinärer Forschung", *GAIA* 4 (1995): 302–314; Checkland and Scholes, *Soft Systems Methodology in Action*.

29 For example, Hirsch Hadorn et al., *Handbook of Transdisciplinary Research*.

30 Compare recent reevaluations of the work of Paul Feyerabend for a discussion of pluralism as a valuable resource even – or especially – in a time of alternative facts and manufactured disagreement, Jamie Shaw, "Feyerabend and Manufactured Disagreement: Reflections on Expertise, Consensus, and Science Policy", *Synthese* 198 (2021): 6053–6084; Karim Bschir, and Jamie Shaw. *Interpreting Feyerabend. Critical Essays* (Cambridge, UK: Cambridge Univ. Press, 2021).

31 Christoph Kueffer, Flurina Schneider, and Urs Wiesmann, "Addressing Sustainability Challenges with a Broader Concept of Systems, Target, and Transformation Knowledge", *GAIA* 28, no. 4 (2019): 386–388.

32 Franziska Humair, Peter J. Edwards, Michael Siegrist, and Christoph Kueffer, "Understanding Misunderstandings in Invasion Science: Why Experts Don't Agree on Common Concepts and Risk Assessments", *NeoBiota* 20 (2014): 1–30; Christoph Kueffer, and Gertrude Hirsch Hadorn, "How to Achieve Effectiveness in Problem-Oriented Landscape Research – The Example Of Research on Biotic Invasions", *Living Reviews in Landscape Research* 2 (2008): 2.

33 Helga Nowotny, Peter Scott, and Michael Gibbons, *Re-Thinking Science: Knowledge and the Public in an Age of Uncertainty* (Cambridge, UK: Polity Press, 2001).

34 For example, Stephen Bocking, *Nature's Experts: Science, Politics, and the Environment* (New Brunswick, NJ: Rutgers Univ. Press, 2004).

35 Menno Schilthuizen, *Darwin Comes to Town: How the Urban Jungle Drives Evolution* (London: Picador, 2018); Herbert Sukopp, "On the Early History of Urban Ecology in Europe", in *Urban Ecology*, eds. John M. Marzluff, Eric Shulenberger, Wilfried Endlicher, Marina Alberti, Gordon Bradley, Clare Ryan, Ute Simon, and Craig ZumBrunnen. (New York: Springer, 2008), 79–97.

36 Catherine Driscoll, "Sociobiology", in *The Stanford Encyclopedia of Philosophy, Spring 2018*, ed. Edward N. Zalta. (Stanford, CA: Metaphysics Research Lab, Stanford Univ., 2018), https://plato.stanford.edu/archives/spr2018/entries/sociobiology/.

37 Sharon E. Kingsland, *Modeling Nature* (Chicago: Univ. of Chicago Press, 1995).

38 Kingsland, *Modeling Nature*.

39 Ludwig Trepl and Annette Voigt, "The Classical Holism-Reductionism Debate in Ecology", in *Ecology Revisited. Reflecting on Concepts, Advancing Science*, eds. Astrid Schwarz and Kurt Jax (Heidelberg: Springer, 2011), 45–83.

40 Voigt, "The Rise of Systems Theory in Ecology", in *Ecology Revisited. Reflecting on Concepts, Advancing Science*, eds. Astrid Schwarz and Kurt Jax (Heidelberg: Springer, 2011), 183–194.

41 Voigt, "The Rise of Systems Theory in Ecology".

42 Peter J. Taylor, "Technocratic Optimism, H.T. Odum, and the Partial Transformation of Ecological Metaphor after World War II", *Journal of the History of Biology* 21, no. 2 (1988): 213–244.

43 Voigt, "The Rise of Systems Theory in Ecology".

44 Paul N. Edwards, *A Vast Machine. Computer Models, Climate Data, and the Politics of Global Warming* (Cambridge, MA: MIT Press, 2010).

45 Michael E. Soulé and Bruce Wilcox, *Conservation Biology: An Evolutionary-Ecological Perspective* (Sunderland: Sinauer Associates, 1980).

46   A research field later called biodiversity research, e.g. Michel Loreau, S. Naeem, P. Inchausti, J. Bengtsson, J.P. Grime, A. Hector, D.U. Hooper et al., "Biodiversity and Ecosystem Functioning: Current Knowledge and Future Challenges", *Science* 294, no. 5543 (2001): 804–808.

47   Jared Diamond and Robert May, "Island Biogeography and the Design of Natural Reserves", in *Theoretical Ecology: Principles and Applications*, ed. Robert May (Oxford, UK: Blackwell Scientific Publications, 1976), 163–186; Robert H. MacArthur and Edward O. Wilson, *The Theory of Island Biogeography* (New Jersey: Princeton Univ. Press, 1967).

48   James Justus, *The Philosophy of Ecology. An Introduction* (Cambridge, UK: Cambridge Univ. Press, 2021); Fredrik Andersen, Rani Lill Anjum and Elena Rocca, "Philosophy of Biology: Philosophical Bias is the One Bias that Science Cannot Avoid", *Elife* 8 (2019): e44929; Cheryl Lousley, "E.O. Wilson's Biodiversity, Commodity Culture, and Sentimental Globalism", *RCC Perspectives* 9 (2012): 11–16; Mark A. Davis, Matthew K. Chew, Richard J. Hobbs, Ariel E. Lugo, John J. Ewel, Geerat J. Vermeij, James H. Brown et al., "Don't Judge Species on Their Origins", *Nature* 474, no. 7350 (2011): 153–154; Arturo Escobar, "Whose Knowledge, Whose Nature? Biodiversity, Conservation, and the Political Ecology of Social Movements", *Journal of Political Ecology* 5 (1998): 53–82; Craig Loehle, "Hypothesis Testing in Ecology: Psychological Aspects and the Importance of Theory Maturation", *The Quarterly Review of Biology* 62, no. 4 (1987): 397–409; Stephen J. Gould and R.C. Lewontin, "The Spandrels of San Marco and the Panglossian Paradigm: A Critique of the Adaptationist Programme", *Proceedings of the Royal Society of London. Series B, Biological Sciences* 205, no. 1161 (1979): 581–598.

49   Gould and Lewontin, "The Spandrels of San Marco".

50   Justus, *The Philosophy of Ecology*; John Kricher, *The Balance of Nature* (Princeton: Princeton Univ. Press, 2009).

51   Christoph Kueffer, "Plant Sciences for the Anthropocene: What Can we Learn From Research in Urban Areas?", *Plants, People, Planet* 2, no. 4 (2020): 286–289; Christoph Kueffer, "Plant Invasions in the Anthropocene", *Science* 358, no. 6364 (2017): 10–11; Will Steffen, Angelina Sanderson, Peter Tyson, Jill Jäger, Pamela Matson, Berrien Moore, Frank Oldfield et al., *Global Change and the Earth System. A Planet Under Pressure* (Berlin: Springer, 2004).

52   Chunglin Kwa, "Local Ecologies and Global Science. Discourses and Strategies of the International Geosphere-Biosphere Programme", *Social Studies of Science* 35 (2005): 923–950.

53   Lousley, "E.O. Wilson's Biodiversity".

54   Lousley, "E.O. Wilson's Biodiversity"; Escobar, "Whose Knowledge, Whose Nature?".

55   Christoph Kueffer and Gertrude Hirsch Hadorn, "Effectiveness in Problem-Oriented Landscape Research"; Humair et al., "Understanding Misunderstandings in Invasion Science".

56   Mark Van Kleunen, Franz Essl, Jan Pergl, Giuseppe Brundu, Marta Carboni, Stefan Dullinger, Regan Early et al., "The Changing Role of Ornamental Horticulture in Alien Plant Invasions", *Biological Reviews* 93, no. 3 (2018): 1421–1437; Jared Diamond, *Guns, Germs, and Steel: The Fates of Human Societies. 20th Anniversary Edition* (New York: Norton, 2017); Nicole L. Boivin, Melinda A. Zeder, Dorian Q. Fuller, Alison Crowther, Greger Larson, Jon M. Erlandson, Tim Denham, and Michael D. Petraglia, "Ecological Consequences of Human Niche Construction: Examining Long-Term Anthropogenic Shaping of Global Species

Distributions", *Proceedings of the National Academy of Sciences* 113, no. 23 (2016): 6388–6396; Alfred W. Crosby, *Ecological Imperialism. The Biological Expansion of Europe, 900–1900* (Cambridge: Cambridge Univ. Press, 2013).

57 Arthur W. Whistler, *Plants of the Canoe People. An Ethnobotanical Voyage through Polynesi* (Hawaii, USA: National Tropical Botanical Garden, 2009).

58 J. Donald Hughes, "Europe as Consumer of Exotic Biodiversity: Greek and Roman Times", *Landscape Research* 28, no. 1 (2003): 21–31.

59 Diamond, *Guns, Germs, and Steel*, 2017; Crosby, Ecological Imperialism, 2013.

60 Boivin et al., "Ecological Consequences".

61 Van Kleunen et al., "Ornamental Horticulture"; Whistler, *Plants of the Canoe People.*

62 For example, Marcus Hall, "The Native, Naturalized and Exotic – Plants and Animals in Human History", *Landscape Research* 28, no. 1 (2003): 5–9.

63 Sukopp, "On the Early History of Urban Ecology in Europe".

64 Ingo Kowarik and Petr Pyšek, "The First Steps Towards Unifying Concepts in Invasion Ecology Were Made One Hundred Years Ago: Revisiting the Work of the Swiss Botanist Albert Thellung", *Diversity and Distributions* 18 (2012): 1243–1252.

65 Charles S. Elton, *The Ecology of Invasions by Animals and Plants. 2nd edition. With Contributions by Daniel Simberloff and Anthony Ricciardi* (Berlin: Springer, 2020); Daniel Simberloff, "Charles Elton: Neither Founder nor Siren, but Prophet", in *Fifty Years of Invasion Ecology: The Legacy of Charles Elton,* ed. David M. Richardson (New York: Wiley, 2011), 11–24.

66 Elton, *The Ecology of Invasions by Animals and Plants.*

67 "SCOPE", accessed February 5, 2022, https://scope-environment.org/.

68 Simberloff, "Charles Elton: Neither Founder nor Siren, but Prophet"; Daniel Simberloff, "SCOPE Project", in *Encyclopedia of Biological Invasions,* eds. Daniel Simberloff and Marcel Rejmanek (Berkeley: Univ. of California Press, 2011), 617–619; Kueffer and Hirsch Hadorn, "Effectiveness in Problem-Oriented Landscape Research".

69 Humair et al., "Understanding Misunderstandings in Invasion Science"; Kueffer and Hirsch Hadorn, "Effectiveness in Problem-Oriented Landscape Research".

70 For example, Laura Meyerson, Aníbal Pauchard, Giuseppe Brundu, James T. Carlton, José L. Hierro, Christoph Kueffer, Maharaj K. Pandit, Petr Pyšek, David M. Richardson, and Jasmin G. Packer, "Moving Toward Global Strategies for Managing Invasive Alien Species", in *Global Plant Invasions,* eds. David R. Clements, Mahesh K. Upadhyaya, Srijana Joshi and Anil Shrestha, in press (Cham: Springer, 2022); Sarah Brunel, Eladio Fernández-Galiano, Piero Genovesi, Vernon H. Heywood, Christoph Kueffer, and David M. Richardson, "Invasive Alien Species: A Growing but Neglected Threat?", in *Late Lessons From Early Warnings: Science, Precaution, Innovation. EEA Report No 1/2013,* ed. European Environment Agency (Copenhagen, Denmark: EEA, 2013), 518–540.

71 Brunel et al., "Invasive Alien Species".

72 Christoph Kueffer, Petr Pyšek, and David M. Richardson, "Integrative Invasion Science: Model Systems, Multi-Site Studies, Focused Meta-Analysis and Invasion Syndromes", *New Phytologist* 200, no. 3 (2013): 615–633.

73 Simberloff, "Charles Elton: Neither Founder nor Siren, but Prophet".

74 Dov F. Sax, John J. Stachowicz, James H. Brown, John F. Bruno, Michael N. Dawson, Steven D. Gaines, Richard K. Grosberg et al., "Ecological and Evolutionary Insights from Species Invasions", *Trends in Ecology & Evolution* 22, no. 9 (2007): 465–471.

75 Nowotny et al., *Re-Thinking Science.*

76 Humair et al., "Understanding Misunderstandings in Invasion Science"; Brunel et al., "Invasive Alien Species"; Kueffer and Hirsch Hadorn, "Effectiveness in Problem-Oriented Landscape Research".

77 For example, Franz Essl, Stefan Dullinger, Piero Genovesi, Philip E. Hulme, Jonathan M. Jeschke, Stelios Katsanevakis, Ingolf Kühn et al., "A Conceptual Framework for Range-Expanding Species that Track Human-Induced Environmental Change", *BioScience* 69, no. 11 (2019): 908–919; Tina Heger, Maud Bernard-Verdier, Arthur Gessler, Alex D. Greenwood, Hans-Peter Grossart, Monika Hilker, Silvia Keinath et al., "Towards an Integrative, Eco-Evolutionary Understanding of Ecological Novelty: Studying and Communicating Interlinked Effects of Global Change", *BioScience* 69, no. 11 (2019): 888–899; Humair et al., "Understanding Misunderstandings in Invasion Science"; Bruce L. Webber, and John K. Scott, "Rapid Global Change: Implications for Defining Natives and Aliens", *Global Ecology and Biogeography* 21, no. 3 (2012): 305–311.

78 Heger et al., "Towards an Integrative, Eco-Evolutionary Understanding"; Kueffer, "Plant Invasions in the Anthropocene"; Christoph Kueffer, "Ecological Novelty: Towards an Interdisciplinary Understanding of Ecological Change in the Anthropocene", in *Grounding Global Climate Change. Contributions from the Social and Cultural Sciences*, eds. Heike Greschke and Julia Tischler (Heidelberg: Springer, 2015), 19–37.

79 Humair et al., "Understanding Misunderstandings in Invasion Science".

80 Humair et al., "Understanding Misunderstandings in Invasion Science"; Kueffer and Hirsch Hadorn, "Effectiveness in Problem-Oriented Landscape Research".

81 Humair et al., "Understanding Misunderstandings in Invasion Science".

82 Davis et al., "Don't Judge Species on Their Origins".

83 Humair et al., "Understanding Misunderstandings in Invasion Science".

84 For example, Christoph Kueffer, Curtis C. Daehler, Christian W. Torres-Santana, Christophe Lavergne, Jean-Yves Meyer, Rüdiger Otto, and Luís Silva, "A Global Comparison of Plant Invasions on Oceanic Islands", *Perspectives in Plant Ecology, Evolution and Systematics* 12, no. 2 (2010): 145–161; Curtis C. Daehler, "Performance Comparisons of Co-occurring Native and Alien Invasive Plants: Implications for Conservation and Restoration", *Annual Review of Ecology, Evolution, and Systematics* 34, no. 1 (2003): 183–211.

85 For example, Kueffer, Pyšek, and Richardson, "Integrative Invasion Science"; Christoph Kueffer, "The Importance of Collaborative Learning and Research Among Conservationists from Different Oceanic Islands", *Revue d'Ecologie (Terre et Vie)* Suppl. 11 (2012): 125–135.

86 René van der Wal, Anke Fischer, Sebastian Selge and Brendon M. Larson, "Neither the Public nor Experts Judge Species Primarily on Their Origins", *Environmental Conservation* 42, no. 4 (2015): 349–355; Humair et al., "Understanding Misunderstandings in Invasion Science"; Kristin Shrader-Frechette, "Non-Indigenous Species and Ecological Explanation", *Biology & Philosophy* 16, no. 4 (2001): 507–519; Stephen J. Gould, "An Evolutionary Perspective on Strengths, Fallacies, and Confusions in the Concept of Native Plants", *Arnoldia* 58, no. 1 (1998): 2–10.

87 Davis et al., "Don't Judge Species on Their Origins".
88 Daniel Simberloff and Montserrat Vilà, "Non-Natives: 141 Scientists Object", *Nature* 475 (2011): 36.
89 Mark A. Davis, *Invasion Biology* (Oxford: Oxford Univ. Press, 2009).
90 Davis et al., "Don't Judge Species on Their Origins".
91 Simberloff et al., "Non-Natives".
92 Ross T. Shackleton, David M. Richardson, Charlie M. Shackleton, Brett Bennett, Sarah L. Crowley, Katharina Dehnen-Schmutz, Rodrigo A. Estévez et al., "Explaining People's Perceptions of Invasive Alien Species: A Conceptual Framework", *Journal of Environmental Management* 229 (2019): 10–26; Christoph Kueffer, and Brendon M.H. Larson, "Responsible Use of Language in Scientific Writing and Science Communication", *BioScience* 64 (2014): 719–724; Kueffer, Pyšek, and Richardson, "Integrative Invasion Science"; Brendon Larson, *Metaphors for Environmental Sustainability* (Yale: Yale Univ. Press, 2011).
93 Jonathan M. Jeschke, Felicia Keesing, and Richard S. Ostfeld, "Novel Organisms: Comparing Invasive Species, Gmos, and Emerging Pathogens", *Ambio* 42, no. 5 (2013): 541–548.
94 Juliet C. Stromberg, Matthew K. Chew, Pamela L. Nagler, and Edward P. Glenn, "Changing Perceptions of Change: The Role of Scientists in Tamarix and River Management", *Restoration Ecology* 17, no. 2 (2009): 177–186.
95 For example, in the case of the Swiss national invasive species strategy and the subsequent formulation of national legislation that is not yet completed, accessed February 5, 2022, http://www.bafu.admin.ch/gebietsfremde-arten.
96 For example, Giuseppe Brundu, "Global Guidelines for the Sustainable Use of Non-Native Trees to Prevent Tree Invasions and Mitigate Their Negative Impacts", *NeoBiota* 61 (2020): 65–116.
97 For example, Andreas Roloff, Sten Gillner, Rico Kniesel, and Deshun Zhang, "Interesting and New Street Tree Species for European Cities", *Journal of Forest and Landscape Research* 1 (2018): 1–7.
98 Ingo Kowarik, and Leonie K. Fischer, "Alien Plants in Cities: Human-Driven Patterns, Risks and Benefits", in *The Routledge Handbook of Urban Ecology. 2nd edition*, eds. Ian Douglas, David Goode, Michael C. Houck, and Rusong Wang (Routledge), 472–482.
99 "GISP", accessed February 5, 2022, https://www.gisp.org/.
100 Harold A. Mooney, Richard N. Mack, Jeffrey A. McNeely, Laurie E. Neville, Peter Johan Schei, Jeffrey K. Waage, *Invasive Alien Species: A New Synthesis* (Washington: Island Press, 2005).
101 Kueffer and Hirsch Hadorn, "Effectiveness in Problem-Oriented Landscape Research", 2008.
102 Mooney et al., *Invasive Alien Species*; Gregory M. Ruiz and James T. Carlton, *Invasive Species: Vectors and Management Strategies* (Washington: Island Press, 2003).
103 For example, E. Lakshmi, M. Priya and V. Sivanandan Achari, "An Overview on the Treatment of Ballast Water in Ships", *Ocean & Coastal Management* 199 (2021): 105296.
104 For example, Philip E. Hulme, Giuseppe Brundu, Marta Carboni, Katharina Dehnen-Schmutz, Stefan Dullinger, Regan Early, Franz Essl et al., "Integrating Invasive Species Policies Across Ornamental Horticulture Supply Chains to Prevent Plant Invasions", *Journal of Applied Ecology* 55, no. 1 (2018): 92–98; Van Kleunen et al., "Ornamental Horticulture".
105 Lakshmi, Priya and Achari, "An Overview on the Treatment of Ballast Water in Ships".

106 For example, Hulme et al., "Integrating Invasive Species Policies"; Van Kleunen et al., "Ornamental Horticulture"; Franziska Humair, Michael Siegrist, and Christoph Kueffer, "Working With the Horticultural Industry to Limit Invasion Risks: The Swiss Experience", *EPPO Bulletin* 44, no. 2 (2014): 232–238.

107 For example, Anthony Ricciardi, Josephine C. Iacarella, David C. Aldridge, Tim M. Blackburn, James T. Carlton, Jane A. Catford, Jaimie T.A. Dick et al., "Four Priority Areas to Advance Invasion Science in the Face of Rapid Environmental Change", *Environmental Reviews* 29, no. 2 (2021): 119–141; Simberloff et al., "Non-Natives".

108 Kueffer, "Plant Invasions in the Anthropocene"; Kueffer, Pyšek, and Richardson, "Integrative Invasion Science"; Jeffrey A. McNeely, *The Great Reshuffling: Human Dimensions of Invasive Alien Species* (Gland, Switzerland: IUCN, 2001).

109 Ana S. Vaz, Christoph Kueffer, Christian A. Kull, David M. Richardson, Stefan Schindler, A. Jesús Muñoz-Pajares, Joana R. Vicente et al., "The Progress of Interdisciplinarity in Invasion Science", *Ambio* 46 (2017): 428–442.

110 Mark Sagoff, "Do Non-Native Species Threaten The Natural Environment?", *Journal of Agricultural and Environmental Ethics* 18 (2005): 215–236.

111 David I. Theodoropoulos, *Invasion Biology: Critique of a Pseudoscience* (Blythe: Avvar Books, 2003).

112 Matthew K. Chew, "The Monstering of Tamarisk: How Scientists Made a Plant into a Problem", *Journal of the History of Biology* 42 (2009): 231–266.

113 See compilation of critique in Daniel Simberloff, "Confronting Introduced Species: A Form of Xenophobia?", *Biological Invasions* 5, no. 3 (2003): 179–192.

114 Franz Essl, Philip E. Hulme, Jonathan M. Jeschke, Reuben Keller, Petr Pyšek, David M. Richardson, Wolf-Christian Saul et al., "Scientific and Normative Foundations for the Valuation of Alien-Species Impacts: Thirteen Core Principles", *BioScience* 67, no. 2 (2017): 166–178.

115 Roger A. Pielke Jr., *The Honest Broker*.

116 Hansson and Hirsch Hadorn, *The Argumentative Turn*.

117 Stirling, "Keep it Complex".

118 Turnhout, Tuinstra and Halffman, *Environmental Expertise*; Turnhout et al., "Listen to the Voices of Experience"; Kueffer et al., "Enabling Effective Problem-Oriented Research".

119 Hirsch Hadorn et al., *Handbook of Transdisciplinary Research*.

120 Gross, *Ignorance and Surprise*; Folke et al., "Adaptive Governance of Social-Ecological Systems"; Ludwig et al., "The Classical Holism-Reductionism Debate in Ecology"; Walters and Holling, "Large-Scale Management Experiments".

121 Pontius and McIntosh, *Critical Skills for Environmental Professionals*.

122 Heger et al., "Towards an Integrative, Eco-Evolutionary Understanding"; Kueffer, "Ecological Novelty".

123 Petr Pyšek, Philip E. Hulme, Dan Simberloff, Sven Bacher, Tim M. Blackburn, James T. Carlton et al., "Scientists' Warning on Invasive Alien Species", *Biological Reviews* 95, no. 6 (2020): 1511–1534.

124 Emma Marris, *Rambunctious Garden: Saving Nature in a Post-Wild World* (New York: Bloomsbury, 2011); Christoph Kueffer and Christopher N. Kaiser-Bunbury, "Reconciling Conflicting Perspectives for Biodiversity Conservation in the Anthropocene", *Frontiers in Ecology and the Environment* 12, no. 2 (2014): 131–137.

125 Chris D. Thomas, *Inheritors of the Earth: How Nature is Thriving in an Age of Extinction* (New York: Public Affairs, 2017).
126 Richard J. Hobbs, Eric Higgs, Carol M. Hall, Peter Bridgewater, F. Stuart Chapin III, Erle C. Ellis, John J. Ewel et al., "Managing the Whole Landscape: Historical, Hybrid, And Novel Ecosystems", *Frontiers in Ecology and the Environment* 12, no. 10 (2014): 557–564; Kueffer and Kaiser-Bunbury, "Reconciling Conflicting Perspectives.
127 Menno Schilthuizen, *Darwin Comes to Town: How the Urban Jungle Drives Evolution*; Thomas, *Inheritors of the Earth*; Scott P. Carroll, "Conciliation Biology: The Eco-Evolutionary Management of Permanently Invaded Biotic Systems", *Evolutionary Applications* 4, no. 2 (2011): 184–199.
128 Jens-Christian Svenning, Pil B. M. Pedersen, C. Josh Donlan, Rasmus Ejrnæs, Søren Faurby, Mauro Galetti, Dennis M. Hansen et al., "Science for a Wilder Anthropocene: Synthesis and Future Directions for Trophic Rewilding Research", *Proceedings of the National Academy of Sciences* 113, no. 4 (2016): 898–906; Christine J. Griffiths, Dennis M. Hansen, Carl G. Jones, Nicolas Zuël and Stephen Harris, "Resurrecting Extinct Interactions with Extant Substitutes", *Current Biology* 21, no. 9 (2011): 762–765.
129 Nina Hewitt, N. Klenk, A.L. Smith, D.R. Bazely, N. Yan, S. Wood, J.I. MacLellan, C. Lipsig-Mumme and I. Henriques, "Taking Stock of the Assisted Migration Debate", *Biological Conservation* 144, no. 11 (2011): 2560–2572; Jason S. McLachlan, Jessica J. Hellmann and Mark W. Schwartz, "A Framework for Debate of Assisted Migration in an Era of Climate Change", *Conservation Biology* 21, no. 2 (2007): 297–302.
130 Antoinette J. Piaggio, Gernot Segelbacher, Philip J. Seddon, Luke Alphey, Elizabeth L. Bennett, Robert H. Carlson, Robert M. Friedman et al., "Is it Time for Synthetic Biodiversity Conservation?", *Trends in Ecology & Evolution* 32, no. 2 (2017): 97–107; Kent H. Redford, William Adams, and Georgina M. Mace, "Synthetic Biology and Conservation of Nature: Wicked Problems and Wicked Solutions", *PLoS Biology* 11, no. 4 (2013): e1001530.
131 Ana Novoa, David M. Richardson, Petr Pyšek, Laura A. Meyerson, Sven Bacher, Susan Canavan, Jane A. Catford et al., "Invasion Syndromes: A Systematic Approach for Predicting Biological Invasions and Facilitating Effective Management", *Biological Invasions* 22, no. 5 (2020): 1801–1820; Kueffer, Pyšek and Richardson, "Integrative Invasion Science".
132 Kueffer, Schneider and Wiesmann, "Addressing Sustainability Challenges".
133 For example, David Bart, "Integrating Local Ecological Knowledge and Manipulative Experiments to Find the Causes of Environmental Change", *Frontiers in Ecology and the Environment* 4, no. 10 (2006): 541–546.
134 Kueffer, "Plant Invasions in the Anthropocene".

References

Akandil, Cengiz, Sascha A. Ismail, and Christoph Kueffer. "No Green Deal Without a Nature-Based Economy". *GAIA* 4, no. 4 (2021): 281–283.
Andersen, Fredrik, Rani Lill Anjum, and Elena Rocca. "Philosophy of Biology: Philosophical Bias Is the One Bias that Science Cannot Avoid". *Elife* 8 (2019): e44929.
Bart, David. "Integrating Local Ecological Knowledge and Manipulative Experiments to Find the Causes of Environmental Change". *Frontiers in Ecology and the Environment* 4, no. 10 (2006): 541–546.

Blicharska, Malgorzata, Richard J. Smithers, Grzegorz Mikusiński, Patrik Rönnbäck, Paula A. Harrison, Måns Nilsson, and William J. Sutherland. "Biodiversity's Contributions to Sustainable Development". *Nature Sustainability* 2, no. 12 (2019): 1083–1093.

Bocking, Stephen. *Nature's Experts: Science, Politics, and the Environment*. New Brunswick, NJ: Rutgers Univ. Press, 2004.

Boivin, Nicole L., Melinda A. Zeder, Dorian Q. Fuller, Alison Crowther, Greger Larson, Jon M. Erlandson, Tim Denham, and Michael D. Petraglia. "Ecological Consequences of Human Niche Construction: Examining Long-Term Anthropogenic Shaping of Global Species Distributions". *Proceedings of the National Academy of Sciences* 113, no. 23 (2016): 6388–6396.

Brundu, Giuseppe. "Global Guidelines for the Sustainable Use of Non-Native Trees to Prevent Tree Invasions and Mitigate Their Negative Impacts". *NeoBiota* 61 (2020): 65–116.

Brunel, Sarah, Eladio Fernández-Galiano, Piero Genovesi, Vernon H. Heywood, Christoph Kueffer, and David M. Richardson. "Invasive Alien Species: A Growing but Neglected Threat?" In *Late Lessons From Early Warnings: Science, Precaution, Innovation. EEA Report No 1/2013*, edited by European Environment Agency, 518–540. Copenhagen, Denmark: EEA, 2013.

Bschir, Karim, and Jamie Shaw. *Interpreting Feyerabend. Critical Essays*. Cambridge, UK: Cambridge Univ. Press, 2021.

Carroll, Scott P. "Conciliation Biology: The Eco-Evolutionary Management of Permanently Invaded Biotic Systems". *Evolutionary Applications* 4, no. 2 (2011): 184–199.

Checkland, Peter, and Jim Scholes. *Soft Systems Methodology in Action*. Chichester: Wiley, 1990.

Chew, Matthew K. "The Monstering of Tamarisk: How Scientists Made a Plant into a Problem". *Journal of the History of Biology* 42 (2009): 231–266.

Crosby, Alfred W. *Ecological Imperialism. The Biological Expansion of Europe, 900–1900*. Cambridge: Cambridge Univ. Press, 2013.

Daehler, Curtis C. "Performance Comparisons of Co-Occurring Native and Alien Invasive Plants: Implications for Conservation and Restoration". *Annual Review of Ecology, Evolution, and Systematics* 34, no. 1 (2003): 183–211.

Dasgupta, Partha. *The Economics of Biodiversity: The Dasgupta Review*. London: HM Treasury, 2021.

Davis, Mark A., Matthew K. Chew, Richard J. Hobbs, Ariel E. Lugo, John J. Ewel, Geerat J. Vermeij, James H. Brown et al., "Don't Judge Species on Their Origins". *Nature* 474, no. 7350 (2011): 153–154.

Davis, Mark A. *Invasion Biology*. Oxford: Oxford Univ. Press, 2009.

Diamond, Jared. *Guns, Germs, and Steel: The Fates of Human Societies*. 20th anniversary edition. New York: Norton, 2017.

Diamond, Jared, and Robert May. "Island Biogeography and the Design of Natural Reserves". In *Theoretical Ecology: Principles and Applications*, edited by Robert May, 163–186. Oxford, UK: Blackwell Scientific Publications, 1976.

Driscoll, Catherine. "*Sociobiology*". In *The Stanford Encyclopedia of Philosophy, Spring 2018*, edited by Edward N. Zalta. Stanford, CA: Metaphysics Research Lab, Stanford Univ., 2018. https://plato.stanford.edu/archives/spr2018/entries/sociobiology/.

Edwards, Paul N. *A Vast Machine. Computer Models, Climate Data, and the Politics of Global Warming.* Cambridge, MA: MIT Press, 2010.

Eigenbrode, Sanford D., Michael O'Rourke, J. D. Wulfhorst, David M. Althoff, Caren S. Goldberg, Kaylani Merrill, Wayde Morse et al., "Employing Philosophical Dialogue in Collaborative Science". *BioScience* 57, no. 1 (2007): 55–64.

Elton, Charles S. *The Ecology of Invasions by Animals and Plants.* 2nd edition. With Contributions by Daniel Simberloff and Anthony Ricciardi. Berlin: Springer, 2020.

Escobar, Arturo. "Whose Knowledge, Whose Nature? Biodiversity, Conservation, and the Political Ecology of Social Movements". *Journal of Political Ecology* 5 (1998): 53–82.

Essl, Franz, Stefan Dullinger, Piero Genovesi, Philip E. Hulme, Jonathan M. Jeschke, Stelios Katsanevakis, Ingolf Kühn et al. "A Conceptual Framework for Range-Expanding Species that Track Human-Induced Environmental Change". *BioScience* 69, no. 11 (2019): 908–919.

Essl, Franz, Philip E. Hulme, Jonathan M. Jeschke, Reuben Keller, Petr Pyšek, David M. Richardson, Wolf-Christian Saul et al. "Scientific and Normative Foundations for the Valuation of Alien-Species Impacts: Thirteen Core Principles". *BioScience* 67, no. 2 (2017): 166–178.

Folke, Carl, Thomas Hahn, Per Olsson, and Jon Norberg. "Adaptive Governance of Social-Ecological Systems". *Annual Review of Environment and Resources* 30 (2005): 441–473.

Frawley, Jodie, and Iain McCalman. *Rethinking Invasion Ecologies from the Environmental Humanities.* London: Routledge, 2014.

Foster, Charles. *Being a Beast: Adventures across the Species Divide.* London: Profile Books, 2016.

Funtowicz, Silvio O., and Jerome R. Ravetz. "Science for the Post-Normal Age". *Futures* 25, no. 7 (1993): 739–755.

Future Earth. *Our Future on Earth.* Future Earth, 2020. https://futureearth.org/publications/our-future-on-earth/.

Gould, Stephen J. "An Evolutionary Perspective on Strengths, Fallacies, and Confusions in the Concept of Native Plants". *Arnoldia* 58, no. 1 (1998): 2–10.

Gould, Stephen J., and R.C. Lewontin. "The Spandrels of San Marco and the Panglossian Paradigm: A Critique of the Adaptationist Programme". *Proceedings of the Royal Society of London. Series B, Biological Sciences* 205, no. 1161 (1979): 581–598.

Green, Lesley J.F. "'Indigenous Knowledge' and 'Science': Reframing the Debate on Knowledge Diversity". *Archaeologies* 4, no. 1 (2008): 144–163.

Griffiths, Christine J., Dennis M. Hansen, Carl G. Jones, Nicolas Zuël, and Stephen Harris. "Resurrecting Extinct Interactions with Extant Substitutes". *Current Biology* 21, no. 9 (2011): 762–765.

Gross, Matthias. *Ignorance and Surprise. Science, Society, and Ecological Design.* Cambridge, MA: MIT Press, 2010.

Haila, Yrjö, and Peter Taylor. "The Philosophical Dullness of Classical Ecology, and a Levinsian Alternative". *Biology and Philosophy* 16 (2001): 93–102.

Hall, Marcus. "The Native, Naturalized and Exotic – Plants and Animals in Human History". *Landscape Research* 28, no. 1 (2003): 5–9.

Hansson, Sven Ove, and Gertrude Hirsch Hadorn. *The Argumentative Turn in Policy Analysis. Reasoning about Uncertainty.* Heidelberg: Springer, 2016.

Harremoes, Paul, David Gee, Malcom MacGarvin, Andy Stirling, Jane Keys, Brian Wynne, and Sofia Guedes Vaz. *The Precautionary Principle in the 20ᵗʰ Century: Late Lessons from Early Warnings.* London: Routledge, 2013.

Head, Leslie. "Living in a Weedy Future", In *Rethinking Invasion Ecologies from the Environmental Humanities,* edited by Frawley Jodie, and Iain McCalman, 87–99. London: Routledge, 2014.

Heger, Tina, Maud Bernard-Verdier, Arthur Gessler, Alex D. Greenwood, Hans-Peter Grossart, Monika Hilker, Silvia Keinath et al., "Towards an Integrative, Eco-Evolutionary Understanding of Ecological Novelty: Studying and Communicating Interlinked Effects of Global Change". *BioScience* 69, no. 11 (2019): 888–899.

Hewitt, Nina, N. Klenk, A.L. Smith, D.R. Bazely, N. Yan, S. Wood, J.I. MacLellan, C. Lipsig-Mumme, and I. Henriques. "Taking Stock of the Assisted Migration Debate". *Biological Conservation* 144, no. 11 (2011): 2560–2572.

Hirsch Hadorn, Gertrude, Holger Hoffmann-Riem, Susette Biber-Klemm, Walter Grossenbacher-Mansuy, Dominique Joye, Christian Pohl, Urs Wiesmann, and Elisabeth Zemp. *Handbook of Transdisciplinary Research.* Heidelberg: Springer, 2008.

Hirsch, Gertrude. "Beziehungen zwischen Umweltforschung und disziplinärer Forschung". *GAIA* 4 (1995): 302–314.

Hobbs, Richard J., Eric Higgs, Carol M. Hall, Peter Bridgewater, F. Stuart Chapin III, Erle C. Ellis, John J. Ewel et al. "Managing the Whole Landscape: Historical, Hybrid, and Novel Ecosystems". *Frontiers in Ecology and the Environment* 12, no. 10 (2014): 557–564.

Hughes, J. Donald. "Europe as Consumer of Exotic Biodiversity: Greek and Roman Times". *Landscape Research* 28, no. 1 (2003): 21–31.

Hulme, Philip E., Giuseppe Brundu, Marta Carboni, Katharina Dehnen-Schmutz, Stefan Dullinger, Regan Early, Franz Essl et al. "Integrating Invasive Species Policies Across Ornamental Horticulture Supply Chains to Prevent Plant Invasions". *Journal of Applied Ecology* 55, no. 1 (2018): 92–98.

Humair, Franziska, Peter J. Edwards, Michael Siegrist, and Christoph Kueffer. "Understanding Misunderstandings in Invasion Science: Why Experts Don't Agree on Common Concepts and Risk Assessments". *NeoBiota* 20 (2014): 1–30.

Humair, Franziska, Michael Siegrist, and Christoph Kueffer. "Working With the Horticultural Industry to Limit Invasion Risks: The Swiss Experience". *EPPO Bulletin* 44, no. 2 (2014): 232–238.

Ismail, Sascha A., Robin Pouteau, Mark van Kleunen, Noëlie Maurel, and Christoph Kueffer. "Horticultural Plant Use as a So-Far Neglected Pillar of Ex Situ Conservation". *Conservation Letters* 14, no. 5 (2021): e12825.

Jeschke, Jonathan M., Felicia Keesing, and Richard S. Ostfeld. "Novel Organisms: Comparing Invasive Species, Gmos, and Emerging Pathogens". *Ambio* 42, no. 5 (2013): 541–548.

Johnson, Kristin. "Natural History as Stamp Collecting: A Brief History". *Archives of Natural History* 34, no. 2 (2007): 244–258.

Justus, James. *The Philosophy of Ecology. An Introduction.* Cambridge, UK: Cambridge Univ. Press, 2021.

Kealiikanakaoleohaililani, Kekuhi, and Christian P. Giardina. "Embracing the Sacred: An Indigenous Framework for Tomorrow's Sustainability Science". *Sustainability Science* 11, no. 1 (2016): 57–67.

Kingsland, Sharon E. *Modeling Nature*. Chicago: Univ. of Chicago Press, 1995.

Kowarik, Ingo, and Leonie K. Fischer. "Alien Plants in Cities: Human-Driven Patterns, Risks and Benefits". In *The Routledge Handbook of Urban Ecology*. 2nd edition, edited by Ian Douglas, David Goode, Michael C. Houck, and Rusong Wang, 472–482, London: Routledge.

Kowarik, Ingo, and Petr Pyšek. "The First Steps Towards Unifying Concepts in Invasion Ecology Were Made One Hundred Years Ago: Revisiting the Work of the Swiss Botanist Albert Thellung". *Diversity and Distributions* 18 (2012): 1243–1252.

Kricher, John. *The Balance of Nature*. Princeton, NJ: Princeton Univ. Press, 2009.

Kueffer, Christoph. "Plant Sciences for the Anthropocene: What Can We Learn From Research in Urban Areas?". *Plants, People, Planet* 2, no. 4 (2020): 286–289.

Kueffer, Christoph, Manuela di Giulio, Kathrin Hauser, and Caroline Wiedmer. "Time for a Biodiversity Turn in Sustainability Science". *GAIA* 29, no. 4 (2020): 272–274.

Kueffer, Christoph, Flurina Schneider, and Urs Wiesmann. "Dressing Sustainability Challenges with a Broader Concept of Systems, Target, and Transformation Knowledge". *GAIA* 28, no. 4 (2019): 386–388.

Kueffer, Christoph. "Plant Invasions in the Anthropocene". *Science* 358, no. 6364 (2017): 10–11.

Kueffer, Christoph. "Ecological Novelty: Towards an Interdisciplinary Understanding of Ecological Change in the Anthropocene". In *Grounding Global Climate Change. Contributions from the Social and Cultural Sciences*, edited by Heike Greschke, and Julia Tischler, 19–37. Heidelberg: Springer, 2015.

Kueffer, Christoph, and Brendon M.H. Larson. "Responsible Use of Language in Scientific Writing and Science Communication". *BioScience* 64 (2014): 719–724.

Kueffer, Christoph, and Christopher N. Kaiser-Bunbury. "Reconciling Conflicting Perspectives for Biodiversity Conservation in the Anthropocene". *Frontiers in Ecology and the Environment* 12, no. 2 (2014): 131–137.

Kueffer, Christoph, Petr Pyšek, and David M. Richardson. "Integrative Invasion Science: Model Systems, Multi-Site Studies, Focused Meta-Analysis and Invasion Syndromes". *New Phytologist* 200, no. 3 (2013): 615–633.

Kueffer, Christoph. Evelyn Underwood, Gertrude Hirsch Hadorn, Rolf Holderegger, Michael Lehning, Christian Pohl, Mario Schirmer et al. "Enabling Effective Problem-Oriented Research for Sustainable Development". *Ecology and Society* 17, no. 4 (2012): 8.

Kueffer, Christoph. "The Importance of Collaborative Learning and Research Among Conservationists from Different Oceanic Islands". *Revue d'Ecologie (Terre et Vie)* Suppl. 11 (2012): 125–135.

Kueffer, Christoph, Curtis C. Daehler, Christian W. Torres-Santana, Christophe Lavergne, Jean-Yves Meyer, Rüdiger Otto, and Luís Silva. "A Global Comparison of Plant Invasions on Oceanic Islands", *Perspectives in Plant Ecology". Evolution and Systematics* 12, no. 2 (2010): 145–161.

Kueffer, Christoph, and Gertrude Hirsch Hadorn. "How to Achieve Effectiveness in Problem-Oriented Landscape Research – The Example Of Research on Biotic Invasions". *Living Reviews in Landscape Research* 2 (2008): 2.

Kuhn, Thomas S. *The Structure of Scientific Revolutions*. Chicago, IL: Univ. of Chicago Press, 1962.

Kwa, Chunglin. "Local Ecologies and Global Science. Discourses and Strategies of the International Geosphere-Biosphere Programme". *Social Studies of Science* 35 (2005): 923–950.

Lakshmi, E., M. Priya, and V. Sivanandan Achari. "An Overview on the Treatment of Ballast Water in Ships". *Ocean & Coastal Management* 199 (2021): 105296.

Larson, Brendon. *Metaphors for Environmental Sustainability*. Yale: Yale Univ. Press, 2011.

Loehle, Craig. "Hypothesis Testing in Ecology: Psychological Aspects and the Importance of Theory Maturation". *The Quarterly Review of Biology* 62, no. 4 (1987): 397–409.

Loreau, Michel, S. Naeem, P. Inchausti, J. Bengtsson, J.P. Grime, A. Hector, D.U. Hooper et al., "Biodiversity and Ecosystem Functioning: Current Knowledge and Future Challenges". *Science* 294, no. 5543 (2001): 804–808.

Lousley, Cheryl, "E.O. Wilson's Biodiversity, Commodity Culture, and Sentimental Globalism". *RCC Perspectives* 9 (2012): 11–16.

Ludwig, Donald, Marc Mangel, and Brent Haddad. "Ecology, Conservation, and Public Policy". *Annual Review of Ecology and Systematics* 32 (2001): 481–517.

MacArthur, Robert H., and Edward O. Wilson. *The Theory of Island Biogeography*. New Jersey: Princeton Univ. Press, 1967.

Mace, Georgina M. "Whose Conservation?". *Science* 345, no. 6204 (2014): 1558–1560.

Marris, Emma. *Rambunctious Garden: Saving Nature in a Post-Wild World*. New York: Bloomsbury, 2011.

McLachlan, Jason S., Jessica J. Hellmann, and Mark W. Schwartz. "A Framework for Debate of Assisted Migration in an Era Of Climate Change". *Conservation Biology* 21, no. 2 (2007): 297–302.

McNeely, Jeffrey A. *The Great Reshuffling: Human Dimensions of Invasive Alien Species*. Gland, Switzerland: IUCN, 2001.

Meyerson, Laura, Aníbal Pauchard, Giuseppe Brundu, James T. Carlton, José L. Hierro, Christoph Kueffer, Maharaj K. Pandit et al. "Moving Toward Global Strategies for Managing Invasive Alien Species", In *Global Plant Invasions*, edited by David R. Clements, Mahesh K. Upadhyaya, Srijana Joshi, and Anil Shrestha, 331–360, Cham: Springer, 2022.

Mooney, Harold A., Richard N. Mack, Jeffrey A. McNeely, Laurie E. Neville, Peter Johan Schei, and Jeffrey K. Waage. *Invasive Alien Species: A New Synthesis*. Washington, DC: Island Press, 2005.

Moore, Jason W. "The End of Cheap Nature, Or: How I Learned to Stop Worrying About 'the' Environment and Love the Crisis of Capitalism". In *Structures of the World Political Economy and the Future of Global Conflict and Cooperation*, edited by Christian Suter, and Christopher Chase-Dunn, 285–314. Berlin: LIT, 2014.

Nixon, Rob. *Slow Violence and the Environmentalism of the Poor*. Cambridge, MA: Harvard Univ. Press, 2011.

Novoa, Ana, David M. Richardson, Petr Pyšek, Laura A. Meyerson, Sven Bacher, Susan Canavan, Jane A. Catford et al. "Invasion Syndromes: A Systematic Approach for Predicting Biological Invasions and Facilitating Effective Management". *Biological Invasions* 22, no. 5 (2020): 1801–1820.

Nowotny, Helga, Peter Scott, and Michael Gibbons. *Re-Thinking Science: Knowledge and the Public in an Age of Uncertainty*. Cambridge, UK: Polity Press, 2001.

OECD. *Beyond Growth: Towards a New Economic Approach*. *New Approaches to Economic Challenges*, Paris: OECD Publishing, 2020.

Parrique, Timothée, Jonathan Barth, François Briens, Christian Kerschner, Alejo Kraus-Polk, Anna Kuokkanen, and Joachim H. Spangenberg. *Decoupling Debunked: Evidence and Arguments against Green Growth as a Sole Strategy for Sustainability*. Brussels: European Environmental Bureau, 2019.

Piaggio, Antoinette J., Gernot Segelbacher, Philip J. Seddon, Luke Alphey, Elizabeth L. Bennett, Robert H. Carlson, Robert M. Friedman et al. "Is It Time for Synthetic Biodiversity Conservation?". *Trends in Ecology & Evolution* 32, no. 2 (2017): 97–107.

Pickett, Steward T.A., Jurek Kolasa, and Clive G. Jones. *Ecological Understanding: The Nature of Theory and the Theory of Nature*. Amsterdam: Elsevier, 2010.

Pielke, Roger A. Jr. *The Honest Broker: Making Sense of Science in Policy and Politics*. Cambridge, UK: Cambridge Univ. Press, 2007.

Pontius, Jennifer, and Alan McIntosh. *Critical Skills for Environmental Professionals. Putting Knowledge into Practice*. Berlin: Springer, 2020.

Potgieter, Luke J., Mirijam Gaertner, Christoph Kueffer, Brendon M. H. Larson, Stuart W. Livingstone, Patrick J. O'Farrell, and David M. Richardson. "Alien Plants as Mediators of Ecosystem Services and Disservices in Urban Systems: A Global Review". *Biological Invasions* 19, no. 12 (2017): 3571–3588.

Pyšek, Petr, Philip E. Hulme, Dan Simberloff, Sven Bacher, Tim M. Blackburn, and James T. Carlton et al. "Scientists' Warning on Invasive Alien Species". *Biological Reviews* 95, no. 6 (2020): 1511–1534.

Raworth, Kate. *Doughnut Economics, Seven Ways to Think Like a 21st-Century Economist*. London: Random House Business, 2017.

Rayner, Steve, and Daniel Sarewitz. "Policy Making in the Post-Truth World. On the Limits of Science and the Rise of Inappropriate Expertise". *The Breakthrough Institute Blog*, March 1, 2021. https://thebreakthrough.org/journal/no-13-winter-2021/policy-making-in-the-post-truth-world.

Redford, Kent H., William Adams, and Georgina M. Mace. "Synthetic Biology and Conservation of Nature: Wicked Problems and Wicked Solutions". *PLoS Biology* 11, no. 4 (2013): e1001530.

Ricciardi, Anthony, Josephine C. Iacarella, David C. Aldridge, Tim M. Blackburn, James T. Carlton, Jane A. Catford, Jaimie T.A. Dick et al. "Four Priority Areas to Advance Invasion Science in the Face of Rapid Environmental Change". *Environmental Reviews* 29, no.2 (2021): 119–141.

Rittel, Horst W.J., and Melvin M. Webber. "Dilemmas in a General Theory of Planning". *Policy Sciences* 4, no. 2 (1973): 155–169.

Roloff, Andreas, Sten Gillner, Rico Kniesel, and Deshun Zhang. "Interesting and New Street Tree Species for European Cities". *Journal of Forest and Landscape Research* 1 (2018): 1–7.

Ruiz, Gregory M., and James T. Carlton. *Invasive Species: Vectors and Management Strategies*. Washington, DC: Island Press, 2003.

Sagoff, Mark. "Do Non-Native Species Threaten The Natural Environment?". *Journal of Agricultural and Environmental Ethics* 18 (2005): 215–236.

Sarewitz, Daniel. "How Science Makes Environmental Controversies Worse". *Environmental Science & Policy* 7 (2004): 385–403.

Sax, Dov F., John J. Stachowicz, James H. Brown, John F. Bruno, Michael N. Dawson, Steven D. Gaines, and Richard K. Grosberg et al. "Ecological and Evolutionary Insights from Species Invasions". *Trends in Ecology & Evolution* 22, no. 9 (2007): 465–471.

Schilthuizen, Menno. *Darwin Comes to Town: How the Urban Jungle Drives Evolution.* London: Picador, 2018.

Shackleton, Ross T., David M. Richardson, Charlie M. Shackleton, Brett Bennett, Sarah L. Crowley, Katharina Dehnen-Schmutz, Rodrigo A. Estévez et al. "Explaining People's Perceptions of Invasive Alien Species: A Conceptual Framework". *Journal of Environmental Management* 229 (2019): 10–26.

Shaw, Jamie, "Feyerabend and Manufactured Disagreement: Reflections on Expertise, Consensus, and Science Policy". *Synthese* 198 (2021): 6053–6084.

Shrader-Frechette, Kristin. "Non-Indigenous Species and Ecological Explanation". *Biology & Philosophy* 16, no. 4 (2001): 507–519.

Simberloff, Daniel. "Charles Elton: Neither Founder nor Siren, but Prophet", In *Fifty Years of Invasion Ecology: The Legacy of Charles Elton*, edited by David M. Richardson, 11–24, New York: Wiley, 2011.

Simberloff, Daniel. "SCOPE Project", In *Encyclopedia of Biological Invasions*, edited by Simberloff, Daniel and Marcel Rejmanek, 617–619, Berkeley: Univ. of California Press, 2011.

Simberloff, Daniel, and Montserrat Vilà. "Non-Natives: 141 Scientists Object". *Nature* 475 (2011): 36.

Simberloff, Daniel. "Confronting Introduced Species: A Form of Xenophobia?. *Biological Invasions* 5, no. 3 (2003): 179–192.

Soulé, Michael E., and Bruce Wilcox. *Conservation Biology: An Evolutionary-Ecological Perspective.* Sunderland, MA: Sinauer Associates, 1980.

Steffen, Will, Angelina Sanderson, Peter Tyson, Jill Jäger, Pamela Matson, Berrien Moore, Frank Oldfield et al. *Global Change and the Earth System. A Planet Under Pressure.* Berlin: Springer, 2004.

Stiglitz, Joseph E., Amartya Sen, and Jean-Paul Fitoussi. *Mismeasuring Our Lives: Why GDP Doesn't Add Up.* New York: The New Press, 2010.

Stirling, Andy. "Keep It Complex". *Nature* 468 (2010): 1029–1031.

Stromberg, Juliet C., Matthew K. Chew, Pamela L. Nagler, and Edward P. Glenn. "Changing Perceptions of Change: The Role of Scientists in Tamarix and River Management". *Restoration Ecology* 17, no. 2 (2009): 177–186.

Sukopp, Herbert. "On the Early History of Urban Ecology in Europe". In *Urban Ecology*, edited by John M. Marzluff, Eric Shulenberger, Wilfried Endlicher, Marina Alberti, Gordon Bradley, Clare Ryan, Ute Simon, and Craig ZumBrunnen, 79–97. New York: Springer, 2008.

Svenning, Jens-Christian, Pil B. M. Pedersen, C. Josh Donlan, Rasmus Ejrnæs, Søren Faurby, Mauro Galetti, Dennis M. Hansen et al. "Science for a Wilder Anthropocene: Synthesis and Future Directions for Trophic Rewilding Research". *Proceedings of the National Academy of Sciences* 113, no. 4 (2016): 898–906.

Taylor, Peter J. "Technocratic Optimism, H.T. Odum, and the Partial Transformation of Ecological Metaphor after World War II". *Journal of the History of Biology* 21, no. 2 (1988): 213–244.

Theodoropoulos, David I. *Invasion Biology: Critique of a Pseudoscience*. Blythe, CA: Avvar Books, 2003.

Thomas, Chris D. *Inheritors of the Earth: How Nature Is Thriving in an Age of Extinction*. New York: Public Affairs, 2017.

Todd, Zoe. "Indigenizing the Anthropocene". In *Art in the Anthropocene: Encounters Among Aesthetics, Politics, Environments and Epistemologies*, edited by Heather Davis, and Etienne Turpin, 241–254, London: Open Humanities Press, 2015.

Trepl, Ludwig, and Annette Voigt. "The Classical Holism-Reductionism Debate in *Ecology*". In *Ecology Revisited. Reflecting on Concepts, Advancing Science*, edited by Astrid Schwarz, and Kurt Jax, 45–83. Heidelberg: Springer, 2011.

Tsing, Anna. "A Threat to Holocene Resurgence Is a Threat to Livability". In *The Anthropology of Sustainability. Beyond Development and Progress*, edited by Marc Brightman, and Jerome Lewis, 51–65, London: Palgrave Macmillan, 2017.

Turnhout, Esther, Willemijn Tuinstra, and Willem Halffman. *Environmental Expertise. Connecting Science, Policy, and Society*. Cambridge, UK: Cambridge Univ. Press, 2019.

Turnhout, Esther, Bob Bloomfield, Mike Hulme, Johannes Vogel, and Brian Wynne. "Listen to the Voices of Experience". *Nature* 488 (2012): 454–455.

UN. *Global Sustainable Development Report 2019: The Future is Now – Science for Achieving Sustainable Development*. New York: United Nations, 2019.

Van der Wal, René, Anke Fischer, Sebastian Selge, and Brendon M. Larson. "Neither the Public nor Experts Judge Species Primarily on Their Origins". *Environmental Conservation* 42, no. 4 (2015): 349–355.

Van Kleunen, Mark, Franz Essl, Jan Pergl, Giuseppe Brundu, Marta Carboni, Stefan Dullinger, and Regan Early et al. "The Changing Role of Ornamental Horticulture in Alien Plant Invasions". *Biological Reviews* 93, no. 3 (2018): 1421–1437.

Vaz, Ana S., Christoph Kueffer, Christian A. Kull, David M. Richardson, Joana R. Vicente, Ingolf Kühn, Matthias Schröter, Jennifer Hauck, Aletta Bonn, and João P. Honrado, "Integrating Ecosystem Services and Disservices: Insights from Plant Invasions". *Ecosystem Services* 23 (2017): 94–107.

Vaz, Ana S., Christoph Kueffer, Christian A. Kull, David M. Richardson, Stefan Schindler, A. Jesús Muñoz-Pajares, and Joana R. Vicente et al. "The Progress of Interdisciplinarity in Invasion Science". *Ambio* 46 (2017): 428–442.

Voigt, Annette. "The Rise of Systems Theory in Ecology". In *Ecology Revisited. Reflecting on Concepts, Advancing Science*, edited by Astrid Schwarz, and Kurt Jax, 183–194. Heidelberg: Springer, 2011.

Walters, Carl J., and Crawford Stanley Holling. "Large-Scale Management Experiments and Learning by Doing". *Ecology* 71 (1990): 2060–2068.

WBGU. *World in Transition. A Social Contract for Sustainability*. Berlin: WBGU, 2011.

Webber, Bruce L., and John K. Scott. "Rapid Global Change: Implications for Defining Natives and Aliens". *Global Ecology and Biogeography* 21, no. 3 (2012): 305–311.

Weizsäcker, von Ernst U., and Anders Wijkman. *Come On! Capitalism, Short-Termism, Population and the Destruction of the Planet*. Heidelberg: Springer, 2018.

Whistler, Arthur W. *Plants of the Canoe People. An Ethnobotanical Voyage through Polynesia.* Hawaii, USA: National Tropical Botanical Garden, 2009.

Wynne, Brian. "Uncertainty and Environmental Learning: Reconceiving Science and Policy in the Preventive Paradigm". *Global Environmental Change* 2, no. 2 (1992): 111–127.

Part III

Questioning the Criteria for Evidence

Health Sciences

5 Surgical Caps and Trouble with Evidence

Epistemology and Ethics of Perioperative Hygiene Measures

Saana Jukola and Mariacarla Gadebusch Bondio

In this chapter, we review the debate concerning the hygienic function of protective clothing in surgical operating theaters. In particular, we analyze the so-called bouffant scandal,[1] a debate which erupted in 2014 after the Association of periOperative Registered Nurses (AORN) issued new guidelines for surgical attire in the perioperative setting. The guideline's recommendation concerning head covers sparked a fierce debate between the AORN, the American College of Surgeons (ACS) and clinicians about how to best protect surgical patients from infections. A central question in this debate was what type and quality of evidence is needed for issuing new policies and practices in the clinical hygienic setting. We illustrate how different epistemic values (e.g., accuracy, internal validity and scope) and non-epistemic (e.g., political and ethical) values and background assumptions can lead to disagreements concerning evidence and guidelines.[2] In this way, we demonstrate one way that de-stabilization and questioning the relevance of evidence can take place. Our aim is to use this case to show how philosophical analysis can help to pinpoint the roles that calls for evidence play in scientific controversies – not only in medicine but also more broadly.

Surgical site infections (SSIs) are associated with significant mortality and morbidity and, consequently, increased healthcare costs and impaired quality of life.[3] Different kinds of interventions are used in healthcare facilities for preventing infections, ranging from hand hygiene, environmental cleaning and disinfection to prophylactic use of antibiotics.[4] Since Evidence-based Medicine (EBM) has become the dominant approach to biomedical research and practice, there have been, on the one hand, an increased demand for high-quality evidence for the effectiveness and safety of hygiene measures in surgery and perioperative care, and, on the other hand, passionate discussions about where the threshold of sufficient evidence should be in this context.[5] The debate on surgical hats demonstrates how references to the EBM standards of evidence can function both as a regulatory epistemic ideal against which empirical bases of recommended measures are assessed and as a rhetorical tool for undermining an opponent's position.

DOI: 10.4324/9781003273509-9

The controversy that became known as the bouffant scandal began in November 2014 after the publication of the AORN's guidelines for surgical attire in the perioperative setting. The main purpose of the guidelines was to reduce the rate of SSIs and shield patients from exposure to microorganisms.[6] They gave directions, for example, as to which fabrics should be used for scrubs, how arms should be covered completely in operating rooms and how jewelry should either be covered or removed during surgical operations.[7] However, the part of the publication that attracted most attention was the section on head covers. It was namely stated that any person who enters an operating room should wear "a clean surgical head cover (e.g., bouffant cap) or hood that confines all hair and completely covers the ears, scalp, sideburns, and nape of the neck".[8] This suggestion caused concern among many members of the surgical community because the recommendation would effectively imply banning the use of the traditional surgical headwear, the skull cap (see Figure 5.1 (A) for a picture of the bouffant hat and Figure 5.1 (B) for the skull cap). The skull cap is a close-fitting cap leaving the ears largely uncovered as well as some of the hair, especially in the back of the head. The surgeons' concerns about losing the skull cap that is "symbolic of the surgical profession"[9] came true – albeit temporarily[10] – when the AORN recommendations were enforced as regulations by accrediting bodies.[11] The Joint Commission on Accreditation of Healthcare Organizations (JCAHO), which is responsible for accrediting hospitals in the US, as well as the leadership of many hospitals, banned skull caps from operating rooms.[12] This meant that healthcare workers, including surgeons, had to wear either an astro bonnet (see Figure 5.1 (C)) or a bouffant hat, which is a loose cap that is secured around the hair with an elastic band.

The publication of the AORN guidelines was followed by an immediate and rather heated response from the ACS and the surgical community.[13] In the debate, the evidential basis of the recommendation to cover all hair for hygiene reasons was questioned and new studies on the effectiveness of different hat types and materials in preventing infections

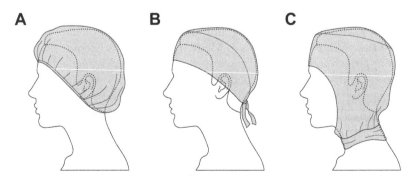

Figure 5.1 (A) A bouffant hat. (B) A skull cap. (C) An astro bonnet. Illustration by Guido I. Prieto.

were conducted. In addition to criticizing the evidential basis of the new recommendation, the opponents of the AORN guidelines raised concerns about cost effectiveness and the environmental impacts of requiring the use of bouffant hats. The American College of Surgeons also emphasized the symbolic nature of the skull cap.

In the following sections, we will explore some central questions that arose in this debate and reflect on their philosophical underpinnings and implications. The analysis will not only help to identify argumentative patterns in this particular debate but also in similarly framed disputes over evidence in future. We first review the evidence that was referred to by the AORN and their critics by examining research articles and commentaries published on the debate. We then analyze the philosophical background assumptions that contributed to the emergence of the disagreement. In particular, by drawing on previous accounts in philosophy of medicine, we will show that the participants in the discussion, on the one hand, had divergent ideals concerning adequate evidence and how to establish claims about causation and, on the other hand, relied on different understandings of the goal of the preventative guidelines, which motivated conducting studies with different endpoints. After this, we briefly discuss the role of different non-epistemic concerns in the debate about the AORN guidelines. The aim of this section is to show how evidence and epistemic concerns alone do not determine what recommendations should be given – practical, ethical and political factors contribute to the content of the guidelines and can spark controversies. In the last substantial section, we broaden the perspective to Germany, where the surgical community has accepted the introduction of new hygienic measures with pragmatism.[14] This comparison allows us to identify and highlight the specifics of the discussion that was sparked in the US before we come to a conclusion.

Evidence and Epistemic Background Assumptions in the Bouffant Scandal

The aim of this section is to articulate how different non-empirical assumptions contributed to the debate between the AORN and the opponents of enforcing the guidelines.[15] After providing an overview of how evidence was debated, we will first discuss the ideals of evidence that influenced the discussion. We finish the section by arguing that distinct understandings of the aims of hygiene control and conceptualizations of the problem at hand added to the disagreement.

Evidence on the Role of Surgical Hats in Infection Control

The debate about surgical headwear centered around the question of what could be inferred from studies that the AORN cited. As the authors of the guidelines acknowledged, the available evidence on surgical wear was limited. They also admitted that no randomized controlled trials

(RCTs) or systematic reviews show that there is a "direct causal relationship between surgical attire and surgical site infections (SSI)".[16] Instead of experimental or epidemiological studies on the connection between different hat types worn in the operating room and the number of infections, the guidelines referred to studies showing that the primary source of bacteria in air in operating rooms is the skin and hair of healthcare workers.[17] These studies were taken to imply that bacteria could get into open wounds of patients after becoming airborne when skin squames are rubbed off.[18] In addition to these studies, the guidelines cited a case report in which psoriatic scalp lesions of an operating room technician were linked to 20 postoperative wound infections.[19] This study of a local infection outbreak demonstrated that pathogen transmission from an infected healthcare worker to a patient was indeed possible.

Many of the commentaries made the critics' disparaging attitude toward this evidential basis of the AORN guidelines evident. For example, Matthew Bartek, Francys Verdial and E. Patchen Dellinger use scare quotes when referring to the studies quoted in the AORN guidelines: "Interestingly, the only 'evidence' cited by the AORN for covering the ears and every last bit of hair is that they contain bacteria that might fall into the surgical wound".[20] Similarly, surgeon and blogger David Gorski commented that "despite invoking 'evidence-based' definitions, notice that AORN didn't cite any actual...oh, you know...evidence to show that wearing surgical skull caps is associated with an increased SSI rate".[21] The core criticism was that evidence of hair and skin carrying bacteria and a case report linking a healthcare worker to infections was not enough to justify a recommendation that would change an established practice. Instead, evidence showing an association between covering all hair and *reduced SSIs rates* would be needed before a recommendation to cover all hair could be given. Consequently, the critics of the AORN guidelines conducted several observational studies to investigate whether there is a connection between the kind of headwear and SSI rates – in other words, if wearing a bouffant hat is more effective in preventing SSIs than wearing a skull cap.[22] For example, Hussain Shallwani et al. analyzed SSI data from 13 months before and 13 months after the mandatory use of bouffant hats was issued to see whether the policy change had an impact on the infection rates. The outcome of this and other published studies was that the type of hat did not correlate with SSI rates.

In addition to the observational studies, a quasi-experimental study on different hats and hat materials was carried out. Troy A. Markel et al. examined how the type of headwear worn in a one hour long mock surgery affects air particle counts.[23] The study assessed both airborne particulate contamination, microbial contamination (by active air samples and passive settle plate assessment) and the physical attributes (permeability, penetrability, thickness and porosity) of different hats. Results from this study indicated that disposable bouffant hats were not better in "regard to particle or actively sampled microbial contamination" than disposable or cloth skull caps.[24] Markel et al. acknowledged that their study had several limitations, such as neither being blinded nor randomized and not comparing headwear in an actual operation. Yet they took its outcomes to contradict the claim that the bouffant hat would prevent particle contamination more effectively than the skull cap.[25] The results from this study were later cited in articles arguing that there is no evidence to support the recommendation to cover all hair as a means to prevent SSIs. For instance, by citing the results of Markel et al., David N. Naumann et al. state that the use of bouffant hats comes with "a potential *increased* risk of SSIs".[26]

Thus, the parties in the debate both referred to empirical evidence, yet they disagreed on what could be inferred from the available evidence. For the AORN, the established fact that scalp and hair are carriers of bacteria means that there is a risk of potential transmission to open wounds. The case report of an incident where SSIs had been connected to the presence of an infection-carrying staff member supported the conclusion that contamination could indeed happen in this way. This possibility, in turn, justifies the need for covering all hair. At the same time, critics of the AORN guidelines argued that the evidence referred to by the AORN did not warrant the claim that covering all hair could help to reduce SSIs. The evidence cited by the AORN is evidence of a *possible* mechanism of contamination *potentially* leading to SSIs. What is missing, according to the critics, is support for the claim that covering all hair actually correlates with reduced rates of SSIs. Consequently, according to the skeptics, the cited evidence could not warrant the recommendation to use headwear covering all hair.[27]

The Role of Epistemic Background Assumptions in Evidence Evaluation

On the face of it, the debate about the surgical hats appears to be a typical scientific controversy about prevention measures, in which weaknesses in the empirical evidence presented by one side of the dispute are criticized by the other side. What is interesting is that in this case the critics did not question the outcomes of the studies referred to by the AORN. What they disagreed about were the conclusions that could be drawn on the basis of these outcomes. What were the reasons for this disagreement? A philosophical analysis will help to reveal them: The parties' evidential reasoning relied on partially conflicting assumptions. In what follows, we focus on two differences in the background assumptions of the discussants: The first tension concerns implicit epistemological views of how a causal connection can be established and more explicit positions concerning what the standards of adequate evidence are. The second disagreement is related to how the problem of infection control should be understood and, relatedly, whether the studies should measure infection rates or pathogen counts. We discuss these in turn.

The AORN's recommendation to cover all hair in the operating room was based on studies reporting hair and skin of the healthcare workers being a main source of pathogens. It is noteworthy that the critics of the AORN guidelines did not question these studies or try to deny the case report about infections being linked to the presence of a staff member with psoriasis. What they disagreed on was what kind of conclusions could be drawn on the basis of this evidence. In this sense, the criticism of the evidential basis of the AORN recommendation (i.e. of the studies showing hair and skin being colonized with bacteria and pathogens becoming airborne on skin squames, as well as the case reports of incidents in which the presence of an infected staff member could be linked to a number of SSIs) follows the lines of argument that are typical in the epistemological framework of EBM.[28] From time to time, appeals to the ideals of EBM are explicitly made in the debate. For example, when Naumann et al. criticized the enforcement of the AORN guidelines as a regulation, they declared that "[t]hose with such authority [to enforce regulation] have a responsibility to lead the way with true evidence-based policies".[29] However, our analysis of the statements that were made in the debate shows that the EBM ideals were appealed to also more implicitly.

According to the tenets of EBM, clinical practice should be based as far as possible on systematic scientific evidence. However, not all types of evidence are taken to be equal. The quality of the available scientific evidence is evaluated according to the ranking that is represented by the evidence hierarchy. In this hierarchy, RCTs, systematic reviews and meta-analyses of RCTs are placed at the top. Observational studies (e.g., cohort studies and case-control studies) are in the middle of the hierarchy, while laboratory research and anecdotal evidence (e.g., case reports) are thought to provide the weakest evidence. When the available evidence relevant for solving a question (concerning, for example, the efficacy of a drug) is conflicting, the evidence originating from study types placed higher on the hierarchy trumps the lower level evidence.[30]

It is worth noting that these EBM ideals of evaluating evidence have become common in other fields outside clinical context. For example, nutrition research is often criticized because RCTs are seldom conducted in the field.[31] In the area of public decision-making, evidence-based policy refers to the idea that decisions should be based on empirical evidence, preferably data from RCTs.[32]

But what is the justification for ranking evidence according to the evidence hierarchy? It has been pointed out that this hierarchical ranking of study types is based on an epistemological premise, namely that internal validity is the most important epistemic value.[33] Internal validity refers to the degree to which the results of the study are accurate to the subjects included in the study.[34] In RCTs, controlling for the effects of biasing factors is considered to be possible as long as studies are properly randomized and blinded.[35] Contrarily, making claims on the basis of study types lower in the hierarchy comes with the risk of confounding and bias of different types. Consequently, on the basis of this criterion, RCTs are taken to provide better evidence than study types lower in the hierarchy.

As scholars applying insights from the philosophy of causality on medical methodology have shown, the EBM ranking of evidence is related to a particular view about how much we need to know about the process that connects interventions to their claimed effects in order to make causal claims.[36] According to these scholars, laboratory and case study evidence is discounted because mechanistic evidence, i.e. evidence about "entities and activities organized in such a way that they are responsible for the phenomenon",[37] in general is discounted in EBM. According to the EBM framework, if an RCT shows that an intervention has an effect X, we do not need information about *why* X happens in order to state that the intervention truly has an effect X. Consequently, the study types highest on the evidence hierarchy produce statistical evidence about average effects in populations without necessarily increasing understanding of why the effect happens. In this framing, evidence of a *possible* mechanism without statistical evidence is simply not enough to warrant causal claims.[38]

The debate raised by the publication of the AORN guidelines is thus an example of a clash of two epistemological frameworks and sets of epistemological assertations about what evidence is sufficient for making claims about causation. The evidence that the AORN based their recommendation on was mainly mechanistic: The studies indicated a potential mechanism that is a part of a causal path through which SSIs might occur: Bacteria becoming airborne and being transferred to a patient via skin squames and thus infecting the patient. On the other hand, the criticism raised by the ACS, Bartek, Verdial and Dellinger and other commentators shows adherence to the EBM framework, in which statistical evidence about the relation between the intervention and outcome is required.[39] Similarly, individual case reports are thought to provide only anecdotal evidence, which cannot be extrapolated to new cases.[40] Consequently, these critics devalue the relevance of the evidence the AORN guidelines cited in support of the policy to cover all hair in operating rooms.

Adherence to the EBM ideals is also apparent in how the critics of the AORN guidelines responded by conducting their own epidemiological studies. Many proponents of the tenets of EBM are generally suspicious of evidence from non-experimental epidemiological studies, given that they cannot control for confounding factors. However, the studies on the association between the type of surgical headwear and the rate of SSIs (conducted, e.g., by Ivy N. Haskins et al. and Hussain Shallwani et al.) are taken to trump the evidence presented by the AORN, because observational studies are ranked higher in the evidence hierarchy than laboratory evidence (in this case showing, e.g., that hair is colonized with bacteria) and case reports of infection outbreaks.

In addition to differences in the criteria for adequate types of evidence for making causal claims, an analysis of the articles published during and after the bouffant scandal shows differences also in how the *purpose* of hygiene guidelines – and thus the whole debate – was framed. When describing the need of guidelines for hygiene measures in perioperative context, Leslie Bourdon writes:

> Although research related to surgical attire is limited and no randomized controlled trials or systematic reviews show a direct causal relationship between surgical attire and surgical site infections (SSIs), increased numbers of microorganisms in the perioperative setting increase the patient's risk of SSI. Thus, health care workers should make efforts to reduce the patient's exposure to microorganisms by following the RP [recommended practices] for surgical attire.[41]

As this quote shows, the AORN was not only interested in the prevention of SSI. Also the increase in the number of pathogens in the perioperative setting was considered to be problematic – because the presence of pathogens in the environment was taken to be a link in the causal chain leading to infections. From this perspective, studies reporting the main source of bacteria in the operating room being the skin and hair of healthcare workers were relevant. For their part, the critics of the guidelines were focused on the number of SSIs and assessing the claim that there is an association between the type of the surgical wear and infection rates. This emphasis on measuring infection rate – not the number or the spread of bacteria – is distinctive of EBM, in which the importance of assessing the effect of different healthcare practices on clinical events (e.g., morbidity, mortality) is emphasized.[42] This shift in the focus on what is being measured becomes evident when surveying the outcomes used in the studies critical of the AORN guidelines. For example, Shallwani et al. compared the numbers of SSIs before and after the policy change.[43] Similarly, Haskins et al., Bradley W. Wills et al. as well as Arturo J. Rios-Diaz et al. had the SSI rate as their main outcome measure.[44] The study by Markel et al. was an exception to this rule as it assessed the claim that the hat type influences particulate and bacterial contamination in the environment.[45] Apart from this one study, the focus of critics of the AORN guidelines was on a different question than that which the AORN sought to address.

The analysis of explicit criteria and implicit suppositions in the debate concerning surgical hats shows how evidential reasoning relies on diverse background assumptions, some of which may be epistemological and conceptual in nature.[46] In the light of different background assumptions, it is possible for individuals to agree on facts (e.g., hair being colonized by bacteria) but disagree on what inferences should be drawn from those facts.[47] The criticism of the AORN guidelines was partially based on epistemological premises central to EBM, namely that internal validity is the most important epistemic value and that statistical evidence trumps evidence about possible mechanisms. Likewise, the critics' framing of the problem of hygiene control differed from that of the AORN: While the AORN saw preventing the spread of pathogens as a central task, the focus of the critics was on SSI reduction. Consequently, the available case reports (the empirical results of which according to the EBM understanding cannot be generalized) and the evidence concerning a *possible* mechanism of pathogen transmission were discounted.

Now that we have outlined the epistemological sources of the disagreement, it is worth considering briefly on what evidence guidelines for hygiene measures should – and could – be based on. In our analysis of EBM criteria, randomized testing of the effectiveness of preventive interventions presents a particular challenge because of the complexity of the transmission pathways and the number of possible agents. Yet the need to develop applicable guidelines that combine available evidence with expert knowledge is ethically imperative in the interest of patient safety. The particular difficulty of developing just such guidelines for the prevention of SSI in the surgical context was also highlighted by the WHO in 2016.[48] Observational studies, especially retrospective studies, in turn, are vulnerable to confounding and bias caused by, for example, the choice of controls in the study.[49] Taking note of the fact that all study types have their weaknesses (either in general or in particular contexts), philosophers have argued for a more pluralistic and broader notion of evidence.[50] Different methods can provide valuable information about the relation of an intervention, its (un)suspected effect or *how* the effect takes place. This implies that a more context-sensitive approach to evidence evaluation is needed. Instead of categorically labeling evidence from certain study types as conclusive or too weak, it would be important to consider *what* the relevant questions are and which methods could best provide answers to these particular questions in a given context. Moreover, given that all methods have weaknesses, acquiring evidence from different sources and different methods to inform practical decision-making would be recommendable. For example, mechanistic evidence from laboratory studies together with evidence from observational studies or RCTs can help to determine whether an intervention works more reliably than evidence from an RCT alone.[51] A potential tool for evaluating and amalgamating evidence from different sources is a framework developed by Jürgen Landes, Barbara Osimani and Roland Poellinger.[52] By drawing on Austin Bradford Hill's views on causality and the Bayesian net approach, their method permits utilizing diverse sources of evidence for assessing causal claims.[53] The tool enables calculating the probability of a causal hypothesis on the basis of available data and updating the estimate as new evidence becomes available.[54] These innovative approaches seem to offer a viable and promising means of evidence evaluation, particularly for the area of hygienic prevention measures and their validation.

Non-Epistemic Concerns in the Bouffant Scandal

In the previous section, we outlined how different epistemic assumptions influenced evidence production and evaluation in the so-called bouffant scandal. In this section, we explain briefly how non-epistemic concerns influenced the debate. Issuing guidelines that are expected to affect practice is not only a matter of following the evidence.[55] As has been argued before, evidence and epistemic concerns alone cannot determine what normative rules, such as guidelines or policies, should be established in the clinical or other practical settings.[56] If there is "no ought from is", as famously argued by David Hume, it is the case that even when the strongest evidence demonstrates the efficacy of an intervention, we need additional normative premises to come to the conclusion that the intervention should be recommended or used.[57] These premises can include, for instance, the desirability of mitigating the symptoms of or preventing a disease. A further reason for stating that non-epistemic concerns are integral in guideline production is the observation that aiming at the clinical and ethical goals of healthcare (cure, care, justice, etc.) under current limited resources requires balancing possibly achievable health-related goods with other societal goods.[58] For example, when guidelines aiming at preventing SSIs are issued, the experts have to consider whether a certain policy – even if the available evidence supports the claim of its efficacy – would require too many resources and hinder implementing other policies considered important. Moreover, the social acceptability and practicality of a given recommendation have to be considered.[59] As a rather amusing example of this, Bartek, Verdial and Dellinger note that despite there being evidence indicating that women shed less bacteria than men and that naked men shed less bacteria than men wearing scrubs, no one has suggested that (clothed) men should be banned from entering operating rooms.[60] Giving a practical recommendation thus involves balancing different practical and ethical concerns with evidential considerations. These non-epistemic concerns, then, can cause debate in the same way as the evidence used.

In the debate concerning surgical hats, critics of the AORN guidelines appealed to different non-epistemic concerns, which, according to them, spoke against enforcing the new bouffant policy. First of these was cost efficiency. This is understandable in the light of the dual role that clinical guidelines serve: On the one hand, it is hoped that grounding practices in scientific evidence is expected to work as means of quality control. On the other hand, demanding that practitioners follow guidelines will save resources.[61] Consequently, for instance, Radwan Dipp Ramos and Kamal Itani point out the increased costs of following the AORN recommendation to wear long-sleeved surgical jackets.[62] In a similar manner, Naumann et al. criticize the bouffant policy for increasing costs.[63] It is noteworthy that the concerns about the increasing costs do not appear to be well-founded. For example, a question left open by Naumann et al. is whether the costs of bouffant hats exceed the costs of skull caps. Moreover, in their retrospective study, Wills et al. report bouffant hats to have been "57.14% less expensive than surgical skull caps".[64]

In addition to the costs, environmental concerns were also raised in the debate. For instance, Naumann et al. write: "[the new] policy has the unintended consequence of environmental damage because these caps are not biodegradable".[65] However, as these authors do not compare the environmental impact of the bouffant hats with that of the skull cap, it remains an open question whether the environmental harms of bouffant hats really are greater than those of the skull caps.

A third non-epistemic concern that was explicitly invoked in the debate was professional autonomy. Requiring new practices (e.g., wearing a bouffant hat instead of a classical surgical skull cap) was claimed to burden the professionals[66] and lead to the disappearance of a hat "symbolic of the surgical profession".[67] In these commentaries, the enforcement of the AORN guidelines and the following ban of skull cap is presented as a disruption of tradition and taking away the agency of a professional group.[68] Similar resistant reactions to guidelines have been reported in previous literature. For instance, Zhicheng Wang, Lisa Bero and Quinn Grundy discuss how experienced violation of clinical and professional autonomy contributed to Canadian radiologists resisting the implementation of new breast cancer screening guidelines in 2018.[69] According to them, fierce debates about guidelines are often political in nature in the sense that professional groups want their group interests to be reflected in the guidelines.[70] A failure to take these interests into consideration when the guidelines are formulated can lead to resistance. The statement by the ACS emphasizing the status of the skull cap as "the symbol" of the professional highlights how the bouffant debate was not only about the most effective means of preventing infections but also about the means that surgeons have for presenting themselves as a group with a shared identity.[71]

Reactions to preventive recommendations, guidelines or orders cannot be characterized solely in terms of their epistemic vs. non-epistemic content. Cultural factors, political, ethical and professional frameworks also shape responses within the scientific community and the professional groups concerned. In the bouffant debate, non-epistemic considerations, for example, cost efficiency, environmental burden and professional autonomy of the surgeons, are not presented as challenging evidential considerations but invoked as additional reasons for opposing the enforcement of the guidelines. If, for example, ethically significant aspects, such as responsibility toward vulnerable individuals in the operating theater or the duty to provide the highest level of protection, had been considered, different weightings could have resulted. This highlights a point previously argued for by numerous scholars: Evidence-based decision-making and guidance are permeated by non-epistemic concerns and evidence alone does not determine which course of action should be taken.[72] Consequently, the disagreements concerning guidelines often cannot be reduced to disagreements about (empirical) evidence. This can be observed also in other contexts. For example, during the COVID-19 pandemic, the recommendations concerning masks have varied between countries and jurisdictions despite the existing evidence being globally available.[73] Similarly, regional differences in adherence and resistance to these policies have varied, and a connection between cultural values and mask use has been observed.[74] In a like manner, experts preparing population-level nutrition guidelines have to consider, for example, local customs and agriculture in addition to the existing science. Consequently, the debates about dietary recommendations are not restricted to evidentiary questions but include arguments that are ethical and political in nature.[75]

Comparing the US and German Guidelines

In the previous two sections, we have pinpointed reasons that contributed to the emergence of the bouffant debate. Next, we contrast this debate with the German discussion related to the role of surgical hats in infection control. This comparison clarifies how differences in institutional settings and participating groups can shape discourses even when the available evidence is the same.

It is noteworthy that in Germany there has been no debate about surgical headwear even after the bouffant scandal took place in the US. Moreover, many of the views of the German surgical community are more in line with the AORN than the ACS. An article published in the *Deutsches Ärzteblatt* in 2019 gets to the heart of the matter: "If one knows that the ears are the most likely to harbor bacteria, then the only consequence can be a general obligation to use astro-bonnets [headwear covering all hair and ears]".[76] The statement shows a confident attitude toward the evidential basis of preventive-hygienic measures in the operating room and simple pragmatism with regard to the problem of effective headgear. The lack of interest in the question of the right hat in the operating room can be better understood by looking at the current positions German experts developed and are developing in hospital hygiene.

In the German material, the direct references to the question of the correct head covering in the operating room can be found in the guidelines for hospital and doctors' office hygiene and the German guidelines for perioperative hygiene.[77] We present these two sources in the following.

For the new edition of the guidelines for hospital and doctors' office hygiene the hygienist Axel Kramer et al. have introduced supplementary chapters on the subject of protective clothing and the updated existing ones.[78] At the beginning of the section "Personal hygiene of nursing and medical staff" (5.2.), the authors state that hygienic misconduct in hospitals is caused by insufficient knowledge about infection transmission. They also recognize that changes in dress code always generate debate due to the lack of "epidemiological evidence" for wearing professional attire and because clothing is often viewed as an expression of personality. This is then followed by a clear positioning on SSI and preventive measures in the perioperative area. According to the interdisciplinary group of authors headed by leading hygienists in Germany, the hair, ears, sideburns and neck must be covered. They cite the AORN guidelines when they state that uncovered hair can become a source of SSIs.[79] The evidence of cross-contamination is not questioned and is considered a premise for compliance with the AORN guidelines: Kramer et al. write that rationales for wearing occupational clothing include the persistence of pathogens on reusable textiles, evidence of pathogens on occupational clothing and evidence of cross-contamination. Accordingly, for the authors, the prerequisites for the adherence to the existing regulations include basic knowledge of sepsis and antisepsis, of hygiene and infectiology and acknowledging that the regulations are based on research and continuously updated. However, common sense also plays a role in adherence. A comparison is made with an everyday area where hygiene measures to prevent contamination are common practice:

> Hair protection is intended to prevent contamination by hair (e.g., operating room, kitchen). If splashing of excitable material is to be expected, single-use hoods that cover the hair on the head and are impermeable to hair should be worn, just as one would when performing/just as during activities that pose a risk of infection (e.g., large-area dressing changes).[80]

By extending the problem of infection prevention in non-high-risk-areas (such as in kitchens), the authors suggest comparison with everyday situations in which common sense is decisive. In the suggested scenario, there is no doubt that workers in the kitchen are obliged to cover their heads and that restaurant guests have a right to "hair-free" dishes.

The German guidelines for perioperative hygiene issued by the working group "Hospital and Practice Hygiene" of the Association of the

Scientific Medical Societies in Germany (AWMF) exemplify a similar attitude and the principle of dual responsibility for protecting both the patient and the staff:

> The surgical gown [has] to provide an effective pathogen barrier [...] This should prevent the path of infection both from the staff to the wound and from the patient to the wound, but not least also from the patient to the staff. Head hair protection is worn to reduce the risk of contamination of the surgical field by microorganisms originating from the hair or scalp of the surgical team. [...] Head hair protection must be designed to completely cover the hair of the head and beard, if applicable.[81]

What could explain this considerable difference between the discussions in Germany and the US? In contrast to the US debate, in which insistence on the quality of evidence has played a major role, the group of German hygiene experts (and with them the surgical community) has welcomed and followed the AORN guidelines. We suggest that this is because in the bouffant debate, the EBM ideals of evidence were invoked partially as a rhetorical tool. As Trisha Greenhalgh and Jill Russell have argued, calls for particular types of evidence can be used as "moves in a rhetorical argumentation game" when conflicts arise in decision-making.[82] One could propose the hypothesis that the fact that groups of medical experts in Germany act as authors of recommendations and guidelines on perioperative hygiene increases the acceptance among physicians.[83] Given that in Germany there was no conflict between professional groups, there was no need to invoke EBM standards of evidence as a weapon in the debate. The German experts on perioperative hygiene were thus able to argue on the basis of existing evidence, common sense and the duty to protect others and the self.

The comparison of the US and the German discussions suggests that different professional or interprofessional constellations of experts and decision makers may influence how claims about evidence and critique are formulated. In order to understand and explain acceptance or questioning of evidence-based guidelines, it would be important to take a close look at such intraprofessional collaborations and tensions in the future.

Conclusion

In this chapter, we anatomized a dispute centered on evidential criticism. The bouffant debate is an instructive example of how evidence can be destabilized in medicine and more broadly. The analysis of epistemic and non-epistemic content in the debate has allowed us to examine the role of implicit assumptions related to evidence evaluation and application. We have shown how epistemic ideals and non-epistemic concerns influenced the debate about surgical hats and thus also explained how it was possible for the AORN and its critics to reach different conclusions about the same evidence. Through this analysis, we were able to pinpoint what understanding of EBM, what ideals of evidence and what actual methodological weaknesses moved the critics in their criticisms. Finally, we broadened the perspective to include the German context. We suggested that the difference between the US and German debates indicates that references to the EBM standards of evidence also had a rhetorical role in the conflict between the AORN and its critics. Our reflection offers a philosophical contribution to elucidating this debate, in which clinical, normative and rhetorical elements were intertwined.

In the bouffant scandal, no one disputed that the risk of infection must be minimized in the perioperative area. On the face of it, the debate appeared to be about empirical evidence. However, our analysis shows that the disagreement actually originates from conflicting epistemic and non-epistemic assumptions. The position of the critics of the AORN guidelines was based on epistemological assertions typical of EBM, namely the demand for correlational evidence and studies focusing on clinical outcomes. This gave them reasons to discount the mechanistic evidence and the case report that formed the basis of the recommendation to cover all hair. In addition to the evidential reasons, the critics referred to the financial and environmental concerns in their opposition to the new policy. Moreover, in the German context, the same evidence rejected by the critics of the AORN in the US has been accepted by the hygiene experts and also by the surgical community. That is, the proven presence of bacteria on the scalp, hair and ears constitutes sufficient evidence that these regions must be effectively covered to form a barrier that protects against transmission of pathogens.[84] This highlights the fact that evidence alone does not help to determine what good practice in a particular situation is.

Philosophy can help us better understand the sources of disagreement in science, but how can it help to overcome them? For instance, can philosophy offer resources for experts who are in the business of evaluating evidence for hygiene guidelines? Philosophers of medicine interested in evidence evaluation have suggested new approaches that aim to go beyond what EBM offers. For example, more contextual approaches to evidence evaluation have been proposed and an evidence-amalgamation framework based on Bradford Hill's work and Bayesian theory have been developed.[85] In their contribution to this book, Fredrik Andersen et al. in turn, suggest that the so-called explanatory approach to weighing evidence can help expert groups in guideline formulation when evidence or the interpretations of evidence are conflicting. These new ways of assessing evidence for a particular purpose can provide tools for overcoming disagreements in situations where following the EBM ideal is practically or ethically difficult.

As Trisha Greenhalgh and co-authors argued during the heated discussion on the preventive effect of masks in the pandemic, "the search for perfect evidence may be the enemy of good policy".[86] A much-discussed question in the philosophy of science has been whether delaying an intervention until there is so-called higher level evidence is ethically acceptable, especially when the suggested measures do not have foreseeably dangerous effects and bring a potential benefit. It has been argued that it should be possible to adjust the criteria of evidence according to the practical requirements of the situation and the risks involved.[87] Finding the right balance between epistemic ideals, ethical principles and practical concerns is thus a challenge that experts applying evidence to issuing good policies need to face. Given that there are no established guidelines for locating where this balance lies, philosophical reflection is needed.

Notes

1 Clayton C. Petro and Michael J. Rosen, "What Surgeons Need to Know about the Bouffant Scandal", *JAMA Surgery* 154, no. 11 (2019): 989–990.
2 The distinction between epistemic and non-epistemic (or cognitive and non-cognitive or constitutive and contextual) values is a debated topic in philosophy of science, see for example Phyllis Rooney, "On Values in Science: Is the Epistemic/Non-Epistemic Distinction Useful?", *PSA: Proceedings of the Biennial Meeting of the Philosophy of Science Association*, 1992, no. 1 (1992): 13–22; Helen E. Longino, "Cognitive and Non-Cognitive Values in Science: Rethinking the Dichotomy", in *Feminism, Science, and the Philosophy of Science*, eds. Lynn Hankinson Nelson and Jack Nelson (Dordrecht: Springer, 1996), 39–58. We will not question this distinction in the context of this chapter. For another approach to analyzing guideline construction, see the contribution by Fredrik Andersen et al. in this volume.
3 Bradley W. Wills, Walter R. Smith, Alexandra M. Arguello, Gerald McGwin, Elie S. Ghanem and Brent A. Ponce, "Association of Surgical Jacket and Bouffant Use With Surgical Site Infection Risk", *JAMA Surgery* 155, no. 4 (2020): 323–328; Arturo J. Rios-Diaz, Guillaume Chevrollier, Hunter Witmer, Christine Schleider, Scott Cowan, Michael J. Pucci and Francesco Palazzo, "The Art and Science of Surgery: Do the Data Support the Banning of Surgical Skull Caps?", *Surgery* 164, no. 5 (2018): 921–925.
4 Benedetta Allegranzi, Peter Bischoff, Stijn de Jonge, N. Zeynep Kubilay, Bassim Zayed, Stacey M. Gomes, Mohamed Abbas et al., "New WHO Recommendations on Preoperative Measures for Surgical Site Infection Prevention: An Evidence-Based Global Perspective", *The Lancet Infectious Diseases* 16, no. 12 (2016): e276–e287; Harsha Siani and Jean-Yves Maillard, "Best Practice in Healthcare Environment Decontamination", *European Journal of Clinical Microbiology & Infectious Diseases* 34, no. 1 (2015): 1–11.
5 Matthew Bartek, Francys Verdial, and E. Patchen Dellinger, "Naked Surgeons? The Debate about What to Wear in the Operating Room", *Clinical Infectious Diseases* 65, no. 9 (2017): 1589–1592.
6 Leslie Bourdon, "RP First Look: New Recommended Practices for Preoperative Patient Skin Antisepsis", *AORN Connections* 100, no. 4 (2014): 1–3.

7 Bourdon, "RP", 1.

8 Bourdon, "RP", 1.

9 "Statement on Operating Room Attire", American College of Surgeons, accessed September 21, 2021, https://bulletin.facs.org/2016/10/statement-on-operating-room-attire/.

10 In 2017, the AORN published a text stating that the "common belief" that it had mandated the use of bouffant hats and urged the elimination of surgeon's skull caps: "The AORN guideline simply recommends *A clean surgical head cover or hood that confines all hair and completely covers the ears, scalp skin, sideburns, and nape of the neck should be worn*. This recommendation is supported by a number of studies showing that hair can be a source of bacterial organisms and potential surgical site infection". "AORN Guideline for Surgical Attire", AORN, accessed September 9, 2021, www.aorn.org/about-aorn/aorn-newsroom/health-policy-news/2017-health-policy-news/aorn-guideline-for-surgical-attire. In 2020, the AORN put an end to the discussion about the type of head covering: "The scalp and hair should be covered when entering the restricted or semi-restricted areas. An interdisciplinary team that includes the surgical team and infection preventionists may determine the type of head covers that will be worn at the healthcare organization. When developing a facility-level policy, keep in mind that AORN's recommendation is to establish and implement a process for managing cloth hats, including the laundering frequency and method. These recommendations are based on recent evidence that does not demonstrate any association between the type of surgical head covering material and the outcome of SSI rates". "2020 Guideline Revisions: 3 Practice Changes to Make Now", AORN, accessed September 9, 2021, https://www.aorn.org/blog/2020-guideline-revisions.

11 Wills et al., "Association", 324.

12 Wills et al., "Association", 324.

13 David N. Naumann, Max E.R. Marsden, Mary L. Brandt and Douglas M. Bowley, "The Bouffant Hat Debate and the Illusion of Quality Improvement", *Annals of Surgery* 271, no. 4 (2020): 635–636; "Statement on Operating Room Attire"; see also David Gorski, "Bouffant Caps Versus Skull Caps in the Operating Room: A No Holds Barred Cage Match", Science-Based Medicine (Blog), accessed August 20, 2018, https://sciencebasedmedicine.org/bouffant-caps-versus-skull-caps-in-the-operating-room-a-no-holds-barred-cage-match/.

14 We have chosen Germany as a comparative case as the German discussion – or the lack thereof – concerning the role of clothing in infection prevention enables us to make some of the peculiarities of the bouffant scandal more explicit. We do not claim that the German discussion is representative of the debate outside the US.

15 Andersen, Anjum and Rocca have referred to these assumptions as *philosophical biases*: "basic implicit assumptions in science about how the world is (ontology), what we can know about it (epistemology), or how science ought to be practiced (norms)". These assumptions influence research practices by, for example, influencing the development of hypotheses and study design. Fredrik Andersen, Rani Lill Anjum and Elena Rocca, "Philosophy of Biology: Philosophical Bias is the One Bias that Science Cannot Avoid", *Elife* 8 (2019): e44929.

16 Bourdon, "RP", 1.

17 William C. Noble, "Dispersal of Skin Microorganisms", *British Journal of Dermatology* 93, no. 4 (1975): 477–485.

18 Bartek, Verdial and Dellinger, "Naked", 1591.

19 Timothy D. Mastro, Thomas A. Farley, John A. Elliott, Richard R. Fack-
    lam, Janet R. Perks, James L. Hadler, Robert C. Good and John S. Spika,
    "An Outbreak of Surgical-Wound Infections Due to Group A Streptococ-
    cus Carried on the Scalp", *New England Journal of Medicine* 323, no. 14
    (1990): 968–972.
20 Bartek, Verdial and Dellinger, "Naked", 1591.
21 Gorski, "Bouffant Caps".
22 Ivy N. Haskins, A.S. Prabhu, D.M. Krpata, A.J. Perez, L. Tastaldi, C.
    Tu, S. Rosenblatt, B.K. Poulose and M.J. Rosen, "Is There an Associa-
    tion Between Surgeon Hat Type and 30-Day Wound Events Following
    Ventral Hernia Repair?", *Hernia* 21, no. 4 (2017): 495–503; Hussain
    Shallwani, Hakeem J. Shakir, Ashley M. Aldridge, Maureen T. Donovan,
    Elad I. Levy and Kevin J. Gibbons, "Mandatory Change from Surgical
    Skull Caps to Bouffant Caps Among Operating Room Personnel Does Not
    Reduce Surgical Site Infections in Class I Surgical Cases: A Single-Center
    Experience With More Than 15 000 Patients", *Neurosurgery* 82, no. 4
    (2018): 548–554; Rios-Diaz et al., "The Art".
23 Troy A. Markel, Thomas Gormley, Damon Greeley, John Ostojic, Angie
    Wise, Jonathan Rajala, Rahul Bharadwaj and Jennifer Wagner, "Hats Off:
    A Study of Different Operating Room Headgear Assessed by Environmen-
    tal Quality Indicators", *Journal of the American College of Surgeons* 225,
    no. 5 (2017): 573–581.
24 Markel et al., "Hats", 573.
25 Another question concerning the external validity of the Markel et al.
    study can be raised. The study took place at two different hospitals and
    Markel et al. used disposable bouffant hats and skull caps available in
    those hospitals. They note that there are different kinds of bouffant hats
    on the market and acknowledge that some of them might have performed
    better than the ones tested. However, they do not give information about
    the quality of the hats in comparison to other similar products on the
    market. It is thus possible that the results of the Markel et al. study are
    not applicable to all bouffant hats. These limitations also raise question
    concerning the study's reproducibility.
26 Naumann et al., "The Bouffant", 635. Italics in original.
27 A similar discussion took place in the context of mask-wearing as a meas-
    ure to prevent COVID-19. Arguing against critics who claimed there was
    lack of evidence for the "real" protective effect of masks, Trisha Green-
    halgh and co-authors published an analysis of the topic in BMJ, in which
    they make a case for taking account of "anecdotal evidence". Since the
    only evidence to date is that SARS COV-2 is transmitted by aerosols, and
    that it is unrealistic to test the effect of each measure separately from the
    others, wearing masks indoors is recommended on the basis of the pre-
    cautionary principle. Trisha Greenhalgh, Manuel B. Schmid, Thomas
    Czypionka, Dirk Bassler and Laurence Gruer, "Face Masks for the Public
    during the Covid-19 Crisis", *BMJ* 369 (2020): 1–4. See also Shuo Feng,
    Chen Shen, Nan Xia, Wei Song, Mengzhen Fan and Benjamin J. Cowling,
    "Rational Use of Face Masks in the COVID-19 Pandemic", *The Lancet
    Respiratory Medicine* 8, no. 5 (2020): 434–436.
28 One could consider surgeons appealing to the EBM ideals to be slightly
    odd. After all, surgeons themselves have been criticized for "lagging
    behind their medical colleagues in embracing evidence based [sic] med-
    icine". Abdul-Wahed Nasir Meshikhes, "Evidence-Based Surgery: The
    Obstacles and Solutions", *International Journal of Surgery* 18 (2015):
    159–162, here 159. It is indeed the case that in medical literature, it is

common to bemoan the quality of surgical research: Few surgical interventions have been tested in RCTs, and many existing surgical trials have methodological problems. This is mainly because the human factor and unavoidable variables play a major role in surgery. See Laura E. Bothwell and David S. Jones, "Innovation and Tribulation in the History of Randomized Controlled Trials in Surgery", *Annals of Surgery* 274, no. 6 (2021): e616–e624.

29　Naumann et al., "The Bouffant", 635–636.

30　For arguments against the use of evidence hierarchies, see Jacob Stegenga, "Down with the Hierarchies", *Topoi* 33, no. 2 (2014): 313–322.

31　Saana Jukola, "On the Evidentiary Standards for Nutrition Advice", *Studies in History and Philosophy of Science, Part C: Studies in History and Philosophy of Biological and Biomedical Sciences* 73 (2019): 1–9. See also the contribution by Edoardo Maria Pelli and Jutta Roosen in this volume.

32　Holger Strassheim and Pekka Kettunen, "When Does Evidence-Based Policy Turn Into Policy-Based Evidence? Configurations, Contexts and Mechanisms", *Evidence & Policy: A Journal of Research, Debate and Practice* 10, no. 2 (2014): 259–277.

33　Adam La Caze, "Evidence-Based Medicine Must Be...", *Journal of Medicine and Philosophy* 34, no. 5 (2009): 509–527.

34　La Caze, "Evidence-Based", 510. Philosophers of science have previously pointed out that the priority of internal validity, and relatedly that of RCTs, can be questioned. See La Caze, "Evidence-Based", 525; Rani Lill Anjum, Samantha Copeland, and Elena Rocca, "Medical Scientists and Philosophers Worldwide Appeal to EBM to Expand the Notion of 'Evidence'", *BMJ Evidence-Based Medicine* 25, no. 1 (2020): 6–8.

35　The claim that confounding could be removed by randomization has been criticized. See John Worrall, "What Evidence in Evidence-Based Medicine?", *Philosophy of Science* 69, no. S3 (2002): 316–330.

36　Federica Russo and Jon Williamson, "Interpreting Causality in the Health Sciences", *International Studies in the Philosophy of Science* 21, no. 2 (2007): 157–170; Mark R. Tonelli and Robyn Bluhm, "Teaching Medical Epistemology Within an Evidence-Based Medicine Curriculum", *Teaching and Learning in Medicine* 33, no. 1 (2020): 98–105.

37　Phyllis McKay Illari and Jon Williamson, "What Is a Mechanism? Thinking about Mechanisms across the Sciences", *European Journal for Philosophy of Science* 2, no. 1 (2012): 119.

38　Russo and Williamson, "Interpreting", 159. It has been pointed out that in EBM, much of the discussion has focused on how to gather evidence on (pharmacological) treatment efficacy. Even if it was the case that RCTs were the best method in this context, their Gold Standard status and the redundancy of mechanistic evidence can be more questionable when assessing the efficacy of more complex interventions. See Barbara Osimani, "Until RCT Proven? On the Asymmetry of Evidence Requirements for Risk Assessment", *Journal of Evaluation in Clinical Practice* 19, no. 3 (2013): 454–462; Ambika Satija, Edward Yu, Walter C. Willett and Frank B. Hu, "Understanding Nutritional Epidemiology and Its Role in Policy", *Advances in Nutrition* 6, no. 1 (2015): 5–18.

39　Bartek, Verdial and Dellinger, "Naked", 591; La Caze, "Evidence-Based", 520; Andersen et al., "Philosophy".

40　On the use of case studies in establishing causation, see Rachel A. Ankeny, "The Overlooked Role of Cases in Causal Attribution in Medicine", *Philosophy of Science* 81, no. 5 (2014): 999–1011.

41　Bourdon, "RT", 1.

42 Michael Jefford, Martin R. Stockler and Martin H.N. Tattersall, "Outcomes Research: What is It and Why Does It Matter?", *Internal Medicine Journal* 33, no. 3 (2003): 110–118; Jeremy H. Howick, *The Philosophy of Evidence-Based Medicine* (Chichester: John Wiley & Sons, 2011).
43 Shallwani et al., "Mandatory", 548.
44 Haskins et al., "Surgeon Hat Type", 495; Wills et al., "Association", 323; Rios-Diaz et al., "The Art", 921.
45 Markel et al., "Hats", 573.
46 Helen E. Longino, *Science as Social Knowledge* (Princeton: Princeton Univ. Press, 1990).
47 Longino, "Science", 42–43.
48 Allegranzi et al., "New", e276.
49 Alex Broadbent, *Philosophy of Epidemiology* (London: Palgrave MacMillan, 2013).
50 Osimani, "Until", 461–462; Anjum et al., "Medical", 6; Veli-Pekka Parkkinen, Christian Wallmann, Michael Wilde, Brendan Clarke, Phyllis Illari, Michael P. Kelly, Charles Norell, Federica Russo, Beth Shaw, and Jon Williamson, *Evaluating Evidence of Mechanisms in Medicine: Principles and Procedures* (Cham: Springer Nature, 2018).
51 Russo and Williamson, "Interpreting", 159; Anjum et al., "Medical", 6.
52 Jürgen Landes, Barbara Osimani, and Roland Poellinger, "Epistemology of Causal Inference in Pharmacology", *European Journal for Philosophy of Science* 8, no. 1 (2018): 3–49.
53 Austin Bradford Hill, "The Environment and Disease: Association or Causation?", *Proceedings of the Royal Society of Medicine* 58 (1965): 295–300.
54 For technical details, see Landes et al., "Epistemology".
55 Several scholars have pointed out that different non-epistemic factors can influence the production of evidence and that evidence can thus be value-laden. See for example, Longino, "Science", 38–61. For an introduction to the debate about the role of non-epistemic values in research, see Kevin Christopher Elliot. *A Tapestry of Values: An Introduction to Values in Science* (Oxford: Oxford Univ. Press, 2017).
56 Samuli I. Saarni and Heta A. Gylling, "Evidence Based Medicine Guidelines: A Solution to Rationing or Politics Disguised as Science?", *Journal of Medical Ethics* 30, no. 2 (2004): 171–175; Fredrik Andersen and Elena Rocca, "Underdetermination and Evidence-Based Policy", *Studies in History and Philosophy of Science Part C: Studies in History and Philosophy of Biological and Biomedical Sciences* 84 (2020): 101335.
57 Pekka Louhiala, "But Who Can Say What's Right and Wrong? Medicine as a Moral Enterprise", in *Philosophy for Medicine: Applications in a Clinical Context*, eds. Martyn Evans, Pekka Louhiala and Raimo Puustinen (Oxford: Radcliffe Medical Press, 2004), 135–142.
58 Saarni and Gylling, "Evidence-Based", 173.
59 Justin O. Parkhurst and Sudeepa Abeysinghe, "What Constitutes 'Good' Evidence for Public Health and Social Policy-Making? From Hierarchies to Appropriateness", *Social Epistemology* 30, no. 5–6 (2016): 665–679.
60 Bartek, Verdial and Dellinger, "Naked", 1591.
61 Saarni and Gylling, "Evidence-Based", 171.
62 Radwan Dipp Ramos and Kamal M.F. Itani, "Emotions, Common Sense, and Evidence in Operating Room Attire", *JAMA Surgery* 155, no. 4 (2020): 329.
63 Naumann et al., "The Bouffant", 635.
64 Wills et al., "Association", 325.
65 Naumann et al., "The Bouffant", 635.

66 Naumann et al., "The Bouffant", 635.
67 "Statement on operating room attire".
68 Naumann et al., "The Bouffant", 635–636.
69 Zhicheng Wang, Lisa Bero and Quinn Grundy, "Understanding Professional Stakeholders' Active Resistance to Guideline Implementation: The Case of Canadian Breast Screening Guidelines", *Social Science & Medicine* 269 (2021): 113586, 1–10.
70 Wang, Bero and Grundy, "Understanding", 8.
71 A commentator in the debate, surgeon and blogger David Gorski, has rightly pointed out that the position of the skull cap as the symbol of the professional relies on stereotypical assumptions about the gender – or at least the hairstyle – of surgeons. The hat is mainly worn by short-haired surgeons. Long-haired surgeons have already previously worn different headwear. Unfortunately, the gendered aspects of the bouffant debate cannot be discussed in this chapter due to space limitations. Gorski, "Bouffant Caps". For experiences of female surgeons, see Joan Cassell, "Doing Gender, Doing Surgery: Women Surgeons in a Man's Profession", *Human Organization* 56, no. 1 (1997): 47–52.
72 Louhiala, "But Who", 135; Strassheim and Kettunen, "Evidence-Based Policy", 259; Parkhurst and Abeyshinghe, "'Good' Evidence", 665; Jukola, "On the Evidentiary", 1.
73 Elaine He and Lionel Laurent, "The World is Masking Up. Some Are Opting Out", Bloomberg.com, July 17, 2020, https://www.bloomberg.com/graphics/2020-opinion-coronavirus-global-face-mask-adoption.
74 Jackson G. Lu, Peter Jin and Alexander S. English, "Collectivism Predicts Mask Use during COVID-19", *Proceedings of the National Academy of Sciences* 118, no. 23 (2021): 1–8.
75 Susanne Freidberg, "Wicked Nutrition: The Controversial Greening of Official Dietary Guidance", *Gastronomica* 16, no. 2 (2016): 69–80; Jukola, "On the Evidentiary", 1.
76 Martina Lenzen-Schulte, "Hygienemanagement: Gegen routinierte Nachlässigkeit [Hygiene Management: Against Routine Negligence]", *Deutsches Ärzteblatt* 116 (2019): A1026. All translations by the authors.
77 Axel Kramer, Ojan Assadian, Martin Exner, Nils-Olaf Hübner and Arne Simon, eds., *Krankenhaus- und Praxishygiene. Hygienemanagement und Infektionsprävention in medizinischen und sozialen Einrichtungen [Hospital and Doctors' Office Hygiene. Hygiene Management and Infection Prevention in Medical and Social Institutions]*, 3rd edition (München: Urban & Fischer, 2016); Arbeitsgemeinschaft der Wissenschaftlichen Medizinischen Fachgesellschaften e.V. 2010. *Leitlinien zur Hygiene in Klinik und Praxis [Guidelines for Hygiene at the Clinic and in Doctors' Offices]*, accessed September 21, 2021, https://www.awmf.org/uploads/tx_szleitlinien/029-012l_S1_OP-Kleidung_und_Patienten abdeckung_2019-07.pdf.
78 Axel Kramer, Ojan Assadian, Martin Exner, Nils-Olaf Hübner, and Arne Simon, eds., *Krankenhaus- und Praxishygiene. Hygienemanagement und Infektionsprävention in medizinischen und sozialen Einrichtungen [Hospital and Doctors' Office Hygiene. Hygiene Management and Infection Prevention in Medical and Social Institutions]* (München: Urban & Fischer, 2022). We would like to take this opportunity to thank Axel Kramer for allowing us to use and quote the current version of the standard work before its publication.
79 Kramer et al., "Krankenhaus- und Praxishygiene", 311–314.
80 Kramer et al., "Krankenhaus- und Praxishygiene", 314.

81 Arbeitsgemeinschaft der Wissenschaftlichen Medizinischen Fachge-sellschaften e.V., "Leitlinien", 1–2.
82 Trisha Greenhalgh and Jill Russell, "Reframing Evidence Synthesis as Rhetorical Action in the Policy Making Drama", *Healthcare Policy* 1, no. 2 (2006): 34.
83 In our material, we could not find any reflection on the fact that when the German experts relied on the AORN guidelines, they referenced guide-lines issued by a professional association of nurses.
84 Silvana R. Tridico, Dáithí C. Murray, Jayne Addison, Kenneth P. Kirk-bride and Michael Bunce, "Metagenomic Analyses of Bacteria on Human Hairs: A Qualitative Assessment for Applications in Forensic Science", *Investigative Genetics* 5, no. 1 (2014): 1–13; Kate L. Owers, E. James and G.C. Bannister, "Source of Bacterial Shedding in Laminar Flow Thea-tres", *Journal of Hospital Infection* 58, no. 3 (2004): 230–232.
85 Anjum et al., "Medical", 6–7; Landes et al., "Epistemology", 3.
86 Trisha Greenhalgh et al, "Face Masks", 3–4.
87 Heather Douglas, "Inductive Risk and Values in Science", *Philosophy of Science* 67, no. 4 (2000): 559–579; Matthew Bennett, "Should I Do as I'm Told? Trust, Experts, and COVID-19", *Kennedy Institute of Ethics Journal* 30, no. 3 (2020): 243–263.

References

Allegranzi, Benedetta, Peter Bischoff, Stijn de Jonge, N. Zeynep Kubilay, Bassim Zayed, Stacey M. Gomes, and Mohamed Abbas et al. "New WHO Recommendations on Preoperative Measures for Surgical Site Infection Prevention: An Evidence-Based Global Perspective". *The Lancet Infectious Diseases* 16, no. 12 (2016): e276–e287.
American College of Surgeons. "Statement on Operating Room Attire", Accessed September 21, 2021. https://bulletin.facs.org/2016/10/statement-on-operating-room-attire/.
Andersen, Fredrik, Rani Lill Anjum, and Elena Rocca. "Philosophy of Biology: Philosophical Bias Is the One Bias that Science Cannot Avoid". *Elife* 8 (2019): e44929.
Andersen, Fredrik, and Elena Rocca. "Underdetermination and Evidence-Based Policy". *Studies in History and Philosophy of Science Part C: Studies in History and Philosophy of Biological and Biomedical Sciences* 84 (2020): 101335.
Anjum, Rani Lill, Samantha Copeland, and Elena Rocca. "Medical Scientists and Philosophers Worldwide Appeal to EBM to Expand the Notion of 'Evidence'". *BMJ Evidence-Based Medicine* 25, no. 1 (2020): 6–8.
Ankeny, Rachel A. "The Overlooked Role of Cases in Causal Attribution in Medicine". *Philosophy of Science* 81, no. 5 (2014): 999–1011.
Arbeitsgemeinschaft der Wissenschaftlichen Medizinischen Fachgesellschaften e.V. 2010. "Leitlinien zur Hygiene in Klinik und Praxis [Guidelines for Hygiene at the Clinic and in Doctors' Offices]". Accessed September 21, 2021. https://www.awmf.org/uploads/tx_szleitlinien/029-012l_S1_OP-Kleidung_und_Patienenabdeckung_2019-07.pdf.
Bartek, Matthew, Francys Verdial, and E. Patchen Dellinger. "Naked Surgeons? The Debate About What to Wear in the Operating Room". *Clinical Infectious Diseases* 65, no. 9 (2017): 1589–1592.

Bennett, Matthew. "Should I Do as I'm Told? Trust, Experts, and COVID-19". *Kennedy Institute of Ethics Journal* 30, no. 3 (2020): 243–263.

Bothwell, Laura E., and David S. Jones. "Innovation and Tribulation in the History of Randomized Controlled Trials in Surgery". *Annals of Surgery* 274, no. 6 (2021): e616–e624.

Bourdon, Leslie. "RP First Look: New Recommended Practices for Preoperative Patient Skin Antisepsis". *AORN Connections* 100, no. 4 (2014): 1–3.

Broadbent, Alex. *Philosophy of Epidemiology*. London: Palgrave MacMillan, 2013.

Cassell, Joan. "Doing Gender, Doing Surgery: Women Surgeons in a Man's Profession". *Human Organization* 56, no. 1 (1997): 47–52.

Douglas, Heather. "Inductive Risk and Values in Science". *Philosophy of Science* 67, no. 4 (2000): 559–579.

Elliott, Kevin Christopher. *A Tapestry of Values: An Introduction to Values in Science*. Oxford: Oxford Univ. Press, 2017.

Feng, Shuo, Chen Shen, Nan Xia, Wei Song, Mengzhen Fan, and Benjamin J. Cowling. "Rational Use of Face Masks in the COVID-19 Pandemic". *The Lancet Respiratory Medicine* 8, no. 5 (2020): 434–436.

Freidberg, Susanne. "Wicked Nutrition: the Controversial Greening of Official Dietary Guidance". *Gastronomica* 16, no. 2 (2016): 69–80.

Gorski, David. "Bouffant Caps Versus Skull Caps in the Operating Room: A No Holds Barred Cage Match", *Science-Based Medicine* (Blog), August 20, 2018. https://sciencebasedmedicine.org/bouffant-caps-versus-skull-caps-in-the-operating-room-a-no-holds-barred-cage-match/.

Greenhalgh, Trisha and Jill Russell. "Reframing Evidence Synthesis as Rhetorical Action in the Policy Making Drama". *Healthcare Policy* 1, no. 2 (2006): 34–42.

Greenhalgh, Trisha, Manuel B. Schmid, Thomas Czypionka, Dirk Bassler, and Laurence Gruer. "Face Masks for the Public During the Covid-19 Crisis". *BMJ* 369 (2020): 1–4.

Haskins, I.N., A.S. Prabhu, D.M. Krpata, A.J. Perez, L. Tastaldi, C. Tu, S. Rosenblatt, B.K. Poulose, and M.J. Rosen, "Is There an Association Between Surgeon Hat Type and 30-Day Wound Events Following Ventral Hernia Repair?. *Hernia* 21, no. 4 (2017): 495–503.

He, Elaine and Lionel Laurent. "The World is Masking Up. Some Are Opting Out", Bloomberg.com, July 17, 2020. https://www.bloomberg.com/graphics/2020-opinion-coronavirus-global-face-mask-adoption.

Hicks, D.J. "Epistemological Depth in a GM Crops Controversy". *Studies in History and Philosophy of Science Part C: Studies in History and Philosophy of Biological and Biomedical Sciences* 50 (2015): 1–12.

Hill, Austin Bradford. "The Environment and Disease: Association or Causation?". *Proceedings of the Royal Society of Medicine* 58 (1965): 295–300.

Howick, Jeremy H. *The Philosophy of Evidence-Based Medicine*. Chichester: John Wiley & Sons, 2011.

Illari, Phyllis McKay, and Jon Williamson. "What Is a Mechanism? Thinking About Mechanisms Across the Sciences". *European Journal for Philosophy of Science* 2, no. 1 (2012): 119–135.

Jefford, Michael, Martin R. Stockler, and Martin H.N. Tattersall. "Outcomes Research: What Is It and Why Does It Matter? *Internal Medicine Journal* 33, no. 3 (2003): 110–118.

Jukola, Saana. "On the Evidentiary Standards for Nutrition Advice". *Studies in History and Philosophy of Science Part C: Studies in History and Philosophy of Biological and Biomedical Sciences* 73 (2019): 1–9.

Kramer, Axel, Ojan Assadian, Martin Exner, Nils-Olaf Hübner, and Arne Simon, eds. *Krankenhaus- und Praxishygiene. Hygienemanagement und Infektionsprävention in medizinischen und sozialen Einrichtungen [Hospital and Doctors' Office Hygiene. Hygiene Management and Infection Prevention in Medical and Social Institutions]*. München: Urban & Fischer, 2016.

Kramer, Axel, Ojan Assadian, Martin Exner, Nils-Olaf Hübner, and Arne Simon, eds. *Krankenhaus- und Praxishygiene. Hygienemanagement und Infektionsprävention in medizinischen und sozialen Einrichtungen [Hospital and Doctors' Office Hygiene. Hygiene Management and Infection Prevention in Medical and Social Institutions]*. München: Urban & Fischer, 2022.

La Caze, Adam, "Evidence-Based Medicine Must Be…". *Journal of Medicine and Philosophy* 34, no. 5 (2009): 509–527.

Landes, Jürgen, Barbara Osimani, and Roland Poellinger. "Epistemology of Causal Inference in Pharmacology". *European Journal for Philosophy of Science* 8, no. 1 (2018): 3–49.

Lenzen-Schulte, Martina. "Hygienemanagement: Gegen Routinierte Nachlässigkeit [Hygiene Management: Against Routine Negligence]". *Deutsches Ärzteblatt* 116 (2019): A1026.

Longino, Helen E. *Science as Social Knowledge*. Princeton: Princeton Univ. Press, 2020.

Longino, Helen E. "Cognitive and Non-Cognitive Values in Science: Rethinking the Dichotomy". In *Feminism, Science, and the Philosophy of Science*, edited by Lynn Hankinson Nelson, and Jack Nelson, 39–58. Dordrecht: Springer, 1996.

Louhiala, Pekka. "But Who Can Say What's Right and Wrong? Medicine as a Moral Enterprise". In *Philosophy for Medicine: Applications in a Clinical Context*, edited by Martyn Evans, Pekka Louhiala, and Raimo Puustinen, 135–142. Oxford: Radcliffe Medical Press, 2004.

Lu, Jackson G., Peter Jin, and Alexander S. English. "Collectivism Predicts Mask Use During COVID-19". *Proceedings of the National Academy of Sciences* 118, no. 23 (2021): 1–8.

Markel, Troy A., Thomas Gormley, Damon Greeley, John Ostojic, Angie Wise, Jonathan Rajala, Rahul Bharadwaj, and Jennifer Wagner. "Hats Off: A Study of Different Operating Room Headgear Assessed by Environmental Quality Indicators". *Journal of the American College of Surgeons* 225, no. 5 (2017): 573–581.

Mastro, Timothy D., Thomas A. Farley, John A. Elliott, Richard R. Facklam, Janet R. Perks, James L. Hadler, Robert C. Good, and John S. Spika. "An Outbreak of Surgical-Wound Infections Due to Group A Streptococcus Carried on the Scalp". *New England Journal of Medicine* 323, no. 14 (1990): 968–972.

Meshikhes, Abdul-Wahed Nasir. "Evidence-Based Surgery: The Obstacles and Solutions". *International Journal of Surgery* 18 (2015): 159–162.

Naumann, David N., Max E.R. Marsden, Mary L. Brandt, and Douglas M. Bowley. "The Bouffant Hat Debate and the Illusion of Quality Improvement". *Annals of Surgery* 271, no. 4 (2020): 635–636.

Noble, William C. "Dispersal of Skin Microorganisms". *British Journal of Dermatology* 93, no. 4 (1975): 477–485.

Osimani, Barbara. "Until RCT Proven? On the Asymmetry of Evidence Requirements for Risk Assessment". *Journal of Evaluation in Clinical Practice* 19, no. 3 (2013): 454–462.

Owers, Kate L., E. James, and G.C. Bannister. "Source of Bacterial Shedding in Laminar Flow Theatres". *Journal of Hospital Infection* 58, no. 3 (2004): 230–232.

Parkhurst, Justin O., and Sudeepa Abeysinghe. "What Constitutes 'Good' Evidence for Public Health and Social Policy-Making? From Hierarchies to Appropriateness". *Social Epistemology* 30, no. 5–6 (2016): 665–679.

Parkkinen, Veli-Pekka, Christian Wallmann, Michael Wilde, Brendan Clarke, Phyllis Illari, Michael P. Kelly, Charles Norell, Federica Russo, Beth Shaw, and Jon Williamson. *Evaluating Evidence of Mechanisms in Medicine: Principles and Procedures*. Cham: Springer Nature, 2018.

Petro, Clayton C., and Michael J. Rosen. "What Surgeons Need to Know About the Bouffant Scandal". *JAMA Surgery* 154, no. 11 (2019): 989–990.

Ramos, Radwan Dipp, and Kamal M.F. Itani. "Emotions, Common Sense, and Evidence in Operating Room Attire". *JAMA Surgery* 155, no. 4 (2020): 329–329.

Rios-Diaz, Arturo J., Guillaume Chevrollier, Hunter Witmer, Christine Schleider, Scott Cowan, Michael J. Pucci, and Francesco Palazzo. "The Art and Science of Surgery: Do the Data Support the Banning of Surgical Skull Caps?. *Surgery* 164, no. 5 (2018): 921–925.

Rooney, Phyllis. "On Values in Science: Is the Epistemic/Non-Epistemic Distinction Useful?, *PSA: Proceedings of the Biennial Meeting of the Philosophy of Science Association* 1992, no. 1 (1992): 13–22.

Russo, Federica, and Jon Williamson. "Interpreting Causality in the Health Sciences", *International Studies in the Philosophy of Science* 21, no. 2 (2007): 157–170.

Saarni, Samuli I., and Heta A. Gylling. "Evidence Based Medicine Guidelines: A Solution to Rationing or Politics Disguised as Science?". *Journal of Medical Ethics* 30, no. 2 (2004): 171–175.

Satija, Ambika, Edward Yu, Walter C. Willett, and Frank B. Hu. "Understanding Nutritional Epidemiology and Its Role in Policy". *Advances in Nutrition* 6, no. 1 (2015): 5–18.

Shallwani, Hussain, Hakeem J. Shakir, Ashley M. Aldridge, Maureen T. Donovan, Elad I. Levy, and Kevin J. Gibbons. "Mandatory Change from Surgical Skull Caps to Bouffant Caps Among Operating Room Personnel Does Not Reduce Surgical Site Infections in Class I Surgical Cases: A Single-Center Experience With More Than 15 000 Patients". *Neurosurgery* 82, no. 4 (2018): 548–554.

Stegenga, Jacob. "Down With the Hierarchies". *Topoi* 33, no. 2 (2014): 313–322.

Strassheim, Holger, and Pekka Kettunen. "When Does Evidence-Based Policy Turn Into Policy-Based Evidence? Configurations, Contexts and Mechanisms". *Evidence & Policy: A Journal of Research, Debate and Practice* 10, no. 2 (2014): 259–277.

Tonelli, Mark R., and Robyn Bluhm. "Teaching Medical Epistemology Within An Evidence-Based Medicine Curriculum". *Teaching and Learning in Medicine* 33, no. 1 (2020): 98–105.

Tridico, Silvana R., Dáithí C. Murray, Jayne Addison, Kenneth P. Kirkbride, and Michael Bunce. "Metagenomic Analyses of Bacteria on Human Hairs: A Qualitative Assessment for Applications in Forensic Science". *Investigative Genetics* 5, no. 1 (2014): 1–13.

Wang, Zhicheng, Lisa Bero, and Quinn Grundy. "Understanding Professional Stakeholders' Active Resistance to Guideline Implementation: The Case of Canadian Breast Screening Guidelines". *Social Science & Medicine* 269 (2021): 113586.

Wills, Bradley W., Walter R. Smith, Alexandra M. Arguello, Gerald McGwin, Elie S. Ghanem, and Brent A. Ponce. "Association of Surgical Jacket and Bouffant Use With Surgical Site Infection Risk". *JAMA Surgery* 155, no. 4 (2020): 323–328.

Worrall, John. "What Evidence in Evidence-Based Medicine?". *Philosophy of Science* 69, no. S3 (2002): 316–330.

6 Negotiating Consensus for Diverging Evidence

An Application of the Explanatory Approach to Guidelines for the Treatment of Substance Use Disorders and Addiction

Fredrik Andersen, * *Yngve Herikstad,* *
Stefan Sütterlin, and Anders Dechsling

Clinical health practice is increasingly framed within evidence-based guidelines, where these guidelines are intended to ensure quality treatment based on the best current knowledge of the relevant field. Guidelines are intended to ensure that patients receive care reliant on a community of specialists rather than decisions made by individual clinicians. As such, guidelines are meant to harmonize and increase the quality of healthcare. Clinical practice guidelines should be regarded as "statements that include recommendations intended to optimize patient care that are informed by a systematic review of evidence and an assessment of the benefits and harms of alternative care options".[1] Guidelines are typically formulated by expert groups who evaluate relevant existing evidence and, on the basis of this evidence, suggest treatments. Expert group members are selected in a multitude of ways and there is considerable divergence concerning who is selected. For instance, some expert groups consist mainly of methodological specialists who evaluate the strengths of conflicting evidence, other expert groups consist mainly of clinicians, while others still consist mainly of researchers in relevant fields. In effect, differences in ideology, disciplinary background and experience are to be expected both within and between groups.

Expert groups face numerous challenges and it is unclear how one should approach guideline formulation.[2] Evidence relating to the effect of specific treatments also often diverges, and thus it is hard to reach consensus. Furthermore, as different types of treatment are tested using different kinds of methods, the value of evidence is often contested on methodological terms. Expert groups, who evaluate already existing evidence generated by others using multiple methods, must weigh evidence and types of evidence against each other, which can be done in multiple

* Equal contribution.

DOI: 10.4324/9781003273509-10

ways. All strategies for evidence evaluation can be, and often are, contested. Ultimately, decisions made by expert groups concerning these issues can shape treatment policies for decades and thus radically affect the lives of numerous people. It is therefore paramount that guidelines are formulated by expert groups in the best available manner.

We identify four main categories of approaches to weighing evidence: Qualitative rule-based techniques, quantitative algorithmic techniques, social process techniques and the explanatory approach.[3] All approaches aim to remove threats to the quality of the final decision: Qualitative rule-based approaches apply methodological hierarchies in order to avoid decisions made from weak evidence. Quantitative algorithmic techniques apply algorithms in order to avoid expert biases. Social process techniques collect expert and stakeholder opinions in order to maintain contact with clinical reality and avoid exclusion of key evidence. The explanatory approach focuses on how experts explain apparent counter-evidence and evaluates explanations in order to exclude ad-hocness.

The aim of this chapter is to illustrate the benefits of the explanatory approach as compared with social process techniques. As a case in point, we will discuss the Norwegian *National Guideline for the Treatment of Substance Use Disorders and Addiction*, currently written within a social process, and argue that this guideline would benefit from switching to the explanatory approach. We have chosen the guideline for the treatment of substance use disorders and addiction, as this illustrates a wide array of challenges to guideline formulation. The field is diverse both in terms of clinician background and research paradigms. There is wide-spread political debate concerning legality, along with scientific debates concerning how various substances impact health and society. In addition, as there is a wide array of treatment schemes, it is hard to negotiate consensus concerning which treatments should be recommended. In combination, the variety of choices presented to expert groups constructing guidelines for the treatment of substance use disorder and addiction illustrates both the complexity and depth required for good guidelines. It is thus an ideal arena for testing various approaches.

In this chapter, we first describe the most common challenges to guideline formulation, both epistemic and non-epistemic. After presenting evidence evaluation criteria that are suggested in the literature, we give an overview of four different approaches to guideline formulation, with an emphasis on the explanatory approach. We then introduce the topic of substance use disorders and addiction along with the Norwegian *National Guideline for the Treatment of Substance Use Disorders and Addiction* as a case example. Following this, we show how the standard challenges are relevant in the specific case, and how the Norwegian *National Guideline for the Treatment of Substance Use Disorders and Addiction* could be improved through an application of the explanatory approach. Finally, we argue that the explanatory approach is preferable as a general approach to guideline formulation within the health sciences.

Challenges to Guideline Formulation

Guideline formulation faces multiple challenges and demands. For instance, there are epistemic challenges relating to how one should weigh different evidence. The purpose of guidelines is to provide the best treatment available, and in order to decide what the best treatment is, one must look at existing evidence. However, different evidence often points in divergent directions and expert groups must find ways to single out the stronger and more relevant evidence.

There is an asymmetry between the necessary generalization of scientific evidence describing groups of individuals and the individual peculiarities of a single patient. While patients typically have comorbidities and individual manifestations of their underlying problem, the general evidence provided by science cannot address every possible combination of comorbidities and contexts. Since scientific knowledge is general rather than universal, clear-cut deductions for specific cases or between groups do not exist.

The scientific focus on the general poses relevance-related questions for guidelines. Does the success of a treatment on group A imply the future success of that same treatment on group B? Given that the treatment is effective at group level, how can clinicians be confident that their patient is among those who will benefit? These types of translation issues are common and lack clear and univocal answers. However, if we expect guidelines to be implemented in clinical practice, translation issues must somehow be dealt with.

There is a further challenge in relation to healthcare and general scientific evidence. While research aims to influence practice, it has been argued that there is a lack of effective feedback from practice to research, resulting in an inefficient/asymmetric coupling between science and practice.[4] Expert groups should consider how to incorporate data from naturalistic settings within the included evidence.

Expert groups formulating guidelines must also decide how to relate to these issues. When facing contradictory evidence, should we rank methods, and if so, how should methods be ranked? Should we emphasize local evidence, and if so, what constitutes "local" (do we mean gender, age, country, culture etc.)? Should we prefer scientific evidence where clinician generated data are included? There are no obvious answers to these questions and they are therefore settled case by case.

There are also non-epistemic challenges relating to guidelines. By non-epistemic challenges, we include political, ideological, ethical, metaphysical and ontological challenges. For instance, guidelines must consider not only the potential outcomes of a treatment, but also how that treatment is experienced by the patient, whether the patient must have full understanding of the details of the treatment, whether the treatment should be mandatory and how much the treatment will cost. In other words, guidelines must balance societal needs, ethical issues and political issues along with analyses of expected effectiveness. These judgments can be subject to the so-called philosophical BIAS affecting research and policy writing.[5] A philosophical BIAS in this sense is a Basic, Implicit, Assumption in Science. These are assumptions relating, for instance, not only to ethics and politics but also to metaphysical and ontological questions such as what we mean by causality, and whether the "entities" under scrutiny are best understood as processes, assemblies, capacities, powers or substances. The "philosophical BIAS" concept argues that these types of assumptions are necessary and irremovable aspects of any scientific procedure.[6] In relation to guidelines, we should expect a variety in basic assumptions both within research and within the different members of the expert groups. It has been argued, however, that basic assumptions of this kind can affect evaluations of policy and guidelines as well as basic research.[7]

As various authors have emphasized, the distinction between epistemic and non-epistemic values is fuzzy at best.[8] Epistemic and non-epistemic values interact and we should not expect expert groups to construct guidelines on a purely evidential basis. Rather, expert group decisions will be shaped by experience, ideology and expertise, along with positions within ethics, politics and more basic views of reality. A first step in the process of domesticating such issues is to make assumptions explicit and thus increase transparency.

So far, we have outlined the common challenges to guideline formulation. In the next section, we discuss the available tools to overcome such challenges, i.e. the available criteria for evaluating evidence and approaches to guideline formulation.

Approaches to Guideline Formulation

In an attempt to approach challenges systematically, six criteria for evaluating approaches to weighing evidence are presented here.[9] The stated purpose of these criteria is as desiderata against which we can evaluate approaches to evidence weighing, and therefore guideline formulation. They build on Thomas S. Kuhn's criteria for a good scientific theory and are as follows: Completeness, rigor, scope, transparency, communicability and practicality. It is to be emphasized that the totality of these criteria must consider both democratic and epistemic issues.[10]

The criteria for evaluating processes that lead to guidelines frame some of the controversies surrounding guideline formulation for clinical practice. For instance, we see that criticism of the evidence-based medicine (EBM) approach can be formulated as a criticism related to completeness. EBM operates with a clear hierarchy of evidence with systematic reviews and randomized controlled trials (RCTs) rated as the highest quality evidence. If included, expert opinion is typically placed at the bottom level of the hierarchy. Case studies are in between. If an expert group has systematic reviews and RCTs available that show no evidence for the effectiveness of a particular treatment, the expert group is likely to ignore case studies that show effectiveness since case studies are of lower epistemic value than systematic reviews and RCTs within the EBM approach. In effect, relevant data might be excluded, which is in conflict with the completeness criterion. Analyzing approaches to guidelines through the criteria presented in the explanatory approach provides us with a language and an argumentative base from where we can evaluate different strategies.[11]

So how are these criteria met within the most common approaches to guideline formulation? We can distinguish four general approaches to guideline and policy formulation: Qualitative rule-based approaches, quantitative algorithmic techniques, social process approaches and the explanatory approach. Each approach is an attempt at balancing threats connected to completeness, rigor, scope, transparency, communicability and practicality.

Qualitative rule-based approaches focus on removing expert biases while increasing rigor by introducing rules for decision procedures. Typically, rules relate directly to evidence weighing. Phenomena for which there is contradictory evidence are handled by elevating evidence achieved through methods from the top of the hierarchy. Rule-based approaches like this are well adapted to limiting variability within a field, but face some standard challenges. Contestations of qualitative rule-based approaches are most prominently present in the broad criticisms of EBM. For instance, it is argued that EBM operates with a narrow notion of evidence and that it provides no understanding of causal mechanisms.[12] Attempts at reformulations have been made within EBM, such as replacing rule following with expert judgment.[13] On a more foundational issue, approaches like EBM arguably apply the types of judgments they were intended to avoid and thus uncritically adopt judgments that should be under public scrutiny.[14] Here, we wish to emphasize the lack of completeness, and the relationship between hypotheses and methods. It is a well-established norm within science that methods ought to depend on which kind of research question one is asking. However, if we accept a strict hierarchy of methods, we end up incentivizing research related to certain types of hypotheses.

If we are interested in helping people with substance use disorders and we want to investigate why certain people are susceptible to substance use disorders and others less so, we could pose biological, social or political hypotheses. These hypotheses might best be investigated through different methods. For instance, a biological hypothesis, such as the prominent current notion that substance use disorders are chronic brain diseases might best be investigated through mechanistic studies on animal models. Mechanisms established in animal models are then generalized to humans through functional neuroimaging studies.[15] An opposing view, where addiction is viewed as voluntary choice, rather than chronic brain disease, is presented in Gene M. Heyman.[16] Contrary to the mechanistic evidence for addiction as brain disease, Heyman presents evidence from addicts' narratives, psychiatric epidemiology, treatment studies and insights from behavioral economics.[17] If there is empirical support for both hypotheses, and we know the evidential hierarchy, we can already predict the outcome of any competition between proponents of these two hypotheses. The hypothesis connected to the higher ranked methodology wins, and, in effect, we risk "settling" deep questions by definition, rather than by increased understanding of the phenomena. In other words, hypotheses dictate research methods, and if we provide a strict hierarchy of methods, we are already deciding which types of hypotheses we are willing to adopt. All in all, qualitative rule-based approaches emphasize rigor and methodological transparency at the cost of completeness.

A second approach to guideline formulation is to apply quantitative algorithmic techniques, in which the goal is to establish a clear relationship between hypotheses and evidence without any interference from individual experts. Such techniques are intended to avoid subjectivity and biases introduced by individual experts.[18] We know, however, that algorithms also come with biases introduced by whoever is writing them.[19] These biases will remain invisible to the reader, who is typically a non-expert in the quantitative algorithmic techniques involved. In effect, quantitative algorithmic approaches emphasize rigor and completeness, but tend to lack transparency and communicability.

Social process approaches "attempt to avoid the problems related to the subjectivity of 'best professional judgement' of a single expert by combining a larger set of expert judgements".[20] The idea behind a social process approach is to gather a variety of expert judgments and arrive at a preferable position. There are multiple ways to accomplish this. For instance, experts can make individual judgments that are later collected or be placed together for a debate.[21] There are, naturally, many variations imaginable for how one would practically perform a social process approach to guideline formulation, but the common idea is to avoid the weaknesses of algorithmic and rule-based approaches by including expert judgments, and then to avoid the challenges of subjectivity directed toward expert judgments by including a wider variety of experts.

Social process approaches face a particular challenge concerning the selection of experts. Algorithmic or rule-based approaches will typically reach similar results independent of expert group members, provided algorithms and evidential hierarchies are clearly stated. Social process approaches, however, might yield different results depending on which experts' judgments one seeks. Granted, a variety of experts will guarantee a variety of judgments, but it is by no means clear how to determine the degree and kind of variety to be sought. Should we include experts on alternative treatments, or should we stay close to the mainstream? Should we choose experts who are likely to find agreement, or should we choose experts whose ideas differ radically? These are contentious questions and, crucially, they are settled before the actual expert group is chosen. There is, therefore, a lack of transparency built into the social process approach. Even when we have collected an acceptable pool of experts, it is not clear how their judgments should be aggregated, if every voice should carry equal weight, if consensus is demanded and how to treat dissent.[22]

If no consensus is demanded and there are no clear rules concerning which types of decisions the expert group should reach, social process approaches risk becoming impotent in the sense that expert groups give an overview of the field rather than clear decisions concerning what the clinical practice should look like. As such, "Rather than serving an amalgamating purpose, they are perhaps better suited for an exploratory purpose, to ensure a full range of perspectives are considered".[23] All in all, social process approaches have potential for completeness, communicability and practicality, but generally struggle in relation to transparency and rigor.

The explanatory approach starts from the basic premise that scientists will disagree, and that one central reason for disagreement is that different scientists will give different explanations for why the existing evidence looks the way it does.[24] In other words, the issue is not so much that experts disagree on what the evidence is, or what types of evidence are preferable for a given topic but rather that researchers and experts have different ways to deal with evidence which appears to oppose their view. Reformulating the issue in this way helps us disentangle situations where there is disagreement over both specific evidence and standards of evidence more generally. The key is to expect evidence contestation and disagreement rather than consensus. Contestations and disagreements are typically formulated as arguments, and the explanatory approach focuses on the quality of these arguments and how they fare in relation to standard criteria for scientific explanations.

With regard to socially relevant issues, it is rarely the case that all evidence points in a single direction. However, researchers invested in a particular understanding of the world will rarely give up their position unless they have to. A common reaction is therefore to find ways to explain why contradictory evidence only appears contradictory, and explain this appearance from within one's preferred understanding.[25] Furthermore, in cases where evidence supports our position, we give explanations for why this is so. The explanatory approach starts from how proponents of certain theories explain the appearance of confirmatory and contradictory evidence. The quality of these explanations is then rated in order to find the more appropriate one. Theory or treatment supported by the more appropriate explanation should be recommended. The explanatory approach dictates that we (1) gather all competing explanations, (2) assess them for adequacy and (3) single out the most adequate explanation.[26] Pre-clinical studies on mouse models might show that a medicine can cause kidney failure, but clinical studies might not confirm this finding. A proponent of the idea that the medicine causes kidney failure could explain the situation by referring to the experimental set-up of the clinical study, for instance that the study did not include participants predisposed to kidney failure. A proponent of the idea that the medicine does not cause kidney failure might argue that there are deficiencies in the pre-clinical studies, for instance that a too high dose or too simple experimental set-up was used. Both of these explanations are reasonable and call for further investigation. If further studies show that one of the explanations is inadequate, it ceases to function as an explanation and the position fails. In essence, therefore, the explanatory approach is what is known as an inference to best explanation approach. The explanatory approach applies the following adequacy criteria: Internal consistency, empirical competency and predictive potential.[27]

Once the various explanations are tested for adequacy, we can remove inadequate explanations and thus reduce the amount of possible positions. In this way, we can reduce the number of relevant hypotheses and gradually move toward a single recommendation in our guidelines. There will, naturally, be cases where the process of selection leads to more than one adequate explanation, and thus more than one possible treatment. In such cases, we have three main options. We can accept a group of adequate treatments and let clinicians choose freely among them, introduce further adequacy criteria in order to resolve tie-breaks or appeal to institutional responses.[28] Whether one ends up in an institutional response, introduces further criteria or accepts a group of treatments as superior, the explanatory approach separates itself from other approaches to guidelines by mimicking the scientific process. We can, for instance, see that the criteria introduced in the explanatory approach are standard criteria for scientific theorizing and debate. Furthermore, the role of evidence within the explanatory approach is as premise in an overall argument about what the phenomena are, how they should be understood and what we should do about them. This is the typical role for evidence when we attempt to understand something within science.

The main cost of applying the explanatory approach is practicality, as collecting and analyzing all available explanations is an arduous task. The upshot, however, is that the explanatory approach offers potential in terms of completeness, transparency, rigor and communicability. We can visualize the strengths and weaknesses of different approaches in Figure 6.1.

	Completeness	Rigor	Transparency	Communicability	Practicality
Qualitative rule-based	X	X	X	√	√
Quantitative algorithmic	√	√	X	X	√
Social process	√	X	X	√	√
Explanatory approach	√	√	√	√	X

Figure 6.1 Categories of approaches/techniques for weighing evidence and criteria for evaluation. Table by Fredrik Andersen et al.

So far, we have discussed challenges to guideline formulations along with a set of criteria for evaluating evidence and different approaches in light of these criteria. In the following, we present a case example and show how the general debate applies to specific cases. We discuss the topic of substance use disorders and addiction, and more specifically the current Norwegian guideline for the treatment of these issues. The focus of our discussion stays on the approaches to guideline formulations and we avoid discussions concerning which specific treatments one should prefer.

The Norwegian *National Guideline for the Treatment of Substance Use Disorders and Addiction*

Traditionally, treatment of substance use disorders and addiction has relied on charity, religious initiatives and volunteer work. More recently, treatment has come under state supervision and regulation. As such, the treatment of substance use disorders and addiction is considered part of healthcare and is expected to be implemented through evidence-based guidelines. However, the issue of substance use in the broad sense is more politically controversial than most conditions requiring health-care. There are continuous debates concerning the proper legal status of substance use and therefore whether substance use disorders should be treated within the penal or health system.[29] Along with debates over legality, there are academic discussions concerning which substances are more harmful and wider issues relating to social injustice.[30] Against the backdrop of legal, moral and political debates concerning substance use, expert groups constructing guidelines for the treatment of substance use disorders are under constant scrutiny.

Formulating guidelines for treatment of substance use disorders and addiction presents extra challenges to guideline formulation. Research on treatment effectiveness is multidisciplinary in the sense that information about disorders comes from a wide range of fields spanning from psycho-social studies to medical laboratory studies. Additionally, clinical treatment is performed by professionals as diverse as nurses, social educators, psychologists, psychiatrists, medical doctors and environmental therapists, and treatment is often most effective when interdisciplinary. We therefore have variety both at the level of research and clinical practice. On the research end, we face immediate challenges concerning, for instance, whether a guideline should focus on specific disorder-related outcomes such as abstinence over time, or on more general outcomes such as well-being. There is no clear answer as to which type of outcome we should emphasize, and the answer will to some extent rely on common non-epistemic assumptions within the expert group. Here we see an immediate challenge relating to the completeness criterion, as once the focus of the expert group is established, certain evidence is already favored. The issue of evidence relevance also stresses the need for transparency. A guideline focusing on disorder-specific outcomes might, rationally, be ignored by practitioners working toward more general outcomes. A guideline can be thought of as a "how to" for achieving particular outcomes, and if a clinician is presented with a guideline directed at a different outcome, this guideline might simply be ignored.

From a practitioners' perspective, it is paramount that evaluations of relevance are transparent. Otherwise, practitioners with a strong belief that a certain treatment is preferable might assume that their preferred treatment was never considered by the expert group, and thus that the guideline can be ignored. In order for a guideline to improve and update actual practice so that patients receive the best treatment currently available, it should be transparent to the practitioner which types of evidence have been considered, and why. Issues of practicality will also have an impact on any expert group making evaluations concerning which types of evidence and which types of treatment they consider.

The particular multidisciplinarity of practitioners who provide treatment to persons with substance use disorders and addiction constrains the possible ways in which decisions can be communicated. Practitioners with varying backgrounds will have varying degrees of training in reading different types of arguments. This implies that, if one is to communicate guidelines in such a way that they can be implemented, the communication must be more broadly accessible than if one was constructing guidelines for an audience with homogenous backgrounds.

We now examine the Norwegian *Guideline for the Treatment of Substance Use Disorders and Addiction*, determine the approach used and suggest improvements in line with the explanatory approach.[31] In Norway, clinical practice guidelines are systematically developed by the public Norwegian Directorate of Health within various fields of professional healthcare and treatment. Guidelines are developed as a means to prevent unwanted variation as well as securing quality in various health and social services. Clinical practice guidelines have status as national guidelines for clinical practice within specific fields in health-care practices, providing recommendations that function as treatment standardizations for a particular population.[32] In the treatment and rehabilitation of persons with substance use disorders and addiction, several national clinical practice guidelines have been developed in the last decade. The national clinical guideline of opioid substitution treatment was followed by and implemented a national guideline for the assessment, treatment and follow-up of people with co-occurring substance abuse and psychiatric disorders.[33] Most recently, the Norwegian Directorate of Health issued the national clinical guideline for the treatment and rehabilitation of substance use disorders and addiction, which is the focus of the present chapter.[34]

With the stated goal of developing evidence-based, recognized, clinical recommendations for interventions and methods for the rehabilitation and treatment of various substance use disorders, an expert group was established in 2011. The majority of group members (16/19) were experts in the field of substance use disorders and addiction. These are key professionals and experts holding positions in various public and private agencies providing rehabilitation and treatment for persons with substance use disorders. In addition, the group consisted of one researcher from the Norwegian Institute of Public Health, along with two stakeholder representatives. The expert group worked both within a plenum and within sub-groups. Initially, sub-groups were divided between legal and illegal substances, and later a further division was made into groups for coordination, research and external grounding. The groups were then joined for 11 common meetings along with three meetings with a corresponding expert group in Sweden along with conferences and study-trips.[35]

The expert group states no hierarchy of evidence, but points out that in cases where there is little available research, they have emphasized clinical and stakeholder knowledge through internal discussions using the international standardized method Grading of Recommendations, Assessment, Development and Evaluations (GRADE). The GRADE tool is intended to ensure that all relevant issues from cost to feasibility and acceptability are considered. There is no stated, overall strategy for achieving consensus. The guideline is non-exhaustive, and recommendations are presented with various degrees of strength.[36] Recommendations are given within ten overall categories ranging from gender specific care and stakeholder inclusion to mapping strategies and levels and intensity of treatment.

Our focus in this chapter is on part four of the guideline, "Therapeutic approaches to substance use disorders", which evaluates 13 therapeutic approaches. Of the 13 therapeutic approaches, 11 refer to treatment methods, while two recommendations are conditional and thus refer to how therapies should be implemented if they are chosen. Of the 11 treatment methods, two are strongly recommended. These are cognitive behavioral therapies (CBT) and motivational interviews (MI). The remaining nine methods are weakly recommended/suggested.[37]

For each recommendation given, the guideline refers to an evidential base. Pharmaceutical support for persons with alcohol use disorders refers to evidence already graded in England by the National Institute for Health and Care Excellence (NICE). In relation to systematic physical activity, pharmaceutical support for persons with alcohol use disorders, group therapies and couples' therapies, the expert group has performed systematic literature searches. The remaining six recommendations are based on the expert group's clinical knowledge and patient knowledge along with discussions with practitioners. These are consensus decisions for which the expert group provides a selection of references, but no summaries or systematic treatment of the available literature.

The treatment and evaluation of evidence made within the Norwegian *Guideline for the Treatment of Substance Use Disorders and Addiction* illustrates a series of challenges to guideline formulation. In the following, we clarify and discuss these challenges in more detail.

Challenges and Improvements to the Norwegian *National Guideline for the Treatment of Substance Use Disorders and Addiction*

The approach applied by the Norwegian expert group is best categorized as a social process approach to guidelines. The process described within the guideline follows Heather E. Douglas' description fairly accurately as a group of experts and two stakeholder representatives were selected. These are divided into sub-groups focusing on legal and illegal substances, research, coordination and external grounding. There are no explicit rules for evidence weighing, and no algorithmic techniques reported. Furthermore, all recommendations are reported as consensus recommendations, either with or without an evidential base, implying that the group has reached consensus through debates. These are all hallmarks of a social process approach. Furthermore, there is no reference to an overall approach to aggregating diverging positions, which indicates that no general rule-based approach is followed.

As we have mentioned, there are some standard challenges connected with the social process approach and some of these challenges also apply in the case of the Norwegian *Guideline for the Treatment of Substance Use Disorders and Addiction*. The first challenge relates to the choice of experts and stakeholder representatives.

The choice of experts might heavily influence the focus of the expert group and it would be beneficial, if a reader was presented with a rationale for the choice of experts. For instance, why only one researcher and two stakeholder representatives were permanent members of the expert group, while the remaining 16 represent the practical field. Of the 16 representatives of the practical field, only three are at the level of day-to-day contact with patients (two medical doctors and one substance use coordinator). Expert groups with a majority of researchers, a majority of experts with more direct contact with patients or a majority of stakeholder representatives are all possible options here, and such constellations could have yielded different guidelines.

The absence of any description relating to how expert group members were selected is a threat to transparency. In choosing an expert group, one main objective is to ensure that as many treatments as possible are considered. This is how social process approaches can avoid individual and subjective biases and avoid prematurely excluding possible treatment candidates. In the current guideline, this is only done to a limited extent. Under heading 14.1 "methodological approach" the guideline document states that the choices made are non-exhaustively and that other possible approaches exist. It goes on to mention contingency management, stages of treatment and psychoeducative interventions as relevant treatments. The reader is then referred to other guidelines in which these treatments are discussed. However, if we follow up and read the documentation available in the related guidelines, we find little to go on. In the case of stages of treatment, there is a single reference to an exploratory paper from 1989.[38] In the cases of contingency management and psychoeducative interventions, there is no documentation of effect/lack of effect whatsoever. These treatments are simply mentioned in passing and excluded from the current guideline. This exclusion breaks with the previously established criterion of completeness, and it does so without any apparent justification since there is available evidence to discuss. For instance, in the case of contingency management, Danielle R. Davis et al. found 801 reports in relation to substance abuse between 2009 and 2014.[39] Reviews of this kind, along with related reports, could have been included within the current guideline. In the absence of transparent exclusion rules, the exclusion of this and similar evidence appears accidental rather than principled. A further challenge relates to how the expert group has reached consensus concerning the hierarchy of recommendations.

In a highly multidisciplinary field, one would expect a divergence of initial positions and therefore that there would be a need for a process through which these positions are aggregated. In the Norwegian *Guideline for the Treatment of Substance Use Disorder and Addiction*, no such process is presented. This is the case both concerning recommendations decided on the basis of a collection of evidence and recommendations that are decided through debate. This is a threat to transparency since it is sometimes hard to understand how a strong/weak/no recommendation was reached.

Any clinician who finds a hierarchy of recommended treatments will naturally wonder about the basis of the hierarchy. How did the expert group arrive at this particular hierarchy? In the case of the Norwegian *Guideline for the Treatment of Substance Use Disorders and Addiction*, this is not an easy question to answer. There are three strong recommendations. However, there is no summarized research base for any of these. Neither are there explanations concerning why these recommendations are placed at the top of the hierarchy other than a mention of user experience, clinical knowledge and expert group consensus. We agree that user experience and clinical knowledge are central aspects of the evaluation of any treatment and assume that this is a position we share with most clinicians. However, the guideline provides no description of what these user experiences are, and no description of the clinical knowledge.

As knowledge base, the guideline presents some, although limited, literature. For instance, CBT is strongly recommended in the guideline. The knowledge base presented for this recommendation consists of the following four sources: Christian S. Hendershot et al. provide a systematic review of the efficacy of the relapse model for addiction.[40] This review not only shows the relapse model to be effective but also warns against confusing terminologies as relapse prevention "has evolved into an umbrella term synonymous with most cognitive-behavioral skills-based interventions addressing high-risk situations and coping responses".[41] Furthermore, Hendershot et al. call for more research concerning the predictability, scope and applicability of relapse prevention methods.[42] In total, Hendershot et al.[43] put forward a good argument for relapse prevention methods while emphasizing the need for a distinction between relapse prevention and general cognitive behavioral approaches. As such, it is hard to see how their review functions as an explanation for the preference for CBT. An explanation of the rationale of the expert group on this issue would increase transparency along with the efficiency of communication.

Kathryn R. McHugh et al. conduct a systematic review of cognitive behavioral treatments of substance abuse disorders, showing that cognitive behavioral treatments are effective.[44] However, McHugh et al. treat behavioral and cognitive behavioral treatments as a collective umbrella which encompasses MI, contingency management and relapse prevention.[45] This somewhat confuses the taxonomy given in the Norwegian *Guideline for the Treatment of Substance Use Disorders and Addiction*, where CBT is treated as distinct from contingency management and MI. Contingency management is treated more as an afterthought in the current guideline and referred to as a possible treatment along with psychoeducative interventions and stages of treatment. MI, on the other hand, is at the top of the recommendation hierarchy as a distinct form of treatment along with CBT. It is therefore difficult to see how we are to understand McHugh et al. in relation to the Norwegian *Guideline for the Treatment of Substance Use*

Disorders and Addiction.[46] Here it would be helpful for the reader if the expert group had provided explanations for the discrepancies in taxonomy, along with reasons for choosing the current taxonomy, and shown how the recommendations are supported by McHugh et al.[47]

Hallgeir Brumoen's book describes itself as predominantly a practical treatise, which shows how one can effectively apply behavioral therapeutic techniques to central elements of substance use disorders.[48] Brumoen does not provide any basis for comparing behavior analytic approaches to other therapies.[49]

Nina E. Andresen and Kari Lossius offer a handbook for the treatment of substance use disorders, which recommends, among other treatments, CBT, MI and mentalization techniques.[50] Andresen and Lossius do not advance arguments for comparing strategies but rather show how these strategies can be effective and how they should be implemented.[51]

All in all, the four sources provided in support of CBT are relevant and respectable, but it is hard to see how they constitute an evidential base for placing CBT at the top of any treatment hierarchy. One issue is that McHugh et al. introduce a taxonomy that differs from the taxonomy used in the guideline and subsume mentalization techniques, MI and contingency management under the umbrella of CBT.[52] If we were to accept this taxonomy, the guideline actually only strongly recommends one type of treatment: CBT. However, the guideline document apparently does not accept the McHugh et al. taxonomy and treats these therapies as separate types of interventions.[53] Furthermore, the separate interventions are graded radically differently within the guideline document. CBT and MI are recommended, mentalization techniques are suggested and contingency management is neither recommended nor suggested but rather mentioned as an afterthought in the methodology section of the document. All of this could probably be defended. For instance, the guideline authors could argue that McHugh et al.'s taxonomy is erroneous, and that there are differences in effect between the treatments.[54] It could also be argued that CBT and MI are better documented and thus more trustworthy as treatments. Another possible line of argument would be to refer to existing expertise among Norwegian clinicians and claim that CBT is more practical for pragmatic reasons. Finally, one could focus on social or cultural differences between places where the original research was conducted. All of these are potentially reasonable arguments that clinicians could scrutinize, contest or be convinced by. However, no such arguments are presented in the guideline document, which simply states that the evaluation was unanimously decided among the experts. In effect, the entire reasoning behind the taxonomy of treatments as well as their prioritization within the hierarchy is black boxed.

A final challenge to the Norwegian *Guideline for the Treatment of Substance Use Disorder and Addiction* concerns its applicability for clinical treatment. The guideline presents a hierarchy of recommendations and there are two strongly recommended treatments: CBT and MI. As we have seen, MI can also be treated as a subgenre of CBT, such as in the McHugh et al. taxonomy.[55] This implies that one should first and foremost attempt to apply CBT. The guideline document also mentions relevant treatments that were never considered, such as contingency management, stages of treatment and psychoeducative therapies. This implies that the guideline is non-exhaustive and it is unclear how a clinician is to relate to these latter treatments. Can we treat them as strongly recommended along with CBT, or are they at a lower level? One effect of omitting therapies is that the hierarchy of the guideline is weakened. The existence of a hierarchy is necessary since a central role of the guideline is to set a common standard, to harmonize treatment and to avoid unwanted variety. However, if the document does not provide arguments for the hierarchy, but rather refers to consensus, and admits to omitting relevant treatments, it might ultimately be read more as an exploratory document that maps parts of the available treatments. All in all, the guideline suffers from challenges in relation to rigor, transparency and completeness and in effect faces challenges relating to effective communication.

It is our contention that the challenges presented here can be met, and that they have their origin in the type of approach chosen for the expert group. As we have seen, the expert group members are experts in the relevant field and have extensive knowledge of actual practice. However, the expert group was instructed to consider other guidelines, but given no particular instructions concerning how to weigh evidence, how to aggregate diverging positions and which criteria to apply in cases of disagreements.[56] We suggest that clarifying such issues is central not only to the reader, but also to the expert group itself.

Hitherto, we have discussed general challenges to guideline formulation, different approaches to meeting these challenges, and illustrated these approaches through the example of the Norwegian *Guideline for the Treatment of Substance Use Disorders and Addiction*. In the following, we indicate how the guideline could be improved through the explanatory approach.

The explanatory approach applies three main steps: (1) collect the widest possible sample of explanations, (2) assess the adequacy of the explanations, (3) reduce the number of explanations according to the adequacy assessment. The first step might appear counterintuitive given that a central purpose of the guideline is to reduce unwanted variety and instead set a common standard of treatment. However, if we wish to set a common standard, we are obliged to set the best available treatment as standard. In order to ensure we have the best available treatment, we fare better by considering the widest possible variety. In this way, we avoid accidental omissions. We have seen how the Norwegian

Guideline for the Treatment of Substance Use Disorders and Addiction has omitted relevant treatments and that one effect of this is that the suggested hierarchy is weakened. By applying the explanatory approach, and collecting the widest possible variety, this could be avoided and we would better satisfy the completeness criterion.

The second step within the explanatory approach is to assess all explanations according to a set of criteria for adequacy. The explanatory approach includes criteria of internal consistency, empirical competency and predictive potential. A rigorous application of these criteria can be presented along with the recommendations hierarchy. In relation to the Norwegian *Guideline for the Treatment of Substance Use Disorders and Addiction*, such a presentation would have to include explanations for the preferred taxonomy, the relation between available evidence and recommendations and the specifics of stakeholder and clinicians' relevant experiences. In effect, the reader would know how different treatments were assessed prior to any hierarchy of recommendations.

In the final step of an explanatory approach, the expert group would present a recommendations hierarchy and thus limit the amount of strongly recommended treatments. This is done in the presented Norwegian guideline, but within the explanatory approach this hierarchy would now be fully transparent as the reader would have already been presented with an adequacy assessment. For a field as multidisciplinary as substance use disorders and addiction, one might expect that the adequacy assessment will fail to pick out a single treatment as most adequate. As mentioned earlier, this could be resolved by either introducing further criteria or calling for an institutional response. However, it is also possible to maintain a group of strongly recommended treatments provided this group is a clear limitation of the original variability.

In conclusion, it is by no means clear that an explanatory approach would yield different recommendations than those already present in the current guideline. However, by applying the explanatory approach, we could substantially increase the level of completeness, rigor, transparency and communicability of the guideline document without seriously reducing the practicality of the expert group workload.

Conclusion

Treatment for substance abuse and addiction is a notoriously difficult field. Treatment aims at different aspects, such as achieving abstinence or reduced abuse, to prevent or reduce relapse, in addition to improved adaptive functioning.[57] Achieving abstinence is considered an ideal outcome and is related to better long-term outcomes.[58] However, if abstinence is one important intended outcome, contemporary treatment must be considered less than effective both within Norway and internationally. Favorable outcomes, such as achieved abstinence, are related to treatment completeness, and in the treatment of substance use dependent individuals, drop out from treatment is more common than completeness.[59] Seeing how adverse the effects of substance use and addiction can be for patients and their families, we are obliged to strive for the best possible treatments. Guidelines are, in our view, a welcome contribution to boosting quality in this sense. However, guidelines must be comprehensive, transparent and convincing. It is within the nature of scientific research that different groups of researchers and clinicians will prefer certain treatments over others. When these preferences are not met, contestation of evidence and critical evidence evaluation is to be expected. Furthermore, in any scientific field where something of value is at stake, we should expect scrutiny and disagreement. If scrutiny, disagreement and evidence contestation are treated as integral rather than opposed to scientific knowledge, we might better be able to deal with disagreement both within science and in science communication directed at the general public. In our view, the expectation that scientific evidence should always align often stands in the way of meaningful debates concerning reasonable disagreement over evidence as well as policy. We suggest that new guidelines treat evidence as premises within arguments and apply the explanatory approach to evidence-based guidelines in the future.

Importantly, our critique is not directed toward the conclusions drawn in the guideline. It might well be that CBT is superior as treatment for substance use disorders and addiction. Indeed, three of the authors are members of the Norwegian Association for Behavior Analysis, and firm believers in the effectiveness of behavioral interventions. Our concern is with how the guidelines are formulated and the function they are meant to have in order to improve the quality of care given to sufferers of substance abuse and addiction.

Finally, we wish to comment on the generalizability of our argument. The central aim of this chapter is to show that guideline formulation benefits from the explanatory approach. As we have seen, the main challenges facing the Norwegian *Guideline for the Treatment of Substance Use Disorders and Addiction* are standard challenges to any social process approach. There will always be possible controversy concerning the choice of experts and stakeholder representatives. Furthermore, the social process approach struggles with issues of transparency and completeness concerning how certain treatments are evaluated and which treatments are considered to begin with. The explanatory approach provides solutions to these issues and, by focusing on explanation and argumentation, mimics the scientific process in a way that is familiar to practitioners as well as researchers. There is, of course, room for development of the explanatory approach as well. For instance, one could argue that there is a need for more adequacy criteria or for better solutions in cases where multiple treatments appear appropriate. The explanatory approach can also be made flexible in the sense that each expert group can formulate context-dependent criteria for adequacy. Still, provided the criteria are made public and transparent, we contend that the explanatory approach should be the preferred approach to guideline formulation within the health sciences.

Notes

1 Robin Graham, Michelle Mancher, Dianne M. Wolman, Sheldon Greenfield and Earl Steinberg, *Clinical Practice Guidelines We Can Trust* (Washington DC: National Academies Press, 2011), 4.
2 Rani L. Anjum, Samantha Copeland, Roger Kerry and Elena Rocca, "The Guidelines Challenge—Philosophy, Practice, Policy", *Journal of Evaluation in Clinical Practice* 24, no. 5 (July 2018): 1120–1126
3 Heather E. Douglas, "Weighing Complex Evidence in a Democratic Society", *Kennedy Institute of Ethics Journal* 22, no. 2 (June 2012): 144.
4 Anjum et al., "The Guidelines Challenge", 3.
5 Fredrik Andersen, Rani L. Anjum and Elena Rocca, "Philosophical Bias Is the One Bias That Science Cannot Avoid", *ELife* 8 (March 2019): 1.
6 Andersen et al., "Philosophical Bias", 3.
7 See, for instance Elena Rocca and Fredrik Andersen, "How Biological Background Assumptions Influence Scientific Risk Evaluation of Stacked Genetically Modified Plants: An Analysis of Research Hypotheses and Argumentations", *Life Sciences, Society and Policy* 13, no. 1 (August 2017), 1–20, along with: Fredrik Andersen and Elena Rocca, "Underdetermination and Evidence-based Policy", *Studies in History and Philosophy of Science. Part C, Studies in History and Philosophy of Biological and Biomedical Sciences* 84 (December 2020): 1–8.
8 There is a long tradition for this argument within philosophy of science. Recent examples are Andersen et al., "Philosophical Bias", and Heather E. Douglas, "Inductive Risk and Values in Science", *Philosophy of Science* 67, no. 4 (December2000): 560.
9 Douglas, "Weighing Complex Evidence", 141.
10 Thomas S. Kuhn, "Objectivity, Value Judgement, and Theory Choice", in *The Essential Tension: Selected Studies in Scientific Tradition and Change* (Chicago: The University of Chicago Press, 1977), 320 introduces criteria for evaluating a scientific theory. The relevance of these criteria for an open and democratic society is argued in Douglas, "Weighing Complex Evidence", 143.
11 Douglas, "Weighing Complex Evidence", 141.
12 Rani L. Anjum, Samantha Copeland, and Elena Rocca, "Medical Scientists and Philosophers Worldwide Appeal to EBM to Expand the Notion 'Evidence'", *BMJ Evidence-Based Medicine* 25 (November 2018): 6–8.
13 Trisha Greenhalgh, Jeremy Howick and Neal Maskrey, "Evidence Based Medicine: A Movement in Crisis?", *BMJ* 248 (June 2014): 4.
14 Elena Rocca, "The Judgements that Evidence-Based Medicine Adopts", *Journal of Evaluation in Clinical Practice* 24 (October 2018): 2.
15 Marco Leyton, "Conditioned and Sensitized Responses to Stimulant Drugs in Humans", *Progress in Neuro-Psychopharmacology & Biological Psychiatry* 31, no. 8 (August 2007): 1602.
16 Gene M. Heyman, "Addiction and Choice: Theory and New Data", *Frontiers in Psychiatry* 4 (May 2013): 4.
17 Heyman, "Addiction and Choice".

18 Douglas, "Weighing Complex Evidence", 147.
19 Wenlong Sun, Olfa Nasraoui and Patrick Shafto, "Evolution and Impact of Bias in Human and Machine Learning Algorithm Interaction", *PLoS One* 15, no. 8 (August 2020): 30.
20 Douglas, "Weighing Complex Evidence", 149.
21 Douglas, "Weighing Complex Evidence", 149.
22 Douglas, "Weighing Complex Evidence", 151.
23 Douglas, "Weighing Complex Evidence", 151.
24 Douglas, "Weighing Complex Evidence", 151.
25 This strategy is also common in the natural sciences and was famously adopted by Johannes Kepler, Galileo Galilei and Albert Einstein. See, for instance Fredrik Andersen, "Experience and Theory: A Defense of the Kantian A Priori and Kepler's Philosophy of Science in Light of Modern Space-Time Physics" (2017), and Johan A. Myrstad, "The Use of Converse Abduction in Kepler" Foundations of Science 9 (September 2004): 321–338.
26 Douglas, "Weighing Complex Evidence", 152.
27 Douglas, "Weighing Complex Evidence", 153–156.
28 Andersen, and Rocca, "Underdetermination".
29 Joanne Csete, Adeeba Kamarulzaman, Michel Kazatchkine, Frederick Altice, Marek Balicki, Julia Buxton, Javier Cepeda et al., "Public Health and International Drug Policy", *The Lancet (British Edition)* 387, no. 10026 (March 2016): 1427–1480.
30 For an example of diverging assessments of harm, see: David Nutt, Leslie A. King, William Saulsbury, and Colin Blakemore, "Development of a Rational Scale to Assess the Harm of Drugs of Potential Misuse", *The Lancet (British Edition)* 369, no. 9566 (March 2007): 1047. For debates concerning social issues, see for instance: Brian D. Earp, Jonathan Lewis, and Carl L. Hart, "Racial Justice Requires Ending the War on Drugs", *The American Journal of Bioethics* 21, no. 4 (January 2021): 4.
31 Norwegian Directorate of Health, *Nasjonal Faglig Retningslinje for Behandling og Rehabilitering av Rusmiddelproblemer og Avhengighet* (Oslo: Helsedirektoratet, 2017).
32 Norwegian Directorate of Health, *Om Helsedirektoratets Normerende Produkter* (Oslo: Helsedirektoratet 2019).
33 Norwegian Directorate of Health, *Nasjonal Retningslinje for Legemiddelassistert Rehabilitering ved Opioidavhengighet. Nasjonale Faglige Retningslinjer* (Oslo: Helsedirektoratet, 2010).
34 Norwegian Directorate of Health, *Nasjonal Faglig Retningslinje for Behandling.*
35 Norwegian Directorate of Health, *Nasjonal Faglig Retningslinje for Behandling,* Ch. 15.
36 Norwegian Directorate of Health, *Nasjonal Faglig Retningslinje for Behandling,* Ch. 14
37 The nine weakly recommended therapeutic approaches are mentalization based therapies, systematic physical activities, music therapy, 12-step programs, mindfulness techniques, pharmaceutical support for persons with alcohol use disorders, urine sampling and the use of biological markers as motivational factors, group therapies and couples therapies.
38 Fred C. Osher and Lial L. Kofoed, "Treatment of Patients with Psychiatric and Psychoactive Substance Abuse Disorders", *Hospital & Community Psychiatry* 40, no. 10 (April 1989): 1025–1030.
39 Danielle R. Davis, Allison N. Kurti, Joan M. Skelly, Ryan Redner, Thomas J. White, and Stephen T. Higgins, "A Review of the Literature on Contingency Management in the Treatment of Substance Use Disorders, 2009–2014", *Preventive Medicine* 92 (November 2016): 40.

40 Christian S. Hendershot, Katie Witkiewitz, William H. George and G. Alan Marlatt, "Relapse Prevention for Addictive Behaviors", *Substance Abuse Treatment, Prevention and Policy* 6, no. 1 (July 2011): 1–17.
41 Hendershot et al., "Relapse Prevention", 2.
42 Hendershot et al., "Relapse Prevention", 13.
43 Hendershot et al., "Relapse Prevention", 13.
44 Kathryn R. McHugh, Bridget A. Hearon and Michael W. Otto, "Cognitive-Behavioral Therapy for Substance Use Disorders", The Psychiatric Clinics of North America 33, no. 3 (September 2010): 511–525.
45 McHugh et al., "Cognitive-Behavioral Therapy", 511.
46 McHugh et al., "Cognitive-Behavioral Therapy".
47 McHugh et al., "Cognitive-Behavioral Therapy".
48 Hallgeir Brumoen, *Bygging av Mestringstillit: En Metodebok om Mestring av Rusproblemer* (Oslo: Gyldendal Akademisk, 2000).
49 Hallgeir Brumoen, *Bygging av Mestringstillit*.
50 Nina E. Andresen and Kari Lossius. *Håndbok i Rusbehandling: Til Pasienter med Moderat til alvorlig Rusmiddelavhengighet*. 2nd ed. (Oslo: Gyldendal Akademisk, 2012).
51 Andresen et al., *Håndbok i Rusbehandling*.
52 McHugh et al., "Cognitive-Behavioral Therapy", 511.
53 McHugh et al., "Cognitive-Behavioral Therapy".
54 McHugh et al., "Cognitive-Behavioral Therapy".
55 McHugh et al., "Cognitive-Behavioral Therapy".
56 Norwegian Directorate of Health, *Nasjonal Faglig Retningslinje for Behandling*, Ch. 15.
57 See, for instance Hanne H. Brorson, Espen A. Arnevik, Kim Rand-Hendriksen, and Fanny Duckert, "Drop-Out from Addiction Treatment: A Systematic Review of Risk Factors", *Clinical Psychology Review* 33, no. 8 (December 2013): 1011. Or Herbert D. Kleber, Roger D. Weiss, Raymond F. Anton Jr., Tony P. George, Shelly F. Greenfield, Thomas R. Kosten, Charles P. O'Brien et al., "Treatment of Patients with Substance Use Disorders", *The American Journal of Psychiatry* 164, no. 4 (April 2007): 5–123.
58 Kleber et al., "Treatment of Patients with Substance Use Disorders".
59 Brorson et al., "Drop-out".

References

Andersen, Fredrik. "Experience and Theory: A Defense of the Kantian A Priori and Kepler's Philosophy of Science in Light of Modern Space-Time Physics". PhD diss., Norwegian Univ. of Life Sciences, 2017.

Andersen, Fredrik, Rani Lill Anjum, and Elena Rocca, "Philosophical Bias Is the One Bias That Science Cannot Avoid". *ELife* 8 (March 2019): e44929.

Andersen, Fredrik, and Elena Rocca, "Underdetermination and Evidence-Based Policy". *Studies in History and Philosophy of Science. Part C, Studies in History and Philosophy of Biological and Biomedical Sciences* 84 (2020): 101335.

Andresen, Nina Elin, and Kari Lossius. *Håndbok i Rusbehandling: Til Pasienter med Moderat til alvorlig Rusmiddelavhengighet*. 2nd ed. Oslo: Gyldendal Akademisk, 2012.

Anjum, Rani Lill, Samantha Copeland, and Elena Rocca. "Medical Scientists and Philosophers Worldwide Appeal to *EBM* to Expand the Notion of 'Evidence'". *BMJ Evidence-Based Medicine* 25 (November 2018): 6–8.

Anjum, Rani Lill, Samantha Copeland, Roger Kerry, and Elena Rocca. "The Guidelines Challenge—Philosophy, Practice, Policy". *Journal of Evaluation in Clinical Practice* 24, no. 5 (July 2018): 1–7.

Brorson, Hanne H., Espen A. Arnevik, Kim Rand-Hendriksen, and Fanny Duckert, "Drop-Out from Addiction Treatment: A Systematic Review of Risk Factors". *Clinical Psychology Review* 33, no. 8 (December 2013): 1010–1024.

Brumoen, Hallgeir. *Bygging av Mestringstillit: En Metodebok om Mestring av Rusproblemer*. Oslo: Gyldendal Akademisk, 2000.

Csete, Joanne, Adeeba Kamarulzaman, Michel Kazatchkine, Frederick Altice, Marek Balicki, Julia Buxton Javier Cepeda et al. "Public Health and International Drug Policy". *The Lancet (British Edition)* 387, no. 10026 (March 2016): 1427–1480.

Davis, Danielle R., Allison N. Kurti, Joan M. Skelly, Ryan Redner, Thomas J. White, and Stephen T. Higgins. "A Review of the Literature on Contingency Management in the Treatment of Substance Use Disorders, 2009–2014". *Preventive Medicine* 92 (November 2016): 36–46.

Douglas, Heather E. "Inductive Risk and Values in Science". *Philosophy of Science* 67, no. 4 (2000): 559–579.

Douglas, Heather E. "Weighing Complex Evidence in a Democratic Society". *Kennedy Institute of Ethics Journal* 22, no. 2 (June 2012): 139–162.

Earp, Brian D., Jonathan Lewis, and Carl L. Hart, "Racial Justice Requires Ending the War on Drugs". *The American Journal of Bioethics* 21, no. 4 (January 2021): 4–19.

Graham, Robin, Michelle Mancher, Dianne M. Wolman, Sheldon Greenfield, and Earl Steinberg. *Clinical Practice Guidelines We Can Trust*. Washington, DC: National Academies Press, 2011.

Greenhalgh, Trisha, Jeremy Howick, and Neal Maskrey. "Evidence Based Medicine: A Movement in Crisis?". *BMJ 348* (June 2014): g3725.

Hendershot, Christian S., Katie Witkiewitz, William H. George, and G. Alan Marlatt. "Relapse Prevention for Addictive Behaviors". *Substance Abuse Treatment, Prevention and Policy* 6, no. 1 (July 2011): 17.

Heyman, Gene M. "Addiction and Choice: Theory and New Data". *Frontiers in Psychiatry* 4 (May 2013): 31.

Kleber, Herbert D., Roger D. Weiss, Raymond F. Anton Jr, Tony P. George, Shelly F. Greenfield, Thomas R. Kosten, and Charles P. O'Brien et al. "Treatment of Patients With Substance Use Disorders". *The American Journal of Psychiatry* 164, no. 4 (April 2007): 3–123. https://pubmed.ncbi.nlm.nih.gov/17569411/.

Kuhn, Thomas S. "Objectivity, Value Judgement, and Theory Choice". In *The Essential Tension: Selected Studies in Scientific Tradition and Change*, 320–339. Chicago, IL: The University of Chicago Press, 1977.

Leyton, Marco. "Conditioned and Sensitized Responses to Stimulant Drugs in Humans". *Progress in Neuro-Psychopharmacology & Biological Psychiatry* 31, no. 8 (August 2007): 1601–1613.

McHugh, Kathryn R., Bridget A. Hearon, and Michael W. Otto. "Cognitive-Behavioral Therapy for Substance Use Disorders". *The Psychiatric Clinics of North America* 33, no. 3 (September 2010): 511–525.

Myrstad, Johan A. "The Use of Converse Abduction in Kepler". *Foundations of Science* 9 (September 2004): 321–338.

Norwegian Directorate of Health. *Nasjonal Faglig Retningslinje for Behandling og Rehabilitering av Rusmiddelproblemer og Avhengighet*. Oslo: Helsedirektoratet, 2017.

Norwegian Directorate of Health. *Om Helsedirektoratets normerende produkter*. Oslo: Helsedirektoratet, 2019.

Norwegian Directorate of Health. *Nasjonal Retningslinje for Legemiddelassistert Rehabilitering ved Opioidavhengighet. Nasjonale Faglige Retningslinjer*. Oslo: Helsedirektoratet, 2010.

Norwegian Directorate of Health. *Nasjonal Faglig Retningslinje for Utredning, Behandling og Oppfølging av Personer med Samtidig Rus – Og Psykisk Lidelse. Nasjonale Faglige Retningslinjer*. Oslo: Helsedirektoratet, 2011.

Nutt, David, Leslie A. King, William Saulsbury, and Colin Blakemore. "Development of a Rational Scale to Assess the Harm of Drugs of Potential Misuse". *The Lancet (British Edition)* 369, no. 9566 (March 2007): 1047–1053.

Osher, Fred C., and Lial L. Kofoed. "Treatment of Patients with Psychiatric and Psychoactive Substance Abuse Disorders". *Hospital & Community Psychiatry* 40, no. 10 (April 1989): 1025–1030.

Rocca, Elena. "The Judgements That Evidence-Based Medicine Adopts". *Journal of Evaluation in Clinical Practice* 24, no. 5 (October 2018): 1184–1190.

Rocca, Elena, and Fredrik Andersen. "How Biological Background Assumptions Influence Scientific Risk Evaluation of Stacked Genetically Modified Plants: An Analysis of Research Hypotheses and Argumentations". *Life Sciences, Society and Policy* 13, no. 1 (August 2017): 1–20.

Sun, Wenlong, Olfa Nasraoui, and Patrick Shafto. "Evolution and Impact of Bias in Human and Machine Learning Algorithm Interaction". *PLoS One* 15, no. 8 (August 2020): E0235502.

Part IV

Challenging Academic Evidence

Counter Science, Citizen Science and Environmental Activism

7 Evidence against the "Nuclear State"

Contesting Technoscience through *Gegenwissenschaft* in the 1970s and 1980s

Stefan Esselborn and Karin Zachmann

In March 1982, Günter Altner, a professor of Lutheran theology and member of the Enquête-Kommission zur zukünftigen Kernenergiepolitik (Parliamentary Commission of Enquiry on Future Nuclear Energy Policy), witnessed what he described as a scene "fraught with symbolism" for the further development of science and technology in the Federal Republic of Germany (FRG).[1] During a Commission hearing on the potential of catastrophic releases of energy in sodium-cooled fast breeder reactors, "a student in sneakers and a so-called leader of German science, Federal Cross of Merit in his black-rimmed button hole, faced off against each other". Even more remarkable than the visual contrast, Altner thought, was the fact that Richard Donderer, the student who had yet to earn his diploma, more than held his own against Heinz Maier-Leibnitz, a highly decorated atomic physicist, founding director of the country's first operational research reactor and former head of the German Research Foundation (DFG). In fact, the student even seemed to have the better technical arguments. To the enthusiastic Altner, this exchange foreshadowed the "rise of a new generation" of scientists, and even "a new understanding of science and technology" in general.[2]

Both Altner and Donderer were leading exponents of a new phenomenon in West German science and technology in the late 1970s and early 1980s, which contemporaries variously referred to as *Gegenwissenschaft* (counter science), *Gegenforschung* (counter research), *kritische* or *alternative Wissenschaft* (critical/alternative science).[3] These and a number of similar terms were employed to describe a rather heterogeneous group of actors who set out to contest official academic and corporate science and technology – by (mostly) scientific means, but from an explicitly political and value-oriented point of view. By the second half of the 1970s, these counter experts started to form an increasingly coherent and organized scene, with its own independent institutional infrastructure. Despite some resistance from established science, leading protagonists of *Gegenwissenschaft* not only quickly became prominent voices in the West German public discourse but also managed to gain entry into the sphere of scientific policy consultancy, and ultimately even the private sector consulting economy.

DOI: 10.4324/9781003273509-12

In spite of the substantial contemporary attention counter science received both in the media and the scientific community, it has so far attracted relatively limited scholarly and historical interest. For the case of the FRG, only a small number of mostly short sociological accounts exist.[4] This seems all the more surprising as the political and cultural history of the West German New Left and the so-called New Social Movements in general,[5] and of the environmental and anti-nuclear movement in particular,[6] are by now fairly well known. However, the knowledge practices and epistemic basis of these movements have only recently become an explicit focus of research, under the umbrella term of *Gegenwissen* (counter knowledge).[7] Within this spectrum, *Gegenwissenschaft* as we define it here can be seen as the most formalized and systematic part.

This chapter investigates the emergence and development of West German *Gegenwissenschaft* in the 1970s and 1980s as a case of institutionalized contestation of technoscientific evidence practices, with a special focus on those actors working on environmental topics, and in particular nuclear energy technology.[8] While counter science – similar to the New Social Movements, with which it was closely associated – was a thematically extremely broad and heterogeneous field, ranging from radical feminism through "Third World" issues to ecological agriculture, the two above-mentioned issues seem to us of particular relevance because of their prominence in the institutionalization and early development of counter science in the FRG. Besides offering a brief historical sketch of the field, we are mainly interested in two questions: Firstly, what approach did counter scientists take toward technoscientific evidence practices? Which is to say, to what extent did they attempt to challenge not only the prevalent applications and social organization, but also the epistemic-methodological basis of science and technology? Secondly, what role did *Gegenwissenschaft* play in the much-discussed "crisis of confidence" in science and technology in the 1970s and 1980s?[9] At the time, a number of established scientists and contemporary commentators tended to see counter science simply as "anti-science", a danger to the authority of scientific knowledge, or even part of a "rising flood of irrationality".[10] Contrary to this perception, we argue that at least some of the leading protagonists of *Gegenwissenschaft* saw themselves explicitly as a restabilizing force, who did not want to dispose of technoscientific evidence practices but improve them and use them for progressive political ends. Instead of a radical alternative to modern technoscience, *Gegenwissenschaft* can therefore better be understood as an example of Ulrich Beck's observation of science's turn to reflexivity in reaction to the advent of the "risk society". As Beck himself put it, the new contestatory movement stood "not in [...] contradiction to modernity, but is rather an expression of reflexive modernization beyond the outlines of industrial society".[11]

In trying to answer some of these questions, we start with a brief overview of the origins of counter-scientific ideas in the context of the alternative movements of the 1970s, pointing out the particularly strong connection with the anti-nuclear movement in the West German case. Secondly, we survey the institutionalization and structure of the field in the late 1970s and early 1980s. We then turn more specifically to internal discussions of counter scientists' self-conception and their stance on scientific method and epistemology. Lastly, we trace the reactions of the scientific and political establishment, and the process of (re-)integration. In doing so, the chapter draws mainly on the ample published material produced by protagonists and institutions of *Gegenwissenschaft* themselves, along with coverage of and commentary about them in contemporary print media. This empirical basis was supplemented by some archival research in German government archives, as well as a number of oral history interviews with some of the protagonists – some publicly available, some conducted by ourselves.[12]

Technoscience, Counterculture and the Nuclear Evidence Gap

Although *Gegenwissenschaft* was in both form and content in many ways specific to the (West) German context, its emergence was also part of a larger transnational dynamic, which spawned a number of comparable (and often interconnected) movements and initiatives in various countries across Europe and North America. Their common roots lay in the series of political, economic and cultural shifts and fractures confronting Western industrialized societies in the late 1960s and early 1970s, such as the student revolt of the 1960s and the post-materialistic "silent revolution" in societal values,[13] the end of the postwar economic boom,[14] the discovery of the ecological "limits to growth" and the beginning of the modern environmental movement.[15] After a period of unusually strong faith in science and technology, underpinned by the material gains of the "economic miracle" and the spectacular successes of Cold War Big Science, the promise of unlimited technoscientific progress suddenly seemed to lose some of its luster.

In the United States and Great Britain, it was mainly opposition to the Vietnam War and the "military-industrial-academic complex" which triggered the formation of the so-called radical science movement, leading to the foundation of activist groups such as Science for the People or the British Society for Social Responsibility in Science (BSSRS) in the late 1960s.[16] In the FRG, by contrast, an organized intra-scientific opposition was considerably slower to form, particularly in the natural and technical sciences. Although the leftist political ideas, the anti-authoritarian stance and the new contestatory style of the student movement provided important biographical and intellectual reference points for many later counter scientists, the so-called "68ers" did not seem to have much need for specialized technoscientific (counter)experts – perhaps because they tended to focus more on general societal change than specific technical details.[17] This changed over the course of the 1970s, as the counter-cultural scene disintegrated into a kaleidoscope of different alternative milieus and protest campaigns. After the failure of the revolutionary dreams of 1968, the so-called New Social Movements tried to pursue their dreams of an alternative society on a smaller, more practical scale. While sharing a similar habitus, language and political outlook, specific groups and initiatives usually specialized in one particular issue – such as the position of women, environmental degradation, the situation of the "Third World", or the threat of nuclear weapons.[18] Since they meant to offer detailed criticism, concrete suggestions and practical alternatives, acquiring comprehensive and detailed technical knowledge became not only a necessity, but also a constitutive practice and a point of pride for many of these groups. Unavoidably, this had a significant impact on evidence practices. As one protagonist later summarized pointedly: "People no longer discuss 'Das Kapital', but conduct experiments, measure radiation and construct instruments".[19]

Although a number of these topics became central concerns of the various branches of *Gegenwissenschaft*, the most important catalyst in its institutionalization was arguably the opposition to nuclear energy, which grew into one of the largest and most determined alternative movements in the FRG in first half of the 1970s. In the heated public debates of the time, the conflict was often presented as a clash of fundamentally different worldviews, rather than merely an exchange of technical arguments – Robert Jungk's famous warnings of the coming "Nuclear State" being perhaps the most eloquent example.[20] However, although conflicting normative ideas certainly helped to fuel the controversy, given the nature of the issues at stake, scientific expertise and detailed technical knowledge were also an indispensable resource for both sides, if they wanted to be taken seriously in the public discussion.[21]

It was therefore a particular problem for the West German anti-nuclear movement that its knowledge base was initially relatively weak. Early campaigns mostly originated with local, often rural populations, who usually did not have a formal scientific education, or easy access to detailed technical information. Although the movement soon drew in numerous students and a number of qualified scientists, doctors and engineers, most were not specialists on nuclear issues.[22] In contrast to the situation in the United States, where some of the earliest and most important opponents of nuclear energy came from the ranks of nuclear scientists and engineers themselves, very few insiders from within the German nuclear establishment openly voiced any kind of criticism.[23] Figures like Klaus Traube, a former Interatom manager turned critic after having been targeted by the FRG's secret service for his leftist contacts, were exceedingly rare exceptions.[24] While the proponents of nuclear energy could count on the full force of institutionalized science and technology to support their arguments, German anti-nuclear activists either had to rely on the writings of well-known US critics or try to use the little information they could glean from their opponents to their advantage.[25]

This "evidence gap" – to use a favorite metaphor of the time – became most clearly visible in court.[26] According to German law, building and operating nuclear power stations required a state license, for which the prospective operators had to prove that they had taken the "necessary precautions" against potential dangers posed by the installation.[27] Because these licenses could be challenged in court, the courtrooms soon turned into important battlegrounds in the controversy.[28] However, the public hearings with large numbers of expert witnesses also vividly illustrated the highly uneven distribution of expertise. For instance, in the trial concerning the planned reactor at Wyhl in South-West Germany in 1977 – arguably the most spectacular of many similar cases – observers counted 42 experts supporting the position of the nuclear industry. By contrast, the citizens' initiatives had found only nine "critical" scientists to testify on behalf of their cause – already including two persons specially flown in from the United States.[29] Much to the surprise of everyone involved, the Freiburg court ultimately still decided in favor of the anti-nuclear plaintiffs and revoked the license.[30] Nevertheless, the experience of the Wyhl trial convinced a number of participants that the West German movement urgently needed an institutionalized structure to facilitate access to technoscientific expertise and evidence.[31] In the months following the verdict, a small group around the two attorneys representing the citizens' initiatives in the case, Siegfried de Witt and Rainer Beeretz, therefore started to sound out the possibilities of setting up an organization for this purpose, roughly modeled on the US National Resources Defense Council (NRDC). On November 5, 1977, this resulted in the foundation of the Institut für angewandte Ökologie (Institute for Applied Ecology), or Öko-Institut for short.[32]

From Movement to Institute(s): The Institutionalization of *Gegenwissenschaft*

As some of its leading exponents later claimed rather self-confidently, the creation of the Öko-Institut marked the "birth of 'alternative' science" in the FRG.[33] Indeed, although originally intended mainly as a counter expertise brokerage service, aiming to "help citizens get scientific support in court by providing expert opinions and expert witnesses",[34] the institute quickly grew into the largest, most comprehensive and arguably most professional institution in its field – in the words of the environmental historian Jens-Ivo Engels, the "first professional think tank of the environmental movement" in Germany.[35] Uniquely among comparable institutions, the Öko-Institut could draw on a comparatively strong independent financial and organizational base in the form of its members, whose number quickly grew to a couple of thousand.[36] It was led by a ten-member executive board, made up of prominent alternative scientists, environmental activists from the regional and national level and representatives of the institute's employees, while a curatorium featuring an all-star cast of West German and international ecological activists helped to raise public profile.[37] For day-to-day operations, the Öko-Institut employed a small administrative staff and a handful of scientists, supplemented with additional researchers hired on an ad hoc basis or working pro bono. With the addition of more commissioned project work, the number of permanent personnel increased to around 20 by the early 1980s – already enough to raise some eyebrows with activists steeped in the "small is beautiful" ideals of 1970s counterculture.[38] The project focus and relative independence of the scientific staff soon resulted in organizational decentralization. In 1979, branch offices opened in Darmstadt, concentrating mainly on nuclear power and its risks, and in Hanover, focusing on the conflict surrounding the planned nuclear reprocessing facility at nearby Gorleben – the latter breaking off as an independent institution in 1981 as Gruppe Ökologie (Ecology Group, GÖK).[39]

In addition to its role as a broker of counter expertise in court cases, the Öko-Institut quickly managed to make a name for itself with a number of spectacular counter studies, in the absence of costly laboratory equipment often based on a reinterpretation of data gathered by official institutions.[40] Publications such as the so-called *Energie-Wende* (Energy Transition) study of 1980, describing a potential energy path for the FRG based on renewables and reduced consumption instead of nuclear energy, two studies on harmful chemicals in mothers' milk and drinking water in 1981, or a detailed critique of the West German reactor safety study published in 1983 created heated public controversies and therefore press coverage.[41] In addition, a number of popular-science-style publications on topics like household chemicals reached considerable audiences and in some cases even made it onto national best seller lists.[42]

The creation of the Öko-Institut marked the beginning of a whole wave of further foundations of similar grassroots research institutes and organizations. By 1980, there were enough of them to form a dedicated umbrella organization, called Arbeitsgemeinschaft ökologischer Forschungsinstitute (Ecological Research Institutes Working Group, AGÖF).[43] In 1981, it already boasted almost 40 member institutions from West Germany and neighboring German-speaking areas, covering a wide spectrum of topics and approaches.[44] Closest in form and outlook to the Öko-Institut was arguably the Institut für Energie- und

Umweltforschung (Institute for Research on Energy and the Environment, IFEU) Heidelberg, whose origins were also directly related to the 1977 Wyhl trial. When the Tutorium Umweltschutz (Tutorial Environmental Protection), an environmentalist working group at Heidelberg University, had provided the plaintiffs with a study attacking official calculations of radio-ecological dangers posed by the planned reactor, the University tried to shut them down, leading to the reconstitution of the group as an independent institute.[45] By contrast, the formation of the Katalyse Umweltgruppe (Catalysis Environmental Group) in Cologne in 1978 was triggered by an article series on water and air pollution appearing in a local alternative newspaper. With the revenue from its best selling 1982 publication, *Chemie in Lebensmitteln* (Chemicals in Food), the group was able to finance its own laboratory, which subsequently became the center of its activities.[46] Chemical and radiological pollution were equally the main topics of various groups forming in the orbit of the University of Bremen, which thanks to its radical leftist profile had already garnered a reputation as a "center of academic resistance" against nuclear energy in various court cases.[47] In 1979, two Bremen University working groups on nuclear energy and water pollution merged into the Bremer Arbeits- und Umweltschutz-Zentrum (Bremen Centre for Occupational Safety and Environmental Protection, BAUZ).[48] Among its rather controversial cast of members were not only the Bremen physics professors Inge Schmitz-Feuerhake and Jens Scheer (the latter a well-known Maoist and member of the Communist Party), but also the anti-nuclear activist, lay expert and radical right-wing publicist Walter Soyka.[49] Other topics of interest to AGÖF institutes included ecological agriculture, the "Third World", long-term planning and environmental economics, environmentally friendly architecture and small-scale decentralized "soft" technology. The latter two fields were also especially attractive for ecologically oriented engineering firms, which formed a substantial and growing subgroup of their own within AGÖF.[50]

By the turn of the 1980s, a lively, organized and more or less coherent field of *Gegenwissenschaften* had emerged, closely connected to – yet also distinct from – the political environmental movement and the larger countercultural milieu. It had its own umbrella organizations,[51] held regular meetings and conferences at places such as the Evangelische Akademie Loccum, and published an astonishing amount of material. Besides studies, books and pamphlets, various circulars and periodicals, such as the Öko-Institut's *Öko-Mitteilungen,* the AGÖF *Rundbriefe* or the independent alternative quarterly *Wechselwirkung,* provided a forum for internal debates.[52] Thematically, the field had branched out from its origins in anti-nuclear activism into a variety of topics and directions – mostly but not exclusively within the domain of engineering and the natural sciences. It comprised a wide variety of actors: Critical but established voices within academia, formally qualified scientists and engineers working outside of established institutions, as well as a number of political activists, legal professionals and even the occasional self-taught lay expert. At the same time, *Gegenwissenschaft* was deeply divided over a number of fundamental questions – including not only its stance toward established scientific institutions, but also its understanding of what constituted good scientific method, epistemology and evidence practices.

"Counter Science" or "Anti-Science"? *Gegenwissenschaft* and Its Concept of Science

The identity questions of *Gegenwissenschaft* took on a particular salience in the early 1980s, when the basic infrastructure was in place and the field's lofty ambitions and ideals started to clash increasingly visibly with day-to-day reality. In September 1982, AGÖF-researchers set up a "self-conception group", which debated core ideas regarding the premises of ecologically oriented science.[53] In addition, the Öko-Institut devoted a conference in March 1983 to the topic "Ecological Research – Between Reality and Utopia", which was extensively covered and debated on the pages of the institute's journal.[54]

Discussions in both of these contexts dealt notably with the question of how to produce good evidence and for whom. The ideas developed in the unfolding debate contained evidence criteria for ecologically oriented research along two dimensions: the social and the epistemic. The persuasiveness of knowledge in its social dimension depends on factors such as the social trustworthiness of experts, but also the involvement of and relevance to citizens who are affected by expert knowledge. In contrast, the epistemic dimension of evidence rests on its content-related design and its methodological justification.

As counter-cultural movements had been the midwives for the new institutions of counter research, the former's notions of grassroots democracy influenced the latter's ideas about how to do research in order to obtain reliable results. Self-determined agenda setting, non-hierarchical structures within the new institutes and cooperation instead of competition among the researchers were highly valued ideals within AGÖF-circles.[55] Not all of these social evidence criteria, however, proved easily implementable in practice. In spite of its much-professed support for feminist causes, for instance, many *Gegenwissenschaft* institutions tended to reproduce the gender-division of tasks prevalent in science and technology at the time.[56] Another important case in point was the strongly emphasized commitment to doing research that served the citizens' initiatives. The alignment with "those directly affected" (*Betroffene*) by technoscientific developments (e.g. the population living close to a nuclear power plant) was a widely agreed-on and oft-repeated principle.[57] Whether this commitment also entailed the participation of "ordinary citizens" in the research process itself, however, was a contested issue.[58] One faction insisted on the equality of all forms of knowledge and denied any superiority of scientific knowledge over other forms of knowledge such as everyday or experiential knowledge. For them, the involvement of "affected" laypersons in research was a question of justice.[59] There was, however, more support for the view that the best way to serve affected communities was to counsel them and provide expert evidence in court, in hearings or similar contexts.[60]

At the same time, doing research for the citizens (and not with them) in specialized institutes was not uncontroversial either. Critics objected that this would reproduce the traditional science/society divide, depoliticize *Gegenwissenschaft* and betray the counter movements. The physicist and educator Rainer Brämer, for instance, argued that separated institutes would "run the danger of appropriation by the ruling system".[61] He already saw the beginnings of an alliance of counter research institutions with the ruling circles in state, industry and science, initiated by the quest for acceptance and research funds on the part of counter science. This alliance, Brämer warned, would "checkmate the inexpert citizens' initiatives by expert alignment" (*Expertenabgleich*).[62] Instead, Brämer advocated for the political engagement of counter researchers and a relationship of service-oriented subordination to the counter movement. For him, the superior evidence value of *Gegenwissenschaft* rested on its partiality for the environmental movement, whereas the claim of epistemological and social exclusiveness on the part of established science, which derived from ethical norms such as disinterestedness, demonstrated not superiority, but arrogance and hubris.[63]

The debate on what the political commitment of counter research implied went beyond the social dimension of evidence into the epistemic realm. This became obvious in discussions on appropriate forms of counter expertise, which crystalized around the question whether counter research was to improve established science by the latter's own means, or whether counter research would help to develop new epistemic practices and result in a new mode of science. Many actors in the AGÖF's orbit – especially engineering consultancies, but also well-known project groups within the Öko-Institut, such as the reactor-safety group – took pride in working toward the improvement of established science and technology. While they rejected the value-neutrality norm of established science and technology as a denial of responsibility, they were eager not to compromise epistemic standards, but to adhere to them more closely than their opponents. In particular, the protagonists of parallel research devoted themselves to uncovering flaws and to integrating missing perspectives into mainstream research, but did not intend to abandon the established ways of doing science and technology altogether. Instead, these counter researchers were anxious to meet the experts from the established system as peers and on the same footing.[64]

Those calling for a new mode of science, however, hotly contested this position. Peter Pluschke, a chemist and participant in the Öko-Institut's discussion on the reality and ideals of ecological research, accused quantitatively working counter-researchers of an "alternative faith in science", i.e. remaining committed to a simplistic, positivist belief in the authority of scientific methods and results. Pluschke attacked trust in numbers, linear interpretations and the narrow focus on specific numerical values as inadequate. Instead of "counter numbers" and simplistic numerical counter evidence, he called for a "more complex scientific understanding of the world and a more human-oriented (= nature-oriented) practice of research".[65] Similarly, the AGÖF's self-conception group argued in its final statement in 1983 that in particular the researchers in the field of "contesting dead-end technology" (i.e. the part of *Gegenwissenschaft* concerned with criticizing for example nuclear energy) would necessarily lean too much toward established science and thus adopt the latter's faith in science.[66]

The theologian and biologist Günter Altner, one of the founding fathers and a leading figure of the Öko-Institute in various positions, self-confidently called his vision of a new mode of science *Anti-Wissenschaft* (anti-science) – a term defenders of the status quo liked to employ to dismiss critics as irrational and retrograde. For Altner, "anti-science in the sense of ecologically oriented research always implies the question of justice and of social and ecological compatibility of the knowledge increase and the changes it introduces".[67] Thus, defining evidence with respect to its societally responsible use was one characteristic of Altner's vision. As a second criterion, he considered an alternative mode of cognition. According to him, anti-science differed fundamentally from established science in its relationship to nature, since it did not treat nature as an object to be exploited, but as a partner to live with. Altner remained vague about how such partnership changed the mode of cognition, but he felt compelled to add: "Whoever turns away from objectifying science does not fall into the realm of irrationality, but needs to try new patterns of dialogue and new rules".[68] He insisted on the need for theoretical reflections about how science was to approach nature in an appropriate way. This would allow gaining a "nature-appropriate knowledge of nature" (*naturangemessene Naturerkenntnis*).[69]

How exactly this could be achieved was a core topic in the circles of ecologically oriented counter research in general. Many of its actors criticized established science for analytical reductionism, narrow specialization and a controlling attitude toward nature. Instead of such inappropriate ways of examining "external nature", counter researchers shared a commitment to investigate nature in view of its complexity and to develop adequate approaches, such as analysis of open systems, contextualization and symbiotic thinking.[70] A further source of controversy was the question whether it was possible to take up these challenges within the institutions of established science. Well-known activists and theorists of counter research such as Hartmut Bossel,[71] Hariolf Grupp,[72] Gernot Böhme,[73] Otto Ullrich[74] and others saw themselves as critical scientists who, nevertheless, trusted in a re-stabilization of the critiqued system as source of reliable evidence in the future. Ullrich explicitly emphasized that critical scientists opposed the mainstream in their discipline, but not as dropouts, as they cherished their acquired knowledge and skills.[75] Many, but not all of these critical scientists had successful careers in this very system.

However, Altner's vision of anti-science as an alternative mode of cognition also left room for a different interpretation. It could be read not only as a call for *Gegenwissenschaft* as a more complete and thus "more scientific" form of established science, but also as an invitation to leave the boundaries of scientific method altogether in favor of spiritual, esoteric and similar kinds of knowledge generation. This became evident in the contribution of Arnim Bechmann to the Öko-Institut's discussion in early 1983. Bechmann, who was at that time spokesperson of the institute's board, held a chair at TU Berlin in environmental planning, a field that he developed further in his academic career as professor and managing director of the institute of landscape economics. In 1983, together with Bärbel Kraft, who would later defend a doctoral thesis on *Ecological and Anthroposophical Descriptions of Nature*, Bechmann criticized ecology as a "legitimation science" and called for the inclusion not only of Steiner's anthroposophy, but also the para-scientific theories of Wilhelm Reich.[76] He and Kraft faulted "academic ecology" for failing to provide explanations for phenomena such as "the direction of evolution, the developmental dynamic or the behavior of complex systems, the rhythms of natural processes, the practical successes of organic dynamic farming, the functioning of divining rods etc".[77] Although Bechmann had a successful career in academia and supported the environmental movement in very productive ways, for example with his engagement for environmental impact assessment, he remained actively engaged in the esoteric scene. In 1986, he founded the Institut für ökologische Zukunftsperspektiven (Institute for Ecological Perspectives of the Future) in Barsinghausen, which offered seminars and lectures on "post-materialistic natural science", including Reich's so-called Orgone theory or the ethereal aspects of plant breeding.[78]

The vast majority of the *Gegenwissenschaft* community, however, rejected the turn to spiritualism and esotericism – although not necessarily all for the same reasons. Rainer Brämer denounced it as "alternative science on the path to the myths", with which the researchers in his opinion were shirking their political responsibility.[79] Others rejected both political instrumentalization of science and the call for an alternative epistemology. In a 1983 article, Hansjörg Hemminger, a behavioral biologist and psychologist, who later became one of the country's leading experts on pseudoscientific movements, sharply rebuffed any thoughts about alternative modes of cognition such as anthroposophy. At the same time, he also warned his fellow Öko-Institut members not to confuse the epistemic authority of science with political authority, as this would result in scientism and render politics impossible. Instead, he proposed developing new approaches within established science and demanded an alternative research policy with new priorities with regard to subjects, sites and applications of research.[80]

Thus, the new institutes and actors of counter science did not develop a common denominator for their contestation of established science. These internal disagreements were not simply academic in nature: How *Gegenwissenschaft* defined itself was not only decisive for the further development of the field, but also for its position within the larger scientific and societal framework in the FRG.

From Counter Science to Parallel Science: The (Re-)Integration of *Gegenwissenschaft*

From the point of view of the so-called established West German scientific community, the emergence of *Gegenwissenschaft* – whatever its exact ideas on scientific method and evidence practices – inevitably constituted a direct challenge. Unsurprisingly, many scientists and experts deeply resented the (at the very least implicit) accusation of being uncritical stooges of political and economic interests. Their responses were therefore immediate, sharp and not rarely veered into outright personal attacks, as many counter scientists complained.[81]

The first and arguably fiercest reactions came from those institutions which had been the direct targets of *Gegenwissenschaft*, often energetically supported (or even orchestrated) by industry interest groups and pro-nuclear politicians. When the Öko-Institut published its energy transition study in 1980, for instance, the state-financed Kernforschungsanstalt Jülich (Nuclear Research Institute, KFA) immediately declared it "unfounded in facts, erroneous in analysis, and illusionary in intent", leading to a prolonged and public exchange of arguments and invectives.[82] In a hurriedly prepared counter-counter study, the Jülich scientists criticized the supposedly selective and manipulative use of numbers – an accusation that the ecologists themselves had regularly aimed at their opponents.[83] While the KFA tried to quibble with the numerical counter evidence, others resorted to more direct tactics. In 1981, a group of municipal water works – having called the Öko-Institut's study on drinking water a "criminal creation of mass psychoses by way of pseudo-scientific statements" in their customer magazines – had to save themselves from legal trouble by claiming (not very plausibly) that their statement had only ever referred to press coverage of the study.[84]

By the early 1980s, *Gegenwissenschaft* had become irritating enough for some of the leading exponents of the West German scientific community to take a fundamental stance. During the yearly meeting of the influential Max-Planck-Gesellschaft (MPG) in May 1982, MPG-President Reimar Lüst indignantly rejected the idea of including "so-called 'critical' scientists" in the public debate, which in his opinion would only serve to needlessly undermine the authority of "truly competent scientists".[85] Another heavyweight of official German science, Heinz Maier-Leibnitz, attempted to solve the problem of expert disagreement once and for all with a formalized trustworthiness test. After the encounter in the Enquête-Kommission mentioned at the opening of this chapter, he published an article proposing a method to rate the credibility of experts based on a randomly selected sample of their statements, without having to engage with all of their arguments in detail. Probably not entirely coincidentally, his test case found Richard Donderer, his sneaker-wearing opponent at the Commission hearing, so severely lacking in credibility that the finer points of his statements could be safely ignored.[86]

Nevertheless, even in the hard sciences, positivist ideas of scientific objectivity and the selfless pursuit of truth were no longer uncontroversial in the early 1980s. At the same 1982 MPG meeting at which Lüst had proudly proclaimed the "value neutrality" of science, none other than the federal Minister for Research and Technology, Andreas von Bülow, directly contradicted him. According to von Bülow, given the "unconcious value system" necessarily underlying all scientific activities, value neutrality was an illusion. Therefore, to him, *Gegenwissenschaft* was simply a part of German science – and a useful counterweight to those established professors he deemed "too comfortable" to reconsider their own self-conception.[87] Ensuing discussions on the topic on the pages of *Bild der Wissenschaft*, the FRG's leading popular science magazine, revealed some support for this opinion, particularly among younger scientists. The physics professors Harald Fritzsch and Siegfried Penselin, for instance, could see no reason to differentiate fundamentally between "critical" and "established" scientists, while others, such as the biologists Berndt Heydemann and Helmut Zwölfer, even expressed open admiration for "those courageous institutes" picking a fight with the establishment.[88] In a field like ecology, which the German academic system had been very slow to pick up, the environmental economist Udo Ernst Simonis claimed, institutions such as the Öko-Institut were even ahead of universities and other conventional academic institutions in terms of scientific importance.[89]

Perhaps even more important for the development of *Gegenwissenschaft* than the gradual and often grudging acceptance by German academia was the integration into the political expertise system. Despite the often confrontational rhetoric on both sides, this process had already begun as early as the late 1970s. It was based on a strategic convergence of interests on both sides: On the one hand, groups like Öko-Institut or IFEU had an obvious interest not only in making their position heard with those who ultimately made the decisions, but also in the financial and symbolic resources to be gained by cooperation. Improving the chance for state commissions by creating a central point of contact had in fact already been one of the main motives behind the founding of AGÖF.[90] By 1982, the AGÖF was openly embracing its restabilizing function as an argument for more public funding. Since the public "no longer trusted the industry's tailor-made research results", they reasoned, the political feasibility of any large-scale technical projects would henceforth depend on "appropriate counter studies", which only a strong, independent and therefore ideally state-funded *Gegenwissenschaft* could supply.[91]

On the other hand, given the scale of the public controversy around nuclear energy, the administrative and political system was looking for ways to calm the waves of the heated public debates by offering some form of inclusion and participation. This was especially true for members of the ruling Social Democrat Party (SPD), itself deeply split on the nuclear question.[92] A number of prominent SPD politicians, such as Volker Hauff or Rainer Ueberhorst, therefore actively tried to support the "scientific wing" of the movement, with which they thought "a controversial discussion on a high level of expertise" would be possible.[93] As Federal Minister for Research and Technology from 1978 to 1980, Hauff used his influence to secure the Öko-Institut its first substantial state commission, largely against the resistance of his own subordinates. From 1980 to 1983, a group of renegade experts and engineers led by Lothar Hahn and Michael Sailer received 1.6 million Deutsche Mark for a thoroughgoing critique of the *Deutsche Risikostudie Kernkraftwerke* (DRS), which had itself been one of the Ministry's flagship projects.[94] For the Öko-Institut, this was not only a significant financial boost, enabling it to set up a second office in Darmstadt, but also an important step in establishing itself as a bona fide expert institution.[95]

For his part, Ueberhorst, the chairman of the Enquête-Kommission zur zukünftigen Kernenergiepolitik from 1979 to 1980, was instrumental in the addition of a number of "nuclear skeptics" to this high-profile advisory body, among them the Öko-Institut's Günter Altner.[96] In its concluding report, the commission not only referenced a possible scenario for a future without nuclear energy directly based on the Institute's "Energy Transition" study but also recommended including skeptics in crucial advisory bodies such as the Reactor Safety Commission (Reaktorsicherheitskommission, RSK) and the Radiation Safety Commission (Strahlenschutzkommission, SSK).[97] Furthermore, the Enquête called for new studies on the risks associated with the German breeder reactor project to be done as so-called *Parallelforschung* (parallel research): Two groups of scientists – one "established", one "skeptical" – were to work parallel to each other.[98] In practice, the experiment did not end well: The two teams – one under the aegis of the semi-governmental Society for Reactor Safety (*Gesellschaft für Reaktorsicherheit,* GRS), one a specially formed selection of various "critical" scientists, led by the physicist Jochen Benecke – never managed to build mutual trust and, instead of working together, ended up publicly trading recriminations.[99]

Politically, public commissions for exponents of *Gegenwissenschaft* remained a sensitive issue on both sides. Critics within AGÖF worried not only that member institutions were about to sell out their independence and become susceptible to pressure from the state. By increasing professionalization and the necessity to work "with the weapons of the enemy", they feared, the hunt for public funds would lead to a new "counter-expertocracy", a "new discrimination of everyday rationality" and a "loss of democratic impetus".[100] "Will this not serve to refurbish a conception of science, whose dismantlement, historically long overdue, had just been achieved?" Rainer Brämer asked in *Wechselwirkung.*[101] On the opposite side, objections were expressed no less emphatically. Following the Öko-Institut's first state commission, conservative members of the German parliament positively flooded the government with sharply formulated official inquiries concerning the finances of the Institute.[102] On national television, a CDU-politician declared that it was "irresponsible in the times of empty public coffers that opponents of nuclear energy, for instance, receive money from taxpayers for scientifically untenable studies and agitatory slogans".[103]

Nevertheless, even the takeover by the Conservatives at the federal level in 1982/1983 spelled only a partial and provisional end for the collaboration between political system and alternative institutes. Although federal funds dried up for a while, many regional governments continued to work with the AGÖF institutes. In addition, the rise of the Green party – which was elected to the national parliament in 1983 and for the first time joined a regional government in Hessen in 1985 – also created considerable demand for alternative expertise.[104] This demand increased exponentially shortly afterwards, when the fall-out of the nuclear catastrophe at Chernobyl severely damaged not only the credibility of West German political authorities, but also of official scientific experts and institutions. Geiger counters were sold out for months, as many citizens tried to determine for themselves, if their milk was fit to drink and their vegetables safe to eat.[105] Old-school anti-nuclear activists sometimes deprecatingly referred to the new citizens' initiatives arising out of this as the "Becquerel movement", thinking them more interested in decontextualized measuring of radiation than in principled resistance.[106]

However, the established institutions of *Gegenwissenschaft* likewise received a substantial boost to their credibility. As soon as he heard of the catastrophe, Hessen's new Green minister of the environment, Joschka Fischer, set up a "standing connection to the *Öko-Institut* in

Darmstadt", as he was later to claim.[107] Even the staunchly conservative federal Ministry of the Economy now felt the need to commission a study on the feasibility of a nuclear phase-out from the Freiburg ecologists. As the irate letters to the Minister in the corresponding archival file show, this was still a highly controversial proposition. Paying any money to the Institute, "Irrespective of the content of the 'study'", the highly decorated biology professor Hans Mohr fumed, would "make a mockery of science and a travesty of the principle of scientific expertise".[108] In his answer, Minister Martin Bangemann emphasized once more the function of *Gegenwissenschaft* in the political process. The Ministry, he wrote, had wanted to "demonstrate the attempt to hear as wide as possible a spectrum of opinions in this complex and controversial issue", because that was what the German public expected. Of course, this did not necessarily mean that the study would have any influence on his policy decisions, he quickly added.[109]

Conclusion: Re-Stabilization by Contestation?

When Ludwig Trepl, himself a veteran of various AGÖF institutions, tried to take stock of the state of German *Gegenwissenschaft* in the late 1980s, he found its development not devoid of a certain irony. In his view, in its attempts to harness scientific knowledge, the anti-nuclear and environmental movement had ended up creating a "paradoxical situation": "A movement that is genuinely critical of science and in particular opposed to the natural sciences [...] not only has a (relatively speaking) flourishing scientific branch, but this counter science is mostly of just the type that it always criticized: natural science".[110] For Trepl, this was a deeply problematic development: If *Gegenwissenschaft* gave up on its ambition to formulate a fundamental critique of traditional scientific epistemology, he asked, what really was the difference to the kind of science practiced at conventional academic institutions – other than inferior financial resources, infrastructure and personnel? Did counter scientists' rather narrow orientation toward immediate applicability and practical political gains not fall prey to the same misguided logic of short-term efficiency and end-justifying-the-means thinking they had set out to contest? Was *Gegenwissenschaft,* as another analysis appearing the same year asked pointedly, even still trying to be "an emancipated alternative to science", or had it devolved into a mere "repair shop of industrial society"?[111]

The further development of the field seemed to confirm some of these apprehensions. Over the course of the 1990s, institutions such as Öko-Institut or IFEU not only became fully recognized and indispensable members of the official expert circles advising administration and legislation on environmental questions but also increasingly extended a hand toward the private economy. With the emergence of corporate responsibility programs and the rise of eco-consultancy as a market, counter scientists started to consider themselves less as political activists than as purveyors of environmental expertise as a specific commodity.[112] Against this background, even a former nemesis like Hoechst AG, one of the country's biggest chemical corporations, could eventually become a client.[113] On the one hand, this inevitably led to accusations of "selling out" and an increasing detachment from the activist base. When Öko-Institut's director, Michael Sailer, publicly criticized activists opposing nuclear waste transports in 1996, Günter Altner and several other founders resigned in protest, nearly tearing the institute apart.[114] On the other hand, integration also offered increased influence and new opportunities to introduce environmental perspectives. Particularly after the Green party's ascent to federal political power in the first red-green coalition government in 1998, some of the leading exponents of *Gegenwissenschaft* ended up in charge of the established expert organizations they had started out opposing so vehemently.[115]

Measured against some of its earliest, most radical ambitions, German *Gegenwissenschaft* – or at least its most institutionalized part – can indeed be said to have fallen somewhat short in its contestation of technoscientific knowledge production. Epistemically, in spite of many heated internal debates, those openly advocating abandonment of orthodox scientific methods always remained a small minority. In the social dimension, while *Gegenwissenschaft* did in fact open up alternative career paths for some experts with only relatively basic formal academic qualifications, few counter scientists systematically attempted to involve laypersons and ordinary citizens in the production of knowledge.[116] Apart from the fact that methodological experiments would have been directly counterproductive to the goal of producing evidence that could stand in court, in expert committee hearings, or in public discussions in the media, most counter scientists – even if they vigorously opposed the actual practical conduct of mainstream research and development – firmly believed in the basic soundness of scientific methods. By the early 1980s, at least some of them were not only fully aware of the possibility that their contestation might contribute to the restabilization of the authority of science and technology in the FRG but were actively leveraging this position for state support. Ultimately, this pragmatic (if sometimes contradictory) attitude toward "the system" was an essential factor not only for the institutional success, but also the political efficiency of *Gegenwissenschaft*.[117]

In addition to its political importance as an ally of the environmental and anti-nuclear movement, *Gegenwissenschaft* also had a substantial impact on the history of science and technology in the FRG, which has so far been largely overlooked in the literature. This is true in at least two different respects: Firstly, even though much of the counter scientists' work was oriented toward relatively limited practical goals, their role in the production of technoscientific knowledge should not be underestimated. By engaging in public controversies, by challenging research findings, by proposing new approaches and producing empirical evidence, they contributed to the development of disciplines such as the environmental sciences, risk research or even nuclear engineering. Historians of science and technology, used to focusing on state-funded universities and Big Science institutions, would therefore do well not to forget these alternative spaces of knowledge production.

Secondly, *Gegenwissenschaft* can also rightfully claim some important contributions to the general changes in the understanding of science and its role in society in the 1970s and 1980s. Counter scientists' efforts to achieve the representation of different political viewpoints in scientific expert commissions not only laid the foundations for more integrative practices in German science and technology policy; their insistence on the political and value-based dimension of scientific evidence practices also contributed substantially to the public dismantlement of ideas of scientific objectivity and value-neutrality. In this regard, counter science in its various expressions can be seen as a sort of practical political counterpart to the more theoretically interested contemporary critiques written by scholars from the field of science and technology studies, which was emerging around the same time.[118] How exactly both interacted in practice and what influence *Gegenwissenschaft* had in the larger history of reflexive thought on science and technology still remains to be more fully investigated.

Notes

1 Günter Altner, "Lehren für eine Forschungs- und Technologiepolitik der Zukunft: Die Erfahrungen eines Kernenergiekritikers in der Enquête-Kommission", in *Der Kalkar-Report: Der Schnelle Brüter: Unwägbares Risiko mit militärischen Gefahren?* eds. Roland Kollert, Richard Donderer, and Bernd Franke (Frankfurt am Main: Fischer, 1983): 19–20. Unless otherwise indicated all translations by the authors.

2 Altner, "Lehren", 19–20.

3 The closest English-language equivalent would probably be "radical science", which however carries its own, rather different historical connotations (see below). We therefore will use the German term *Gegenwissenschaft* (as well as occasionally the literal translation) throughout this chapter. Compared to alternatives such as *kritische Wissenschaft*, which many "mainstream" scientists found highly offensive because it implied that they were "uncritical", *Gegenwissenschaft* was used as a relatively value-neutral designation by contemporaries. At the same time, it was more comprehensive than for instance *ökologische Wissenschaft* (ecological science), which described only one (if arguably the dominant) branch within this formation.

4 See in particular Dieter Rucht, "Gegenöffentlichkeit und Gegenexperten: Zur Institutionalisierung des Widerspruchs in Politik und Recht", *Zeitschrift für Rechtssoziologie* 9, no. 2 (1988): 293–294; Margit Kautenburger, *Entstehung, aktuelle Lage und Wissenschaftsverständnis*

ökologisch orientierter Wissenschaft in Deutschland am Beispiel der Arbeitsgemeinschaft ökologischer Forschungsinstitute (Dissertation, Universität Hannover, 1989), 99–110; Ansgar Klein, ed., Special Issue "Gegenexperten in der Risikogesellschaft", *Forschungsjournal Neue Soziale Bewegungen* 3, no. 1 (1990). Specifically on the *Öko-Institut*, arguably the most prominent institution in the field, see Jochen Roose, *Made by Öko-Institut: Wissenschaft in einer bewegten Umwelt* (Freiburg: Öko-Institut, 2002); Jens-Ivo Engels, "'Inkorporierung' und 'Normalisierung' einer Protestbewegung am Beispiel der westdeutschen Umweltproteste in den 1980er Jahren", *Mitteilungsblatt des Instituts für soziale Bewegungen* 40 (2008): esp. 94–97.

5 See in particular Sven Reichardt, *Authentizität und Gemeinschaft: Linksalternatives Leben in den siebziger und frühen achtziger Jahren* (Berlin: Suhrkamp, 2014); Sven Reichardt and Detlef Siegfried, eds., *Das alternative Milieu: Antibürgerlicher Lebensstil und linke Politik in der Bundesrepublik Deutschland und Europa, 1968–1983* (Göttingen: Wallstein Verlag, 2010).

6 On the environmental movement in the FRG, see Jens Ivo Engels, *Naturpolitik in der Bundesrepublik: Ideenwelt und politische Verhaltensstile in Naturschutz und Umweltbewegung, 1950–1980* (Paderborn: Schöningh, 2006); Frank Uekötter, *The Greenest Nation? A New History of German Environmentalism* (Cambridge: MIT Press, 2014); on the West German anti-nuclear movement Dolores L. Augustine, *Taking on Technocracy: Nuclear Power in Germany, 1945 to the Present* (New York: Berghahn Books, 2018); Stephen Milder, *Greening Democracy: The Anti-Nuclear Movement and Political Environmentalism in West Germany and Beyond, 1968–1983* (Cambridge: Cambridge Univ. Press, 2017).

7 See Max Stadler, Nils Güttler and Niki Rhyner, eds., *Gegen|Wissen*, cache 01 (Zürich: intercom, 2020); also Nils Güttler, Margarete Pratschke and Max Stadler, eds., *Wissen, ca. 1980* (Zürich: diaphanes, 2016); Alexander von Schwerin, ed., Special Issue "Gegenwissen: Wissensformen an der Schnittstelle von Universität und Gesellschaft", *NTM* 30, no. 4 (2022).

8 For most of the historical actors concerned, this meant first and foremost ecological issues in West Germany itself. For the slightly different evidence practices of activists concerned with ecological issues in the Global South, see Sarah Ehlers, this volume.

9 The exact extent and nature of this crisis is still an object of controversy. For an attempt to map it empirically in the US context, see Gordon Gauchat, "Politicization of Science in the Public Sphere. A Study of Public Trust in the United States, 1974–2010", *American Sociological Review* 77, no. 2 (2012): 167–187; for the FRG Andie Rothenhäusler, "Die Debatte um die Technikfeindlichkeit in der BRD in den 1980er Jahren", *Technikgeschichte* 80, no. 4 (2013): 273–294.

10 For a contemporary critique of this perception, see Helga Nowotny, "Science and Its Critics: Reflections on Anti-Science", in *Counter-Movements in the Sciences: The Sociology of the Alternatives to Big Science*, eds. Helga Nowotny and Hilary Rose (Dordrecht: Reidel, 1979): 1–26.

11 Ulrich Beck, *Risk Society. Towards a New Modernity* (London: SAGE, 1992): 11. However, Beck does not seem to have been especially interested in the existing institutionalized form of the phenomenon he describes here. This extends to the term *Gegenwissenschaft*, which he only uses once in *Risk Society* – curiously in an inversion of the common meaning–to designate industry-funded attempts to disprove ecological criticism (Beck, *Risk Society*, 42).

12 We are particularly grateful to Michael Sailer, who took the time for an extensive interview, which (although quoted sparsely in this text) has greatly contributed to our general understanding of the developments discussed here.

13 Ronald Inglehart, *The Silent Revolution: Changing Values and Political Styles Among Western Publics* (Princeton: Princeton Univ. Press, 1977); see Bernhard Dietz, Christoph Neumaier, and Andreas Rödder, eds., *Gab es den Wertewandel? Neue Forschungen zum gesellschaftlich-kulturellen Wandel seit den 1960er Jahren* (München: Oldenbourg, 2014).

14 On the end of the "economic miracle" as a turning point in West German history, see the literature inspired by Anselm Doering-Manteuffel and Lutz Raphael, *Nach dem Boom: Perspektiven auf die Zeitgeschichte seit 1970* (Göttingen: Vandenhoeck & Ruprecht, 2008).

15 Donella H. Meadows, Dennis Meadows, and Jorgen Randers, *The Limits to Growth: A Report for the Club of Rome's Project on the Predicament of Mankind* (New York: Universe Books, 1972); on the "ecological revolution" of the 1970s in West Germany, see Joachim Radkau, *The Age of Ecology. A Global History* (Cambridge: Polity Press, 2014): 79–113.

16 On the US, see Sarah Bridger, "Anti-Militarism and the Critique of Professional Neutrality in the Origins of Science for the People", *Science as Culture* 25, no. 3 (2016): 373–378; Kelly Moore, *Disrupting Science: Social Movements, American Scientists, and the Politics of the Military, 1945–1975* (Princeton: Princeton Univ. Press, 2008); Sigrid Schmalzer, Daniel S. Chard, and Alyssa Botelho, eds., *Science for the People: Documents from America's Movement of Radical Scientists* (Amherst, Boston: Univ. of Massachusetts Press, 2018). For Great Britain Alice Bell, "The Scientific Revolution That Wasn't: The British Society for Social Responsibility in Science", *Radical History Review* 2017, no. 127 (2017): 149–172; for an illuminating insider's account Gary Werskey, "The Marxist Critique of Capitalist Science: A History in Three Movements?" *Science as Culture* 16, no. 4 (2007): 397–461. Interestingly, France seems to offer many similarities to the West German case; see Sezin Topçu, "Confronting Nuclear Risks: Counter-Expertise as Politics Within the French Nuclear Energy Debate", *Nature and Culture* 3, no. 2 (2008): 225–245.

17 See Rucht, "Gegenöffentlichkeit", 293–294; Kautenburger, *Entstehung*, 99–110.

18 Roland Roth and Dieter Rucht, eds., *Neue soziale Bewegungen in der Bundesrepublik Deutschland* (Bonn: Bundeszentrale für Politische Bildung, 1991); Ansgar Klein, Hans-Josef Legrand, and Thomas Leif, eds., *Neue soziale Bewegungen: Impulse, Bilanzen und Perspektiven* (Wiesbaden: VS Verlag für Sozialwissenschaften, 1999); Reichardt and Siegfried, *Das alternative Milieu*.

19 Ludwig Trepl, "Was ist alternativ an der alternativen Forschung?" *Wechselwirkung* 40, no. 11 (1989): 17.

20 Robert Jungk, *Der Atom-Staat: Vom Fortschritt in die Unmenschlichkeit* (München: Kindler, 1977).

21 Perhaps the best overview over the use of "technical" arguments in the debate can still be found in Joachim Radkau, "Die Kernkraft-Kontroverse im Spiegel der Literatur", in *Das Ende des Atomzeitalters? Eine sachlich-kritische Dokumentation*, ed. Armin Hermann (München: Moos, 1987): 307–334.

22 On the social structure of the German anti-nuclear movement Milder, *Greening Democracy*.

23 Augustine, *Taking on Technocracy*, Ch. 3; Joachim Radkau, *Aufstieg und Krise der deutschen Atomwirtschaft 1945–1975* (Reinbek bei Hamburg: Rowohlt, 1983), 438–455. On the comparison to the US, see also Christian

Joppke, *Mobilizing Against Nuclear Energy: A Comparison of Germany and the United States* (Berkeley: Univ. of California Press, 1993); Michael L. Hughes, "Civil Disobedience in Transnational Perspective: American and West German Anti-Nuclear-Power Protesters, 1975–1982", *Historical Social Research* 39, no. 1 (2014): 236–253.

24 See Christopher Kirchberg and Marcel Schmeer, "The 'Traube Affair': Transparency as a Legitimation and Action Strategy between Security, Surveillance and Privacy", in *Contested Transparencies, Social Movements and the Public Sphere: Multi-Disciplinary Perspectives*, eds. Stefan Berger and Dimitrij Owetschkin (Cham: Palgrave, 2019).

25 Thus, pirated copies of a doctoral defense by the TÜV engineer Karl-Heinz Lindackers, in which he had calculated a maximum number of 1.67 million possible victims for a reactor accident in the FRG, became one of the most circulated documents in anti-nuclear circles in the early 1970s; Paul Laufs, *Reaktorsicherheit für Leistungskernkraftwerke: Die Entwicklung im politischen und technischen Umfeld der Bundesrepublik Deutschland* (Berlin: Springer Vieweg, 2013), 94.

26 See Martin H. Geyer, "'Gaps' and the (Re-)Invention of the Future Social and Demographic Policy in Germany During the 1970s and 1980s", *Social Science History* 39, no. 1 (2015): 39–61.

27 Gesetz über die friedliche Verwendung der Atomenergie (Atomgesetz), December 23, 1959, § 7 (2).

28 For an overview, see Hartmut Albers, *Gerichtsentscheidungen zu Kernkraftwerken* (Villingen-Schwenningen: Neckar-Verlag, 1980).

29 "Geballte Ladung", *Der Spiegel,* February 17, 1977; Hanno Kühnert, "Läßt der Tiger zu viele Haare?" *Die Zeit,* February 11, 1977.

30 The 1977 decision was ultimately overturned on appeal in 1982. Nevertheless, the almost 15 years of delay caused by the legal battles ultimately led to the abandonment of the project by the nuclear industry. For a reconstruction of the timeline, see Laufs, *Reaktorsicherheit,* 118–121.

31 Hanno Kühnert, "Werkstatt der besseren Argumente. Das Öko-Institut – Ein Versuch gegen die etablierte Übermacht", *Die Zeit,* September 28, 1979.

32 Christian Hey and Rainer Grießhammer, "10 Jahre Öko-Institut: Zwischen Anspruch und Wirklichkeit", *Öko-Mitteilungen* 10, no. 3 (1987): 4–10.

33 Hey and Grießhammer, "10 Jahre Öko-Institut", 7.

34 Foundational declaration of the Öko-Institut, reprinted in Roose, *Made by Öko-Institut*, 16–17.

35 Engels, *Naturpolitik*, 335.

36 By 1982, the original number of 40 members had grown to around 4.000; Arnim Bechmann and Leo Pröstler, "Editorial: Fünf Jahre Öko-Institut", *Öko-Mitteilungen* 5, no. 5 (1982): 5–6.

37 Amongst the early members of the board were for instance, the physicist Hans-Georg Otto, the engineer and systems analyst Hartmut Bossel, the environmental economist Arnim Bechmann, the Professor of Lutheran theology Günter Altner, the "environmental Reverend" Werner Beck, or the head of the Association of Citizens' Initatives (BBU) Hans-Günther Schuhmacher. Prominent members of the curatorium included the writer Carl Amery, the publicist and futurologist Robert Jungk, the zoologist and TV presenter Bernhard Grzimek, the "Limits to Growth" authors Dennis and Donnella Meadows or the politicians Herbert Gruhl (CDU) and Ehrhard Eppler (SPD).

38 Roose, *Made by Öko-Institut,* 33–35; Elisabeth Werner and Bernd Speiser, "Alternative Wissenschaft zwischen Anspruch und Wirklichkeit: Das Öko-Institut", *Wechselwirkung* 6, no. 21 (1984): 16–19.

39 Kautenburger, *Entstehung*, 17–18; Roose, *Made by Öko-Institut*, 28–30.

40 For an overview over early projects, see Peter von Gizycki, "Projektliste im Sinne eines Arbeitsberichts der vergangenen fünf Jahre", *Öko-Mitteilungen* 5, no. 5 (1982): 7–12.

41 Florentin Krause, Hartmut Bossel, and Karl F. Müller-Reißmann, *Energie-Wende*: *Wachstum und Wohlstand ohne Erdöl und Uran* (Frankfurt am Main: Fischer, 1980); Uwe Lahl and Barbara Zeschmar, *Wie krank ist unser Wasser? Die Gefährdung des Trinkwassers* (Freiburg: Öko-Institut, 1981); Elke Pröstler, *Stillen trotz verseuchter Umwelt: Die chemische Belastung der Muttermilch* (Freiburg: Öko-Institut, 1981); Öko-Institut Freiburg, *Risikountersuchungen zu Leichtwasserreaktoren*: *Analytische Weiterentwicklung zur 'Deutschen Risikostudie Kernkraftwerke'*, 3 vols. (Freiburg: Öko-Institut, 1983).

42 Rainer Grießhammer, ed., *Chemie im Haushalt* (Reinbek bei Hamburg: Rowohlt, 1984).

43 See Hartmut Bossel and Wolfhart Dürrschmidt, eds., *Ökologische Forschung*: *Wege zur verantworteten Wissenschaft* (Karlsruhe: Müller, 1981). The acronym was a deliberate pun on the activists' arch-nemesis, the association of state-funded big science institutes AGF (*Arbeitsgemeinschaft der Großforschungsinstitute*).

44 A list of participating institutions can be found in Bossel and Dürrschmidt, *Ökologische Forschung*.

45 "Nackt Am Zaun", *Der Spiegel*, August 7, 1978; Kautenburger, *Entstehung*, 17. See also Ulrich Höpfner, Mario Manchini, eds. *20 Jahre ifeu-Institut: Engagement für die Umwelt zwischen Wissenschaft und Politik* (Braunschweig: Viehweg, 1998).

46 Katalyse-Umweltgruppe, ed., *Chemie in Lebensmitteln* (Frankfurt am Main: Zweitausendeins, 1982); Detlef Kutz, "Im Reagenzglas: Die Bundesrepublik", *taz*, October 14, 1988.

47 In the words of Kühnert, "Tiger", 10. See Birte Gräfing, *Tradition Reform*: *Die Universität Bremen 1971–2001* (Bremen: Donat, 2012).

48 "Projekte", *Wechselwirkung* 1, no. 2 (1979): 56–59. The organizational infrastructure also included a *Verein für Arbeits- und Umweltschutz* (VAU) as the institutional parent of the BAUZ.

49 On Scheer see Lilo Weinsheimer, "Held oder Spinner? Professor Scheer gilt als wilder Kommunist", *Die Zeit*, February 27, 1976. This extreme political range was not completely atypical for the West German environmental movement of the time, see Silke Mende, *"Nicht rechts, nicht links, sondern vorn"*: *Eine Geschichte der Gründungsgrünen* (München: Oldenbourg, 2011).

50 Examples of AGÖF member institutes dedicated to the above-mentioned topics include the GÖK Hanover or the Forschungsinstitut für biologischen Landbau (Research Institute for Biological Agriculture), Bernhardsberg (Switzerland) for ecological agriculture; the Gesellschaft für angepasste Technologie in ländlichen Entwicklungsgebieten OEKOTOP (Society for Adapted Technology in Rural Development Areas) or the Netzwerk Entwicklungsfördernder Werkgruppen (Network of Development Working Groups, NEW), both based in Berlin, for "Third World"; the IÖW Berlin and the SYNOPSIS Institut de Recherche Alternative, Lodève (France) for planning and economics; the Österreichisches Institut für Baubiologie (Austrian Institute for Building Biology, IBO), Vienna, for architecture; and the Arbeitsgruppe Angepasste Technologie (Working Group Adapted Technology), Kassel, for "soft" technology. See also

Kautenburger, *Entstehung*, 19. By the late 1990s, this eventually led to the transformation of AGÖF into a sort of trade association for ecologically oriented construction companies.

51 In addition to AGÖF, the less formalized "science shop" movement had its own umbrella organization, the *Arbeitsgemeinschaft Wissenschaftsläden* (AWILA). See "Ist die Wissenschaft noch zu retten? Wissenschaftsläden in der Bundesrepublik", *Wechselwirkung* 1982; Guido Block-Künzler and Dittmar Graf, *Wissenschaft von unten: Zwischenbilanz und Perspektiven der Wissenschaftsladen-Bewegung* (Frankfurt am Main: VAS, 1993).

52 On the latter see Nils Güttler, Margarete Pratschke and Max Stadler, "Before Critique Ran out of Steam: Die Zeitschrift 'Wechselwirkung', 1979–1989. Ein Interview mit Reinhard Behnisch, Barbara Orland und Elvira Scheich", accessed November 21, 2022, https://www.ultrabeige.de/pdf/2016_wechselwirkung.pdf.

53 The final position paper of this group appeared in the *AGÖF Rundbriefe* no. 6 in 1983. An overview on the positions can be found in Kautenburger, *Entstehung*, 66–75.

54 See *Öko-Mitteilungen* 1983, no. 1 and no. 2.

55 Kautenburger, *Entstehung*, 19–23.

56 Retrospectively, Barbara Orland, one of the editors of *Wechselwirkung*, has called the beginnings of the journal an "all-male project", in which women were only present "in the background as family, wives, lovers, daughters etc.", quoted by Güttler, Pratschke, and Stadler, "Critique", 23. A cursory perusal of early numbers of the *Öko-Institut's* journal equally seems to indicate a certain gender-disbalance in authors.

57 Kautenburger, *Entstehung*, 23.

58 Already during the AGÖF's foundation in 1981, the participation of lay-persons was demanded by some and rejected by others. See Bossel and Dürrschmidt, *Ökologische Forschung*, 44, 47.

59 Along these lines, the Öko-Institut invited Native Americans to Freiburg in order to learn more about their way to communicate with nature. See *Öko-Mitteilungen* no. 3, 4 and 6 of 1981.

60 In their founding declaration, the initiators of the *Öko-Institut* emphasized their commitment to serve citizens' initiatives, see Roose, *Made by Öko-Institut*, 17. A very first model on how to involve affected persons and citizens' initiatives in reviews for approval procedures of environmental protection was drafted as "Wiedenfelser Entwurf" (Wiedenfels draft) as early as 1973. It proposed to involve affected persons and citizens to participate in the validation, but not the generation of evidence. Werner Beck, M. Fischer, G. Glienicke, P. Jansen and H.H. Wüstenhagen, "Gutachten und Genehmigungsverfahren. Wiedenfelser Entwurf zur Neugestaltung des Genehmigungsverfahrens im Umweltschutz", in Günter Altner, ed., *Alternativen. Anders denken – anders handeln. Zum Selbstverständnis der Bürgerinitiativenbewegung* (Freiburg: Dreisam-Verlag), 85–100.

61 Rainer Brämer, "Alternative Wissenschaft auf dem Weg zu den Mythen. Was ist ein Endiviensalat?" *Öko-Mitteilungen* 2 (May 1983): 10; similar Brämer, "AGÖF am Scheideweg".

62 Brämer, "Alternative Wissenschaft", 8.

63 Brämer, "Alternative Wissenschaft", 10.

64 Roose, *Made by Öko-Institut*, 88.

65 Peter Pluschke, "Ökologische Forschung. Naturwissenschaftler zwischen Wissenschaftsgläubigkeit und alternativer Wissenschaftsgläubigkeit", *Öko-Mitteilungen* 1 (March 1983): 25–26.

66 AGÖF-Rundbrief 6/1983, 6–7, quoted by Kautenburger, *Entstehung*, 38.

67 Günter Altner, "Ökologisch orientierte Forschung", *Öko-Mitteilungen* 1 (March 1983): 9.

68 Altner, "Ökologisch orientierte Forschung", 9.

69 Altner, "Ökologisch orientierte Forschung", 9.

70 Kautenburger, *Entstehung*, 69–71. On the question of complexity as a central challenge in the field of (academic) ecology, see also Kueffer, this volume.

71 The engineer and systems analyst Hartmut Bossel was one of the co-authors of the *Öko-Institut's* energy transition study. With more than 600 scientific publications and reports, he fundamentally helped to shape the field of environmental science.

72 Hariolf Grupp, an economist and mathematician by training, was one of the founding members of the IFEU Heidelberg, see Höpfner and Manchini, *20 Jahre ifeu-Institut*, 37.

73 Gernot Böhme, a philosopher with a background in mathematics, dealt extensively with problems of the philosophy of nature and the natural sciences, which made him a close ally of counter science. Together with others, he formed the group "Social Natural Science" and participated in the debate on ecologically oriented research. See Gernot Böhme, "Was ist sozial konstituierte Natur?" *Öko-Mitteilungen* 6, no. 1 (March 1983): 27–28; Joachim Grebe, "Entstehung und Entwicklung des Projekts soziale Naturwissenschaft", in Gernot Böhme and Engelbert Schramm, eds., *Soziale Naturwissenschaft. Wege zu einer Erweiterung der Ökologie* (Frankfurt: Fischer Taschenbuchverlag, 1985), 143–160.

74 Otto Ullrich, a technician turned sociologist of technology, was one of the founding members of the Institut für Ökologische Wirtschaftsforschung (Institute for Ecological Research in Economics, IÖW), which was established in 1985.

75 Otto Ullrich in Loccumer Protokolle 1982, quoted by Kautenburger, *Entstehung*, 86.

76 Arnim Bechmann and Bärbel Kraft, "Ökosystemtheorie – Versuch einer Bestandsaufnahme und Plädoyer für eine Erweiterung", *Öko-Mitteilungen* 6, no. 1 (March 1983): 15–18. See also Bärbel Kraft, *Ökologische und anthroposophische Naturbeschreibung: Ein struktureller Theorievergleich* (Berlin: Techn. Univ. Universitätsbibliothek, 1994).

77 Bechmann and Kraft, "Ökosystemtheorie", 18.

78 Arnim Bechmann, *Zwischenstation 1995: Reprint eines Berichts über das Programm, Entwicklung und Arbeit des Instituts für ökologische Zukunftsperspektiven* (Barsinghausen: Edition Zukunft, 2011). On Bechmann see also Kautenburger, *Entstehung*, 88–91.

79 Brämer, "Alternative Wissenschaft", 7–10.

80 Hansjörg Hemminger, "Wissenschaft und Wissenschaftsglaube – die Ökologie zwischen Erkenntnistheorie und Politik", *Öko-Mitteilungen* 6, no. 1 (March 1983): 13–14.

81 Doris Knoblauch and Linda Mederake, "Die Anfänge der nichtstaatlichen Umweltpolitikforschung und -beratung", Working Paper Oral History Project "Entstehung und Entwicklung der wissenschaftlichen Umweltpolitikberatung" (Berlin: Ecologic Institute, 2014), 12–13.

82 Doris Freiberg, "Energie für Besserwisser", *Die Zeit*, June 26, 1981; Peter Christ, "Widersprüche aus Freiburg", *Die Zeit*, October 14, 1983.

83 Kurt Schmitz and Alfred Voß, "Energiewende? Analyse, Fragen und Anmerkungen zu dem vom Öko-Institut vorgelegten 'Alternativ-Bericht'" (KFA Jülich, April 1980), 5–8.

84 Leo Pröstler, "Angriffe gegen das Institut: Im Kreuzfeuer der Kritik", *Öko-Mitteilungen* 5, no. 5 (1982): 30. The same article also offers a list of further controversies and public attacks.

85 Wolfram Huncke, "Parallel-Forschung", *Bild der Wissenschaft* 19, no. 11 (1982): 5.

86 Heinz Maier-Leibnitz, "Stichprobenverfahren zur Klärung wissenschaftlichtechnischer Kontroversen", *Die Naturwissenschaften* 70 (1983): 65–69. The article awarded Donderer and his group 0.5 credibility points out of 10; the Kernforschungszentrum Karlsruhe (KfK), whose study had been at the root of the controversy, was rated at 9.5.

87 "'Ernst nehmen, aber nicht tragisch': Die Zukunft unserer Industriegesellschaft, Teil 1", *Bild der Wissenschaft* 18, no. 11 (1981): 72–83.

88 "Müssen wir der Technik eine neue Richtung geben? Die Zukunft unserer Industriegesellschaft Teil 2", *Bild der Wissenschaft* 18, no. 12 (1981): 146–166; "Streit um Öko-Gutachten: Dürfen die Wissenschaftler parteiisch sein?" *Bild der Wissenschaft* 18, no. 5 (1981): 122–138.

89 "'Verdammt viele faule Professoren'", *Der Spiegel*, May 26, 1985, 22.

90 See the discussion in Bossel and Dürrschmidt, *Ökologische Forschung*, 52–61.

91 Brämer, "AGÖF am Scheideweg".

92 Bernd Faulenbach and Dieter Dowe, *Das sozialdemokratische Jahrzehnt: Von der Reformeuphorie zur neuen Unübersichtlichkeit* (Bonn: Dietz, 2011), 587–593.

93 See the double interview with Volker Hauff and Michael Sailer by Andreas Kraemer, September 19, 2013, accessed November 21, 2022, https://www.ecologic.eu/de/9986.

94 On the DRS, see Stefan Esselborn and Karin Zachmann, "Safety by Numbers: Probabilistic Risk Analysis as an Evidence Practice for Technical Safety in the German Debate on Nuclear Energy", *History and Technology* 36, no. 1 (2020): 129–164.

95 Michael Sailer, interview by Stefan Esselborn, Darmstadt, September 3, 2021.

96 On the Enquête-Kommission – the first scientific advisory body of this kind in the FRG – see Cornelia Altenburg, *Kernenergie und Politikberatung: Die Vermessung einer Kontroverse* (Wiesbaden: Verlag für Sozialwissenschaften, 2010).

97 The latter recommendation, however, was restricted to the majority opinion and vehemently opposed by the minority; Enquête-Kommission zur zukünftigen Kernenergie-Politik, "Bericht über den Stand der Arbeit und die Ergebnisse gemäß Beschluß des Deutschen Bundestages: Drucksache 8/2628" (Deutscher Bundestag, 1980).

98 Enquete-Kommission zur zukünftigen Kernenergie-Politik, "Bericht".

99 Jochen Benecke, "Die kompromittierte Wissenschaft: Erfahrungen bei der Risikoanalyse" and Adolf Birkhofer, "Die kompromittierte Wissenschaft? Eine Entgegnung", both in *AUSgebrütet – Argumente zur Brutreaktorpolitik*, eds. Klaus M. Meyer-Abich and Reinhard Ueberhorst (Basel: Birkhäuser Basel, 1985). On the problems between the groups, see Altenburg, *Kernenergie und Politikberatung*, 248–265.

100 Brämer, "AGÖF am Scheideweg".

101 Brämer, "AGÖF am Scheideweg".

102 Pröstler, "Angriffe gegen das Institut", 31, lists ten such official inquiries in the space of one year.

103 Pröstler, "Angriffe gegen das Institut", 31.

104 Arnim Gleich, Herbert Mehrtens and Joachim Karnath, "Wer hat die Alternative?" *Wechselwirkung* 6, no. 22 (1984): 52–54.

105 "'Die Sache hat uns kalt erwischt'", *Der Spiegel*, May, 11, 1986.
106 "Die Becquerel-Bewegung", *taz*, April 22, 2006.
107 "'Eine wahre Lust, das Regieren!' Aus dem Tagebuch des Grünen-Ministers Joschka Fischer (II)", *Der Spiegel*, February 22, 1987.
108 Bundesarchiv Koblenz (BAK), B 102–339831, Mohr to BMWi Bangemann, September, 8, 1986.
109 BAK B 102–339831, Bangemann to Mohr, October, 16, 1986.
110 Trepl, "Was ist alternativ an der alternativen Forschung?" 17. At the time, Trepl worked at the IÖW, the result of another break-away of a specialized working group from the *Öko-Institut*.
111 Kautenburger, *Entstehung*, 243.
112 Engels, "'Inkorporierung' und 'Normalisierung'", 96.
113 Roose, *Made by Öko-Institut*, 21–23.
114 Roose, *Made by Öko-Institut*, 23–25.
115 For instance, Sailer served as member and at times head of the RSK from 1999 to 2014 and headed (amongst various other capacities) the FRG's commission on nuclear waste (Entsorgungskommission) since 2011. His close collaborator Lothar Hahn, former head of the reactor safety working group of Öko-Institut, not only joined him in the RSK but also went on to become director of the semi-public GRS from 2002 to 2010.
116 This is arguably the biggest difference to Citizen Science projects starting in the 1990s, which in some ways can be seen as successors to (at least or inspired by) *Gegenwissenschaft*; see Bruno Strasser, Jerome Baudry, Dana Mahr, Gabriela Sanchez, and Elise Tancoigne, "'Citizen Science'? Rethinking Science and Public Participation", *Science and Technology Studies* 32, no. 2 (2019): 52–76; Andreas Wenninger and Kevin Altmann, this volume.
117 As Engels, *Naturpolitik, 392–399*, shows, this highly ambivalent attitude toward the state can be seen as a general trait of the West German environmental movement. While environmentalists often rallied around a rhetoric of resistance to a supposedly repressive and "inhuman" state, they also confidently expected (and often succeeded in obtaining) its material and political support.
118 Although both groups shared many similar ideas and often intersected personally, the exact nature of their interactions is not yet fully clear. For some preliminary thoughts regarding the Anglo-American context, see Schmalzer, *Science for the People*, esp. 6–11; Peter J. Taylor and Karin Patzke, "From Radical Science to STS", *Science as Culture* 30, no. 1 (2021): 1–10.

References

Albers, Hartmut. *Gerichtsentscheidungen zu Kernkraftwerken. Argumente in der Energiediskussion*, Vol. 10. Villingen-Schwenningen: Neckar-Verlag, 1980.

Altenburg, Cornelia. *Kernenergie und Politikberatung: Die Vermessung einer Kontroverse*. Wiesbaden: Verlag für Sozialwissenschaften, 2010.

Altner, Günter. "Ökologisch orientierte Forschung". *Öko-Mitteilungen* 6, no. 1 (March 1983): 7–9.

Altner, Günter. "Lehren für eine Forschungs- und Technologiepolitik der Zukunft: Die Erfahrungen eines Kernenergiekritikers in der Enquête-Kommission". In *Der Kalkar-Report: Der Schnelle Brüter: Unwägbares Risiko mit militärischen Gefahren?*, edited by Roland Kollert, Richard Donderer, and Bernd Franke, 19–20. Frankfurt am Main: Fischer, 1983.

Anonymous. "Ist die Wissenschaft noch zu retten? Wissenschaftsläden in der Bundesrepublik". *Wechselwirkung* 4, no. 14 (1982): 51–54.

Anonymous. "Projekte". *Wechselwirkung* 1, no. 2 (1979): 56–59.

Augustine, Dolores L. *Taking on Technocracy: Nuclear Power in Germany, 1945 to the Present.* New York: Berghahn Books, 2018.

Bechmann, Arnim. *Zwischenstation 1995: Reprint eines Berichts über das Programm, Entwicklung und Arbeit des Instituts für ökologische Zukunftsperspektiven.* Barsinghausen: Edition Zukunft, 2011.

Bechmann, Arnim, and Bärbel Kraft. "Ökosystemtheorie – Versuch einer Bestandsaufnahme und Plädoyer für eine Erweiterung". *Öko-Mitteilungen* 6, no. 1 (March 1983): 15–18.

Bechmann, Arnim, and Leo Pröstler. "Editorial: Fünf Jahre Öko-Institut". *Öko-Mitteilungen* 5, no. 5 (1982): 5–6.

Beck, Ulrich. *Risk Society. Towards a New Modernity.* Translated by Mark Ritter. London: SAGE, 1992.

Beck, Werner, M. Fischer, G. Glienicke, P. Jansen, and H.H. Wüstenhagen. "Gutachten und Genehmigungsverfahren. Wiedenfelser Entwurf zur Neugestaltung des Genehmigungsverfahrens im Umweltschutz". In *Alternativen. anders denken – anders handeln. Zum Selbstverständnis der Bürgerinitiativenbewegung*, edited by Günter Altner, 85–100. Freiburg: Dreisam-Verlag, 1978.

Bell, Alice. "The Scientific Revolution That Wasn't: The British Society for Social Responsibility in Science", *Radical History Review* 2017, no. 127 (2017): 149–172.

Benecke, Jochen. "Die kompromittierte Wissenschaft: Erfahrungen bei der Risikoanalyse". In *AUSgebrütet – Argumente zur Brutreaktorpolitik*, edited by Klaus M. Meyer-Abich, and Reinhard Ueberhorst, 259–279. Basel: Birkhäuser Basel, 1985.

Birkhofer, Adolf. "Die kompromittierte Wissenschaft? Eine Entgegnung". In *AUSgebrütet – Argumente zur Brutreaktorpolitik*, edited by Klaus M. Meyer-Abich, and Reinhard Ueberhorst, 280–288. Basel: Birkhäuser, 1985.

Block-Künzler, Guido, and Dittmar Graf. *Wissenschaft von unten: Zwischenbilanz und Perspektiven der Wissenschaftsladen-Bewegung.* Frankfurt am Main: VAS, 1993.

Böhme, Gernot. "Was ist sozial konstituierte Natur?" *Öko-Mitteilungen* 6, no. 1 (March 1983): 27–28.

Bossel, Hartmut, and Wolfhart Dürrschmidt, eds. *Ökologische Forschung: Wege zur verantworteten Wissenschaft.* Karlsruhe: Müller, 1981.

Brämer, Rainer. "AGÖF am Scheideweg: Alternative Wissenschaft zwischen Staat und Basis", *Wechselwirkung* 4, no. 14 (1982): 49–50.

Brämer, Rainer. "Alternative Wissenschaft auf dem Weg zu den Mythen. Was ist ein Endiviensalat?" *Öko-Mitteilungen* 6, no. 2 (May 1983): 7–10.

Bridger, Sarah. "Anti-Militarism and the Critique of Professional Neutrality in the Origins of Science for the People". *Science as Culture* 25, no. 3 (2016): 373–378.

Christ, Peter. "Widersprüche aus Freiburg: Öko-Institut: Wissenschaftler der Umweltschutzbewegung zerpflücken konservative Denkmodelle". *Die Zeit*, October 14, 1983.

"Die Becquerel-Bewegung". *taz*, April 22, 2006.

"Die Sache hat uns kalt erwischt". *Der Spiegel*, May 11, 1986.

Dietz, Bernhard, Christoph Neumaier, and Andreas Rödder, eds. *Gab es den Wertewandel? Neue Forschungen zum gesellschaftlich-kulturellen Wandel seit den 1960er Jahren*. München: Oldenbourg, 2014.

Doering-Manteuffel, Anselm, and Lutz Raphael. *Nach dem Boom: Perspektiven auf die Zeitgeschichte seit 1970*. Göttingen: Vandenhoeck & Ruprecht, 2008.

"'Eine wahre Lust, das Regieren!': Aus dem Tagebuch des Grünen-Ministers Joschka Fischer (II)". *Der Spiegel*, February 22, 1987.

Engels, Jens Ivo. *Naturpolitik in der Bundesrepublik: Ideenwelt und politische Verhaltensstile in Naturschutz und Umweltbewegung, 1950–1980*. Paderborn: Schöningh, 2006.

Engels, Jens Ivo. "'Inkorporierung' und 'Normalisierung' einer Protestbewegung am Beispiel der westdeutschen Umweltproteste in den 1980er Jahren". *Mitteilungsblatt des Instituts für soziale Bewegungen* 40 (2008): 81–100.

Enquête-Kommission zur zukünftigen Kernenergie-Politik. "Bericht über den Stand der Arbeit und die Ergebnisse gemäß Beschluß des Deutschen Bundestages: Drucksache 8/2628", *Deutscher Bundestag*, June 27, 1980.

"'Ernst nehmen, aber nicht tragisch': Die Zukunft unserer Industriegesellschaft, Teil 1". *Bild der Wissenschaft* 18, no. 11 (1981).

Esselborn, Stefan, and Karin Zachmann. "Safety by Numbers: Probabilistic Risk Analysis as an Evidence Practice for Technical Safety in the German Debate on Nuclear Energy". *History and Technology* 36, no. 1 (2020): 129–164.

Faulenbach, Bernd, and Dieter Dowe. *Das sozialdemokratische Jahrzehnt: Von der Reformeuphorie zur Neuen Unübersichtlichkeit. Die SPD 1969–1982*. Bonn: Dietz, 2011.

Freiberg, Doris. "Energie für Besserwisser", *Die Zeit*, June 26, 1981.

Gauchat, Gordon. "Politicization of Science in the Public Sphere: A Study of Public Trust in the United States, 1974–2010". *American Sociological Review* 77, no. 2 (2012): 167–187.

"Geballte Ladung". *Der Spiegel*, February 17, 1977.

Geyer, Martin H. 'Gaps' and the (Re-)Invention of the Future Social and Demographic Policy in Germany During the 1970s and 1980s". *Social Science History* 39, no. 01 (2015): 39–61.

Gizycki, Peter von. "Projektliste im Sinne eines Arbeitsberichts der Vergangenen fünf Jahre". *Öko-Mitteilungen* 5, no. 5 (1982): 7–12.

Gleich, Arnim, Herbert Mehrtens, and Joachim Karnath. "Wer hat die Alternative?" *Wechselwirkung* 6, no. 22 (1984): 52–54.

Gräfing, Birte. *Tradition Reform: Die Universität Bremen 1971–2001*. Bremen: Donat, 2012.

Grebe, Joachim. "Entstehung und Entwicklung des Projekts Soziale Naturwissenschaft". In *Soziale Naturwissenschaft. Wege zu einer Erweiterung der Ökologie*, edited by Gernot Böhme, and Engelbert Schramm, 143–160. Frankfurt am Main: Fischer, 1985.

Grießhammer, Rainer, ed. *Chemie im Haushalt*. Reinbek bei Hamburg: Rowohlt, 1984.

Güttler, Nils, Margarete Pratschke, and Max Stadler. "Before Critique Ran Out of Steam: Die Zeitschrift *Wechselwirkung – Technik, Naturwissenschaft, Gesellschaft*, 1979–1989. Ein Interview mit Reinhard Behnisch, Barbara Orland und Elvira Scheich". https://www.ultrabeige.de/pdf/2016_wechselwirkung.pdf.

Güttler, Nils, Margarete Pratschke, and Max Stadler, eds. *Wissen, ca. 1980*. Nach Feierabend 12.2016. Zürich: diaphanes, 2016.

Hauff, Volker, and Michael Sailer. Interview by Andreas Kraemer. September 19, 2013.

Hemminger, Hansjörg. "Wissenschaft und Wissenschaftsglaube – die Ökologie zwischen Erkenntnistheorie und Politik". *Öko-Mitteilungen* 6, no. 1 (March 1983): 13–14.

Hey, Christian, and Rainer Grießhammer. "10 Jahre Öko-Institut: Zwischen Anspruch und Wirklichkeit". *Öko-Mitteilungen* 10, no. 3 (1987): 4–10.

Höpfner, Ulrich, and Mario Manchini, eds. *20 Jahre ifeu-Institut: Engagement für die Umwelt zwischen Wissenschaft und Politik*. Braunschweig: Viehweg, 1998.

Hughes, Michael L. "Civil Disobedience in Transnational Perspective: American and West German Anti-Nuclear-Power Protesters, 1975–1982". *Historical Social Research* 39, no. 1 (2014): 236–253.

Huncke, Wolfram. "Parallel-Forschung". *Bild der Wissenschaft* 19, no. 11 (1982).

Inglehart, Ronald. *The Silent Revolution: Changing Values and Political Styles Among Western Publics*. Princeton: Princeton Univ. Press, 1977.

Joppke, Christian. *Mobilizing Against Nuclear Energy: A Comparison of Germany and the United States*. Berkeley: Univ. of California Press, 1993.

Jungk, Robert. *Der Atom-Staat: Vom Fortschritt in die Unmenschlichkeit*. München: Kindler, 1977.

Katalyse-Umweltgruppe. *Chemie in Lebensmitteln*. Frankfurt am Main: Zweitausendeins, 1982.

Kautenburger, Margit. *Entstehung, aktuelle Lage und Wissenschaftsverständnis ökologisch orientierter Wissenschaft in Deutschland am Beispiel der Arbeitsgemeinschaft Ökologischer Forschungsinstitute*, Dissertation, Universität Hannover, 1989.

Kirchberg, Christopher, and Marcel Schmeer. "The 'Traube Affair': Transparency as a Legitimation and Action Strategy Between Security, Surveillance and Privacy". In *Contested Transparencies, Social Movements and the Public Sphere: Multi-Disciplinary Perspectives*, edited by Stefan Berger, and Dimitrij Owetschkin, 173–196. Cham: Palgrave, 2019.

Klein, Ansgar, ed., Special Issue "Gegenexperten in der Risikogesellschaft". *Forschungsjournal Neue Soziale Bewegungen* 3, no. 1 (1990).

Klein, Ansgar, Hans-Josef Legrand, and Thomas Leif, eds. *Neue soziale Bewegungen: Impulse, Bilanzen und Perspektiven*. Wiesbaden: VS Verlag für Sozialwissenschaften, 1999.

Knoblauch, Doris, and Linda Mederake. "Die Anfänge der nichtstaatlichen Umweltpolitikforschung und -beratung", Working Paper Oral History Project "Entstehung und Entwicklung der wissenschaftlichen Umweltpolitikberatung". Berlin: Ecologic Institute, 2014.

Kraft, Bärbel. *Ökologische und anthroposophische Naturbeschreibung: Ein struktureller Theorievergleich*. Berlin: Techn. Univ. Universitätsbibliothek, 1994.

Krause, Florentin, Hartmut Bossel, and Karl F. Müller-Reißmann. *Energie-Wende: Wachstum und Wohlstand ohne Erdöl und Uran*. Frankfurt am Main: Fischer, 1980.

Kühnert, Hanno. "Läßt der Tiger zu viele Haare? Die Sachverständigen, die für die Kernkraftwerke votieren, sind in der Mehrzahl", *Die Zeit*, February 11, 1977.

Kühnert, Hanno. "Werkstatt der besseren Argumente: Das Öko-Institut – Ein Versuch gegen die etablierte Übermacht", *Die Zeit*, September 28, 1979.

Kutz, Detlef. "Im Reagenzglas: Die Bundesrepublik", *taz*, October 14, 1988.

Lahl, Uwe, and Barbara Zeschmar. *Wie krank ist unser Wasser? Die Gefährdung des Trinkwassers. Sachstand und Gegenstrategien.* Freiburg: Öko-Institut, 1981.

Laufs, Paul. *Reaktorsicherheit für Leistungskernkraftwerke: Die Entwicklung im politischen und technischen Umfeld der Bundesrepublik Deutschland.* Berlin: Springer Vieweg, 2013.

Maier-Leibnitz, Heinz. "Stichprobenverfahren zur Klärung wissenschaftlich-technischer Kontroversen". *Die Naturwissenschaften* 70 (1983): 65–69.

Meadows, Donella H., Dennis Meadows, and Jorgen Randers. *The Limits to Growth: A Report for the Club of Rome's Project on the Predicament of Mankind.* New York: Universe Books, 1972.

Mende, Silke. *"Nicht rechts, nicht links, sondern vorn": Eine Geschichte der Gründungsgrünen.* München: Oldenbourg, 2011.

Milder, Stephen. *Greening Democracy: The Anti-Nuclear Movement and Political Environmentalism in West Germany and Beyond, 1968–1983.* Cambridge: Cambridge Univ. Press, 2017.

Moore, Kelly. *Disrupting Science: Social Movements, American Scientists, and the Politics of the Military, 1945–1975.* Princeton: Princeton Univ. Press, 2008.

"Müssen wir der Technik eine neue Richtung geben?" Die Zukunft unserer Industriegesellschaft, Teil 2, *Bild der Wissenschaft 18*, no. 12 (1981): 146–166.

"Nackt am Zaun". *Der Spiegel*, August 7, 1978.

Nowotny, Helga, and Hilary Rose, eds. *Counter-Movements in the Sciences: The Sociology of the Alternatives to Big Science.* Dordrecht: Reidel, 1979.

Öko-Institut. *Risikountersuchungen zu Leichtwasserreaktoren: Analytische Weiterentwicklung zur, Deutschen Risikostudie Kernkraftwerke'.* 3 vols. Freiburg: Öko-Institut, 1983.

Pluschke, Peter. "Ökologische Forschung. Naturwissenschaftler zwischen Wissenschaftsgläubigkeit und alternativer Wissenschaftsgläubigkeit". *Öko-Mitteilungen* 6, no. 1 (March 1983): 25–26.

Pröstler, Elke. *Stillen trotz verseuchter Umwelt: Die chemische Belastung der Muttermilch. Ursachen, Hintergründe und politische Forderungen.* Freiburg: Öko-Institut, 1981.

Pröstler, Leo. "Angriffe gegen das Institut: Im Kreuzfeuer der Kritik". *Öko-Mitteilungen* 5, no. 5 (1982): 29–32.

Radkau, Joachim. *Aufstieg und Krise der deutschen Atomwirtschaft, 1945–1975. Verdrängte Alternativen in der Kerntechnik und der Ursprung der nuklearen Kontroverse.* Reinbek bei Hamburg: Rowohlt, 1983.

Radkau, Joachim. "Die Kernkraft-Kontroverse im Spiegel der Literatur". In *Das Ende des Atomzeitalters? Eine sachlich-kritische Dokumentation*, edited by Armin Hermann, 307–334. München: Moos, 1987.

Radkau, Joachim. *The Age of Ecology: A Global History.* Cambridge: Polity Press, 2014.

Reichardt, Sven. *Authentizität und Gemeinschaft: Linksalternatives Leben in den siebziger und frühen achtziger Jahren.* Berlin: Suhrkamp, 2014.

Reichardt, Sven, and Detlef Siegfried, eds. *Das alternative Milieu: Antibürgerlicher Lebensstil und linke Politik in der Bundesrepublik Deutschland und Europa; 1968–1983.* Göttingen: Wallstein Verlag, 2010.

Roose, Jochen. *Made by Öko-Institut: Wissenschaft in einer bewegten Umwelt.* Freiburg Breisgau: Öko-Institut, 2002.

Roth, Roland, and Dieter Rucht, eds. *Neue soziale Bewegungen in der Bundesrepublik Deutschland*. Bonn: Bundeszentrale für Politische Bildung, 1991.

Rothenhäusler, Andie. "Die Debatte um die Technikfeindlichkeit in der BRD in den 1980er Jahren". *Technikgeschichte* 80, no. 4 (2013): 273–294.

Rucht, Dieter. "Gegenöffentlichkeit und Gegenexperten: Zur Institutionalisierung des Widerspruchs in Politik und Recht", *Zeitschrift für Rechtssoziologie* 9, no. 2 (1988): 290–305.

Schmalzer, Sigrid, Daniel S. Chard, and Alyssa Botelho, eds. *Science for the People: Documents from America's Movement of Radical Scientists*. Amherst, Boston: Univ. of Massachusetts Press, 2018.

Schmitz, Kurt, and Alfred Voß. *Energiewende? Analyse, Fragen und Anmerkungen zu dem vom Öko-Institut vorgelegten 'Alternativ-Bericht'. Spezielle Berichte der Kernforschungsanlage Jülich*, vol. 73, Jülich: Kernforschungsanlage, 1980.

Stadler, Max, Nils Güttler, Niki Rhyner, Mathias Grote, Fabian Grütter, Tobias Scheidegger, Martina Schlünder et al. *Gegen|Wissen*. Cache 01. Zürich: intercom, 2020.

Strasser, Bruno, Jerome Baudry, Dana Mahr, Gabriela Sanchez, and Elise Tancoigne, "'Citizen Science'? Rethinking Science and Public Participation". *Science and Technology Studies* 32, no. 2 (2019): 52–76.

"Streit um Öko-Gutachten: Dürfen die Wissenschaftler parteiisch sein?" *Bild der Wissenschaft* 18, no. 5 (1981): 122–138.

Topçu, Sezin, "Confronting Nuclear Risks: Counter-Expertise as Politics within the French Nuclear Energy Debate". *Nature and Culture* 3, no. 2 (2008): 225–245.

Trepl, Ludwig, "Was ist alternativ an der alternativen Forschung?" *Wechselwirkung* 40, no. 11 (1989): 15–20.

Uekötter, Frank. *The Greenest Nation? A New History of German Environmentalism*. Cambridge: MIT Press, 2014.

"'Verdammt viele faule Professoren'". *Der Spiegel*, May 26, 1985.

Weinsheimer, Lilo. "Held oder Spinner? Professor Scheer gilt als wilder Kommunist", *Die Zeit*, February 27, 1976.

Werner, Elisabeth, and Bernd Speiser. "Alternative Wissenschaft zwischen Anspruch und Wirklichkeit: Das Öko-Institut". *Wechselwirkung* 6, no. 21 (1984): 16–19.

Werskey, Gary. "The Marxist Critique of Capitalist Science: A History in Three Movements?" *Science as Culture* 16, no. 4 (2007): 397–461.

8 Appropriating Evidence

Scientific Criticism and Environmental Activism in the Global Pesticide Controversy during the 1970s and 1980s

Sarah Ehlers

> We urgently need an end to these false assurances, to the sugar coating of unpalatable facts. It is the public that is being asked to assume the risks that the insect controllers calculate. The public must decide whether it wishes to continue on the present road, and it can do so only when in full possession of the facts.
>
> Rachel Carson, *Silent Spring*, 1962.[1]

In 1962, Rachel Carson's *Silent Spring* brought environmental concerns about the dangers of synthetic pesticide use to the American public. Often credited with starting an environmental movement, *Silent Spring* led to a decade of fierce political conflict and intense scientific research on the hazards of pesticide use, which eventually spurred a reversal in the United States' national pesticide policy. It took a full decade before, in 1972, the newly founded US Environmental Protection Agency (EPA) banned DDT for virtually all but emergency uses in the United States because of its persistence in the environment and accumulation in the food chain. By 1974, most industrial countries had followed suit and banned many persistent pesticides.[2] The pesticide industry, however, continued its growth. Shifting their sales of persistent pesticides to the developing world, chemical companies even boosted their output after the publication of *Silent Spring*. By the mid-1970s, the debate about hazardous pesticide use had also shifted, alerting the public to the pesticide problem in the poor countries of the South. When the social and ecological conditions in many parts of Latin America, Sub-Saharan Africa and South Asia magnified the ill effects of hazardous substances, pesticide use in the developing world became a matter of contention.[3] As Rachel Carson had done in 1962, scientists, journalists and activists researched and published about the risks of persistent pesticides, now in the context of developing countries, and urged the public to consider "the facts".

In this chapter, I will examine the variety and competition of forms of knowledge about pesticide use in the Global South during the 1970s and 1980s. Drawing on anti-pesticide material such as activists' publications

DOI: 10.4324/9781003273509-13

and pamphlets as well as journalistic documents and documentaries, I will outline strategies to raise awareness on environmental destruction and health hazards in developing countries. Thereby, this chapter gives an account of how contestations of evidence shaped a controversy that was both academic and political in nature. Evidence, philosopher of science Nancy Cartwright reminds us, is never an abstract phenomenon: "What a claim means in the context in which it is first justified may be very different from what it means in the different contexts in which it will be put to use".[4] Against the backdrop of the increasing use of scientific knowledge as a legitimatory resource in political discourse, it traces the development of evidence criticism and its political implications in the debate on global pesticide use. This has important implications for negotiating social consensus today: Studying how critics made use of different forms of evidence will help, I argue, to understand not only the role of expertise in political conflicts, but also the contingent and dynamic nature of evidence in modern knowledge societies.[5]

To this end, I firstly discuss how a pesticide export industry developed after the ban of DDT and other persistent pesticides in industrial countries. Since the most lucrative market for pesticides was in agriculture and, in particular, in the application for crops to be exported to the US and Europe, debating pesticide use in the South became a matter of environmentalism as well as consumerism. While scientists studied the many effects of pesticide use, activists and newly founded transnational initiatives denounced the export of banned substances to the "third world", calling attention to their immediate environmental and health impacts. Secondly, I explore how already precarious environmental knowledge claims were particularly contested in the Global South. How, given this context, could environmental criticism become a destabilizing factor in established risk assessments? Exploring the link between North-South solidarity and fears about food security, I outline new strategies developed by organizations concerned about environmental destruction. Thirdly, my chapter will focus specifically on the evidence these activists produced: The way they presented and published scientific results, images or voices from the South, and the ways they countered evidence for pesticide use by the chemical industries. Illuminating how activist groups expanded their work beyond national borders to challenge multinational corporations, this chapter emphasizes the tension-laden relationship between global entanglement and environmental regulation differences during the 1970s and 1980s.

Going South: Pesticide Export after *Silent Spring*

Although the EPA hearings had exposed the long-range risks in the continued use of DDT to the public, and most developed countries had banned DDT's and other persistent organic pollutants' agricultural use during the 1970s, they still allowed for its continued manufacture and export. This inconsistency created a paradoxical set of issues that not only complicated the application of environmental regulation for the Western chemical industries. It also posed a moral challenge for development policy. International development aid and the WHO's disease control programs continued to rely heavily on the use of DDT and other hazardous substances in the Global South.[6] While there was some justification for this, mainly in the control of Malaria and other insect-borne diseases, disease vector control was only a small fraction of the pesticide export. Most of the exported pesticides were used in agriculture. More importantly, a rapid increase in observed insecticide resistance created doubts about the overall usefulness of DDT and other organochlorides even in disease control.[7] Yet, in spite of openly debated health and environmental risks, pesticide use in developing countries climbed 23 percent between 1971 and 1973.[8] Moreover, despite these numbers, the pre-dominantly right-wing myth that the DDT ban undermined programs against malaria around the world and therefore condemned millions of people to death from the tropical disease gained popularity and continues to re-surface in political debate until today.[9]

For the pesticide industry, the process of finding new pesticides was tedious and costly. Instead, many companies looked south for expansion. Clearly, it was only because of exports that the period of pronounced growth in pesticide sales continued during the 1970s. Sales dropped briefly during the early 1980s but then began to increase again.[10] From 1972 to 1985, imports of pesticides increased by 261 percent in Asia, 95 percent in Africa and 48 percent in South America.[11] Especially in the poorer markets, companies frequently marketed older pesticides, which were broader in spectrum and cheaper.[12] Conversely, for importing nations, deploying these pesticides for crop production not only meant a fundamental shift in how they produced food and fiber crops, it also created enormous health risks, in particular for farm workers handling the pesticides. While the long-term consequences for human health and the environment are still not fully known, in the years to come, the direct human health impacts of the explosion of the global pesticide market would become apparent in the rising number of pesticide poisonings.[13] In 1972, the WHO estimated there were approximately 500,000 cases of accidental pesticide poisoning, and by 1987, the estimates had risen to a million victims per year.[14] Still, the number of unreported cases exceeded the number of reported cases. Reporting for the *New York Times* in 1977, Alan Riding interviewed a nurse at a local clinic in Guatemala. Treating about 30 or 40 people a day for pesticide poisoning, she explained: "The farmers often tell the peasants to give another reason for their sickness, but you can smell the pesticide in their clothes".[15]

During the 1970s, pesticide harms abroad became a matter of public interest. Following the UN Stockholm Conference in 1972, which had signaled the rapid broadening of the environmental movement, NGOs and activists campaigned for reforming development policies to incorporate ecological considerations. Whereas development and environment were no longer seen as incompatible, the conference exposed the challenges of seeking national solutions for global problems as well as bringing together economic growth with ecological concerns.[16] The new environmental slogan, "sustainable development", also revealed pesticide use in a different light. Compared to the debates following *Silent Spring*, weighing the pros and cons seemed even more complicated. As deeply embedded in global economy as in development policy and in the environment, debating pesticide use wove together a wide range of topics. Therefore, presenting a comprehensive look at the export of banned substances and its direct and long-term consequences for the public created a challenge for activists, journalists and policymakers alike. High profile western media organizations, such as the New York Times, PBS and the Center for Investigative Reporting, published award-winning material on global pesticide issues.[17] Yet, despite garnering considerable public attention, exporting banned hazardous pesticides continued to be legal. Neither national nor international environmental or foreign trade legislation prevented the widespread environmental contamination and the documented poisoning of people.

Challenges of Environmental Knowledge

Although DDT and other hazardous pesticide use was a much-debated issue during the 1970s, for environmental activists, calling out the global pesticide market created a double challenge. Firstly, this was due to the epistemic specificities of environmental knowledge, in the words of historian of science Dominique Pestre, "a weaker form of knowledge".[18] Secondly, arguments against pesticide use tended to appear negligible compared to those for economic growth or agricultural production, notably in the poor countries of the South. Generally, evidence for environmental destruction and claims about environmental hazards are complicated to put forward. Documenting effects from chemical exposure, pollution or radiation is difficult as it takes place gradually and often invisibly. For example, a 1987 scientific publication called the effects of hazardous chemicals in the developing world "Bhopal in Slow Motion", referring to the 1984 industrial disaster at the Union Carbide India Limited (UCIL) pesticide plant in Bhopal, India.[19] Describing the strategic and representational challenges posed by this kind of ecological degradation, environmental scholar Rob Nixon coined the term "slow violence": "Falling bodies, burning towers, exploding heads, avalanches, volcanoes, and tsunamis have a visceral, eye-catching and page-turning power that tales of slow violence, unfolding over years, decades, even centuries, cannot match".[20] Moreover, scientifically, "the environment" is extremely complex as an object of inquiry and implies a large variety of studies, tools and techniques. Environmental sciences therefore only exist as an interdisciplinary academic field, and as such, it mobilizes a set of studies and tools that are not hierarchized and produces data that is varied in nature and complicated to assemble and to assess.[21] Both in the eyes of specialized literature as in the popular imagination, the heterogeneous nature of environmental science therefore produces seemingly inconsistent and less convincing results – especially compared to disciplines such as physics, biology or chemistry.[22]

In the pesticide controversy, disciplinary and epistemic differences meant that science and scientific evidence appeared to some actors to have been weaponized for political ends rather than relating to objective realities.[23] Drawing on ecology, biochemistry, agronomy, toxicology and many more disciplines, both proponents and critics of pesticides sustained their claims with results from a broad range of scientific studies. For non-experts, the debate touched off by *Silent Spring* increasingly brought to light the inconsistencies and contradictory nature of the scientific discourse on pesticide risks. Scientific authority was at center stage early on when, for example, some scientists called Rachel Carson's work "unscientific" or "overwrought" because *Silent Spring* also included emphatic ethical statements and arguments.[24] In a similar vein, in 1971, DDT proponent Thomas Jukes claimed in a letter to the *New York Times*: "If the environmentalists win on DDT, they will achieve, and probably retain in other environmental issues, a level of authority they have never had before. In a sense then, much more is at stake than DDT".[25] After the ban in industrial countries, the controversy developed into a struggle over the benefit-cost ratio of hazardous pesticides in the developing world. While proponents declared that pesticides saved lives, critics argued that they produced unwarranted environmental, economic and social catastrophes in the South.[26]

For pesticide critics, claims of validity, especially concerning long-term health costs, were hard to establish. In 1978, the *New York Times* quoted Dr. Samuel S. Epstein on the health harms of pesticide use: "It's comparable to the relationship between cigarette smoking and lung cancer—it took 20 years to prove".[27] Over the course of the debate, each side even advanced their views about technology and ecology. Scientific disputes on biodiversity, toxic residues and carcinogenics became battlegrounds in the broader collision of environmental ethics and human ethics.[28] For the public, the distinction between the two could be hard to discern. Hence, instead of negotiating consensus, competing forms of knowledge – including political and moral judgments as well as internal scientific differences – were a main driver for the divisiveness of these debates.

Generally, evidence for environmental destruction depends on the political, social or institutional space in which debates take place. The pesticide debates during the 1970s unfolded when norms, regulations and economic instruments were seen to be the solutions of choice and economists as the ultimate experts on environmental protection.[29] Yet, being too complex for a so-called scientific or technological fix, the global pesticide problem did not fit into any of the existing categories.[30] This was particularly relevant for the dominant policy tool of cost-benefit analyses. Although often depicted as neutral or scientific, cost-benefit analyses structurally optimized quantifiable material benefits while the number and design of their parameters was prone to manipulation.[31] Moreover, proposed solutions to environmental hazards were heavily susceptible to political and economic interests. In 1971, the *Programme of Work for the OECD* made this abundantly clear, stating "governmental interest in maintaining an acceptable human environment must now be developed in the framework of economic growth".[32] This way of handling environmental challenges put the pesticide industry in a position of power. While monetary policies that encouraged cash crops for exports increased pesticide sales, only a small number of transnational corporations dominated the market.[33] Given the dogma that population explosion, poverty and hunger had created the need for production increases, which only pesticides could deliver, the pesticide industry was seen as essential to development policy. Yet despite this setting of structurally "weak" environmental knowledge and strong economic interests, professionalized organizations became a critical part of activism and successfully confronted corporate power.

Breaking the Circle of Poison: North-South Solidarity and Global Activism

Breaking the "circle of poison" became a slogan for the environmentalist movement after the publication of David Weir's and Mark Schapiro's *Circle of Poison: Pesticides and People in a Hungry World* in 1981. With their investigative study, the two journalists demonstrated how chemicals banned in the United States would return through food imports.[34] Powerfully written, Weir and Schapiro's circle not only linked the Global North and the Global South through pointing to the chemical industry's profits but also through a notion of shared victimhood:

> Every minute, someone in the Third World becomes a victim of pesticide poisoning. And we are victims too. Illegal levels of pesticides turn up in ten percent of the food shipments that arrive at our borders; yet many of these contaminated foods still reach supermarket shelves, completing the circle.[35]

Rife with tables, graphs and statistics, *Circle of Poison* made plenty of use of scientific studies. At least 25 percent of US pesticide exports, they stated, were products that were banned, heavily restricted or have never been registered for use in the United States. As a result, about ten percent of the food items imported into the United States contained higher levels of pesticides than permitted. For instance, almost 50 percent of the coffee beans imported to the US showed traces of banned pesticides such as DDT or Dieldrin. In developing countries, the amount of toxic residues was even more alarming: For example, the average DDT levels in cow's milk in Guatemala were 90 times as high as allowed in the United States. People in Nicaragua and Guatemala carried 31 times more DDT in their blood than people in the United States. Combining numbers with personal stories and engaging images, the authors exposed the nexus between pesticide exports, food production and public health to the American public. In the wake of its publication, documentaries and publications such as "Pesticides: For Export only" or "The Pesticide Boomerang" took up the issue.[36] Moreover, during the 1990s, a "Circle of Poison Bill" was introduced repeatedly to the American legislation, but failed every time.[37]

While the *Circle of Poison* argued that this regulatory loophole of pesticide export had been a disaster both for industrial countries and for the developing world, it stated clearly that the impact was far worse for the poor countries in the South. Yet in response to being unequally affected, Weir and Schapiro asked for global solidarity and activism. In order to break the circle of poison, North and South had to work together, they argued: "We must begin to see third world people not as a burden or a threat, but as allies".[38] Indeed, their urge to control the export of hazardous substances was backed up by a chorus of voices from the developing world. In 1977, for example, Dr. J.C. Kiano, the Kenyan Minister for Water Development, demanded at a meeting of United Nations Environment Programme (UNEP) that "unless a product has been adequately tested, certified, and widely used in the countries of origin, it should not be used for export".[39] In 1978, the Central American Non-Governmental Conservation Societies Conference adopted a resolution demanding President Carter "extend this protection [the ban of hazardous pesticides] to the rest of the humans of our planet".[40] Frequently quoted in Western activist publications, statements like these debunked the claim that poor countries would welcome imports of banned pesticides or not care about the consequences.

Responding to global corporate power with global activism was also the guiding principle of the Pesticide Action Network (PAN). Founded in 1982 at a meeting in Malaysia by activists from around the world, the network was planned as an international coalition of NGOs, scientists and activists, many of them in the Global South. In 1984, they went public with the so-called dirty dozen campaign as an attempt to ban certain pesticides not only in the North, but worldwide.[41] Confidence in their ability to change global politics was crucial for this new movement.[42] When, for example, Anwar Fazal, one of their founding members, was asked in an interview about the power relationship between multinational corporations and activist groups in the years to come, he declared boldly, "there will be major changes in the pesticide industry. They have no choice".[43] Activists saw the reason for this global power of pesticide criticism in the newfound global public. The way that *Silent Spring* had galvanized the American public into an awareness of the dangers of hazardous chemicals had to be translated to a global level. If industries were acting across borders, Fazal declared, so was environmental criticism: "We have now got muscle globally to deal with them in a way that we never had before: power to organize globally, to organize boycotts, direct actions, shareholder actions, power to embarrass them for engaging in unconscionable activities".[44]

Evidencing Pesticide Criticism

Although the pesticide controversy centered on facts about the risks and benefits of hazardous substances, self-produced scientific knowledge only played a minor role. Instead, most of the pesticide criticism during the 1970s and 1980s relied on data from secondary sources. Compared to other alternative movements from the 1960s onward that produced critical and alternative forms of knowledge at the intersection of academia and activism, in particular the campaign against nuclear energy, structures of knowledge production against hazardous pesticide use were less institutionalized and less well funded.[45] Only rarely did pesticide critics themselves measure, for example, the extent of water pollution or soil contamination. In activists' publications, references to smaller self-funded research projects or scientific investigations into the pesticide problem sponsored by NGOs such as Oxfam were the exception rather than the rule.[46] Instead, statistics, tables, diagrams and calculations from government agencies or international organizations were quoted and referenced in numerous publications. While some of the institutions were newly founded in order to deal with environmental challenges, such as the US Environmental Protection Agency (EPA, 1970) or the United Nations Environment Programme (UNEP, 1972), institutions such as the FAO, the WHO and the OECD also increasingly devoted their attention to environmental issues and published extensively on pesticide issues. Compared to their efforts, activists simply did not have enough resources to produce the data necessary to support their claims.[47] Additionally, especially for data on pesticide exports, publications by the chemical industries themselves were an often-referenced source for pesticide criticism. For pesticide activists, producing knowledge mostly meant compiling, commenting, interpreting and editing externally obtained scientific results; ironically, often produced by institutions that were the very targets of their protest. In this regard, pesticide activism during the 1970s and 1980s was less about questioning scientific results

or official findings, than about negotiating their consequences, and ultimately, about the position of environmental knowledge. Activists would
publish findings produced by government agencies in order to demand
that their respective governments take these findings seriously.

In particular, with regard to publications by the chemical industries,
turning external sources into evidence against pesticides also implied
reading them against the grain. For instance, the German branch of the
Pesticide Action Network stated "we are not aiming at new calculations on pesticide effects [...] Rather, we will analyze their [IPS, Industry
Association for Crop Protection] data and dissect it with our questions".[48]
The chemical industries responded to this activist criticism by turning
the accusation around: "Critical publications on this topic use the wrong
data, or: use data wrongly".[49] Indeed, much of the pesticide controversy
was about how to interpret data on pesticide use and crop productivity
correctly. Many publications from the 1970s and 1980s attacked the
pesticide lobby for spreading misinformation by publishing misleading
data and comparisons.[50] Environmental campaigner Jürgen Knirsch,
for example, accused the German industry association of "number-
juggling" for repeatedly disguising the difference between quantitative
and monetary consumption figures. Their comparison between the monetary consumption of pesticides in industrial and developing countries
was all but useless, he argued, as long as they were referring to different
pesticides. A volume of 33 kilograms DDT, for example, cost the same
as one kilogram of the insecticide Cypermethrin, which was newer and
safer and – unlike DDT – used in industrialized countries.[51] In a similar
vein, Oxfam author and campaigner David Bull criticized the company
Velsicol for advertising their product Chlordane as "comparatively [...]
the safest insecticide": In fact, he stated, Chlordane's toxicity was comparable to the highly toxic chemicals Heptachlor, Aldrin and Dieldrin,
which were banned for almost all uses in the US and Europe.[52]

In addition to critically commenting on published evidence on pesticide
use, environmental critics called for new factors to be included in calculating risks and safety. Being specific to the application of pesticides in
the Global South, many factors did not feature in the political economy
of environmental questions. Yet, critics argued, they were at the heart
of the global pesticide problem. In many ways, dangerous accidents and
intoxications were caused by a lack of safety precautions. Pesticides in
the developing world were often used without any protective gear, such
as boots, coveralls, respirators, chemical-resistant gloves and aprons and
protective eyewear. To illustrate, many activists' publications on pesticides in the developing world included picture material of farm workers
handling dangerous substances wearing only light clothing such as shorts
and a T-Shirt (See Figure 8.1).[53] Besides the lack of protective equipment,
most agricultural use of pesticides in the South also lacked appropriate
information material. Poor literacy in most developing countries made

Figure 8.1 Book Cover, David Bull, A Growing Problem: Pesticides and the Third World Poor (Oxford: OXFAM, 1982), reproduced with the permission of Oxfam, www.oxfam.org.uk.[62]

it impossible to read or follow complex label instructions. Moreover, activists frequently pointed out that hazardous chemicals were sold without any warning or application manuals. As a result, pesticide handlers would simply not know how to follow safety precautions and, for example, not to smoke or chew while spraying or what to do in case of skin or eye contact.[54] However, even if workers knew about the dangers, they were in a structurally weak position. Journalists Shapiro and Weir called attention to the socioeconomic context of pesticide export by asking rhetorically: "Can pesticides – poisons, by definition – be used safely in societies where workers have no right to organize, no right to strike, no right to refuse to carry the pesticides into the fields"?[55] Given the lack of any potent government agency overseeing regulation, the strikingly obvious consequence of pesticide use in poor countries was a complete lack of safety procedures.

Apart from their nefarious role as a hazard factor in the Global South, critics also highlighted the use of highly potent pesticides. Being cheap and easily available, pesticides often replaced other chemical substances in the home or in medical applications. For instance, agricultural pesticides were often used for treating head lice, on parasites on domestic animals, or termites and other insects in houses, for food preservation and fishing.[56] In a collection of statements from German development aid workers published by the German section of PAN, a Niger-based official reported that after a pestbird control spraying campaign killing over 50,000 songbirds, the local population would collect the dead birds in order to eat them. Sold by untrained dealers, different pesticides were frequently mixed together to make them more effective. "One wonders if people really do not know what they are doing", PAN quoted a development worker from the Ivory Coast.[57] Combined with a lack of medical facilities, antidotes and poison treatment centers, as well as the confusion of symptoms of pesticide poisoning with common illness, the wrong application and inappropriate mixing of pesticides made subsequent medical management of poisoned patients particularly complicated.[58]

Another major problem in the South was pesticide storage, disposal and waste. For instance, the Pesticides Trust reported that in 1985, the inhabitants of the city Kalaa Seghira, Tunisia, signed a motion urging the Ministry of Agriculture to deal with "600 tons of HCH and 70 barrels of malathion which are deteriorating, and whose fumes are intoxicating the neighbourhood making people sick".[59] Indeed, during the 1970s, the industrial countries' exports of banned pesticides, frequently already beyond expiry, effectively turned many areas of the Global South into a dumping site for pesticide waste.[60] As a result, unregulated pesticide disposal fostered the increasing spread of toxic materials into the environment, including the contamination of ground water. Poor household storage and disposal, as well as the reuse of poorly cleansed pesticide bottles, barrels or cans as water containers or cisterns further exacerbated serious risks to the environment and human health. In order to call attention to the problem of hazardous pesticide waste in the Global South, pamphlets and articles frequently featured images of leaking pesticide barrels or unsafe reuse of pesticide containers.

Taken together, for PAN-activists and other environmentalists, all these factors resulted in the regular and widespread incidence of poisoning as well as continuing environmental degradation in the Global South. For this reason, any calculation of pesticide safety levels had to focus on the real risks to health. Describing the risks of pesticides only as if they were used sparsely, handled correctly and supplied with appropriate protection did not produce reliable data. Neither did ignoring long-term factors such as pesticide disposal and water contamination. In short, pesticide activists were calling for "real-world evidence", to borrow a term from the medical sciences, in evaluating the benefits and the costs of pesticides.[61] For them, risk-benefit analyses concluding that local benefits derived from the use of pesticides exceeded the risks to the public were obviously operating with false parameters. This critique not only challenged established certainties, such as the dominant policy tool of cost-benefit analyses, but pictures of barefoot children spraying pesticides, people drinking from old pesticide cans or masses of discarded pesticide containers also developed a visual narrative of environmental hazards in poor countries that is still universally understood.

Conclusion

What can we learn from the ways environmental activists tried to establish knowledge claims during the pesticide controversy? Why do their "practices of evidence criticism" matter for broader historical and sociopolitical inquiries? Could they change the precarious status of environmental knowledge? Following historian of science Moritz Epple, perceptions of weakness concerning knowledge are functionally related to historical change.[63] Regarding the development of late modern knowledge societies, the editors of this volume emphasize how dynamics of destabilizing evidence for established certainties and attempts to re-stabilize them determined the way in which societies decided on key issues in the late 20th and early 21st century.[64] With respect to the research on discourses of weakness and resource regimes in the history of science, pesticide criticism showed several features of "weak knowledge". Indeed, it combined the dimensions of "epistemic weakness", as it was difficult to prove, and "practical weakness", as it was seen in opposition to agricultural progress and economic growth. Regarding the dimension of social and cultural weakness, however, a more nuanced picture emerges: Although pesticide criticism started with a weak institutional anchoring, its supporters became more numerous, passionate and powerful and therefore developed a strong cultural embedding.[65]

As such, pesticide criticism never developed an independent institution-alized structure of knowledge production, yet it had a transformative role in environmental discourse. This was due to the close relationship between established science and international environmental activism on the one hand, and the political and normative framing of the pesticide problem on the other.

Protests against pesticide exports to the developing world during the 1970s and 1980s produced a specific form of lay criticism by appro-priating evidence from diverse backgrounds. If we understand evidence as something that is not stable but constantly constructed and re-constructed in complex negotiation processes between experts and the public, it offers a lens through which we can better understand its polit-ical dimension. This perspective sheds light on evidence as a resource and therefore on interlinks with different forms of power, especially in knowledge-based societies. In this regard, the pesticide controversy illuminates how environmental knowledge could rise to pre-eminence. Real-world evidence for the risks of pesticide use in the South in the form of pictures and alarming numbers would appear on the front pages of newspapers and receive considerable attention from administrators and policymakers. What made evidence practices by environmentalists spe-cific was how they reused official data for their own arguments and questions, thereby turning results from secondary sources – often with opposing agendas – into evidence for their own cause. Combined with dramatic visual evidence, they added a real-world dimension to pesticide use in the South. Contextualizing evidence, recycling it and using it for the opposite purpose points to the portability of evidence as a resource of power that can be deployed by influential industry lobbies as well as by marginalized groups. However, although pesticide activism signifi-cantly changed the discourse on pesticide export and called attention to its devastating consequences in the developing world, regulatory effects were remarkably poor. Pesticide activism and the formation of environ-mental knowledge was not strong enough to ground sustained and last-ing action against pesticide hazards in the Global South.

Nevertheless, the debates on pesticide exports point to a pivotal moment in political and environmental history as they reflect the shifting environmental discourse during the late 1970s and the way it opened up for global questions. During the pesticide controversy, environmental, human rights and health activists challenged established certainties by denouncing national solutions and pointing to the fundamentally global nature of the pesticide problem. A decade after *Silent Spring* destabilized the myth of DDT as a "miracle chemical", environmental contamination in the South showed that Western environmentalism had been contained within American and European borders. For environmental activists from the Western world, demanding bans of dangerous substances in their respective home countries no longer sufficed. International coalitions such as PAN called out how environmental regulation differences created a world of environmental inequality. This shift to a more global perspective, however, implied a closer collaboration between citizens from industrial countries and the developing world. Compared to the "scientific" criticism of *Silent Spring*, the influence and authority of environmental criticism now increasingly relied on access to the South, the ability to present evidence from these places and finally, on personal contact with its inhabitants. Ultimately, though relying on stark power imbalances and asymmetries in exposure to risk, this shift brought an increasing recognition of local knowledge.

Notes

1 Rachel Carson, *Silent Spring* (Boston: Houghton Mifflin, 1962), 13.
2 Thomas R. Dunlap, ed., *DDT, Silent Spring, and the Rise of Environmentalism: Classic Texts*, Weyerhaeuser Environmental Classics (Seattle: Univ. of Washington Press, 2008); Mark H. Lytle, *The Gentle Subversive: Rachel Carson, Silent Spring, and the Rise of the Environmental Movement*, New Narratives in American History (New York: Oxford Univ. Press, 2007); David Kinkela, *DDT and the American Century: Global Health, Environmental Politics, and the Pesticide That Changed the World* (Chapel Hill: Univ. of North Carolina Press, 2011); Frederick Rowe Davis, *Banned: A History of Pesticides and the Science of Toxicology* (New Haven: Yale Univ. Press, 2014). On the rise of pollutants and their regulation see Nathalie Jas and Soraya Boudia, eds., *Toxicants, Health and Regulation since 1945*, Studies for the Society for the Social History Medicine 9 (London: Pickering et Chatto, 2013).
3 Peter Hough, *The Global Politics of Pesticides: Forging Concensus from Conflicting Interests* (Hoboken: Taylor and Francis, 2014).
4 Nancy Cartwright, "Well-Ordered Science: Evidence for Use", *Philosophy of Science* 73, no. 5 (2006): 983, https://philpapers.org/rec/CARWSE.
5 For a praxeological perspective on evidence and the relationship between science, technology and the role of knowledge in society see Sarah Ehlers and Stefan Esselborn, "Introduction: Evidence in Action", in *Evidence in Action Between Science and Society: Constructing, Validating and Contesting Knowledge*, eds. Sarah Ehlers and Stefan Esselborn, Routledge Studies in the History of Science, Technology and Medicine (London, New York: Routledge, 2022); see also Karin Zachmann and Sarah Ehlers, eds., *Wissen und Begründen: Evidenz als umkämpfte Ressource in der Wissensgesellschaft* (Baden-Baden: Nomos, 2019). On the role of expertise and consensus for environmental policy see: Michael Oppenheimer, Dale Jamieson, Jessica O'Reilly, Matthew Shindell, Milena Wazeck and Naomi Oreskes, *Discerning Experts: The Practices of Scientific Assessment for Environmental Policy* (Chicago: Univ. of Chicago Press, 2019).
6 Peter Milius and Dan Morgan, "Hazardous Pesticides Sent as Aid", *Washington Post*, December 8, 1976; Thomas Zimmer, "In the Name of World Health and Development: The World Health Organization and Malaria Eradication in India, 1949–1970", in *International Organizations and Development, 1945–1990*, eds. Marc Frey, Sönke Kunkel and Corinna R. Unger (London: Palgrave Macmillan UK, 2014); Marcus Linear, *Zapping the Third World: The Disaster of Development Aid* (London: Pluto

Press, 1987). On the earlier colonial history of DDT see Sabine Clarke, "Rethinking the Post War Hegemony of DDT: Insecticides Research and the British Colonial Empire", in *Environment, Health and History*, eds. Virginia Berridge and Martin Gorsky (Basingstoke: Palgrave Macmillan, 2012). For a sociological perspective on the current dealings with toxicants see Didier Torny, "Managing an Everlastingly Polluted World. Food Policies and Community Health Actions in the French West Indies", in *Toxicants, Health and Regulation since 1945*, eds. Nathalie Jas and Soraya Boudia, Studies for the Society for the Social History Medicine 9 (London: Pickering et Chatto, 2013).

7 Elena Conis, "Debating the Health Effects of DDT: Thomas Jukes, Charles Wurster, and the Fate of an Environmental Pollutant", *Public Health Reports* 125, no. 2 (2010); Thomas R. Dunlap, *DDT: Scientists, Citizens, and Public Policy* (Princeton: Princeton Univ. Press, 1981).

8 Barbara Dinham, ed., *The Pesticide Hazard: A Global Health and Environmental Audit* (London: Zed Books, 1993).

9 Michael Hiltzik, "Rachel Carson, 'Mass Murderer'? A Right-Wing Myth About 'Silent Spring' Is Poised for a Revival", *Los Angeles Times*, February 6, 2017, https://www.latimes.com/business/hiltzik/la-fi-hiltzik-carson-myth-20170206-story.html.

10 Dinham, *The Pesticide Hazard*, 11.

11 Sandra L. Postel, *Defusing the Toxics Threat: Controlling Pesticides and Industrial Waste*, Worldwatch paper 79 (Washington: Worldwatch Inst., 1987); Kinkela, *DDT and the American Century*, 172.

12 Jürgen Knirsch, ed., *Pestizide, Ex- & Import: Folgen des Pestizidexports in Länder der Dritten Welt. Beiträge einer öffentlichen Anhörung zu den Folgen des Pestizidexports in Länder der Dritten Welt* (Köln: Kölner-Volksblatt-Verl., 1985).

13 David Bull, *A Growing Problem: Pesticides and the Third World Poor* (Oxford: OXFAM, 1982).

14 Jerry Jeyaratnam, Kwok C. Lun. and Wai-On Phoon, "Survey of Acute Pesticide Poisoning Among Agricultural Workers in Four Asian Countries", *Bulletin of the World Health Organization* 65, no. 4 (1987).

15 Alan Riding, "Free Use of Pesticides in Guatemala Takes a Deadly Toll", *New York Times*, November 9, 1977.

16 On sustainable development see Stephen J. Macekura, *Of Limits and Growth: The Rise of Global Sustainable Development in the Twentieth Century*, Global and International History (Cambridge: Cambridge Univ. Press, 2015); Iris Borowy, *Defining Sustainable Development for Our Common Future: A History of the World Commission on Environment and Development (Brundtland Commission)* (London: Routledge, 2014); John McCormick, *The Global Environmental Movement: Reclaiming Paradise* (London: Belhaven Press, 1989), 149–170.

17 Robert Richter, "For Export Only – Pesticides". Aired 1981, on PBS; David Weir and Mark Schapiro, *Circle of Poison: Pesticides and People in a Hungry World* (Oakland: Institute for Food and Development Policy, 1981); Ruth Norris, ed., *Pills, Pesticides & Profits: The International Trade in Toxic Substances* (Croton-on-Hudson: North River Press, 1982); Ed Magnuson, "The Poisoning of America", *Time* 116, no. 12 (1980), 58–66.

18 Dominique Pestre, "A Weaker Form of Knowledge? The Case of Environmental Knowledge and Regulation", in *Weak Knowledge: Forms, Functions, and Dynamics*, Discourses of Weakness and Resource Regimes, vol. 4, eds. Epple, Imhausen, and Müller (New York: Campus Verlag, 2020), 295–320.

19 Barry I. Castleman and Vincente Navarro, "International Mobility of Hazardous Products, Industries, and Wastes", *International Journal of Health Services Planning, Administration, Evaluation* 17, no. 4 (1987): 12.

20 Rob Nixon, *Slow Violence and the Environmentalism of the Poor* (Cambridge, MA: Harvard Univ. Press, 2011), 3; David Arnold, *Toxic Histories: Poison and Pollution in Modern India*, Science in History (Cambridge: Cambridge Univ. Press, 2016).

21 Christoph Kueffer, Flurina Schneider, and Urs Wiesmann, "Addressing Sustainability Challenges with a Broader Concept of Systems, Target, and Transformation Knowledge", *GAIA. Ecological Perspectives for Science and Society* 28, no. 4 (2019), 386–388. See also the contribution by Christoph Kueffer to this volume.

22 Pestre, "A Weaker Form of Knowledge? The Case of Environmental Knowledge and Regulation", 297.

23 On strategic uses of scientific evidence in the tobacco industry see Naomi Oreskes and Erik M. Conway, *Merchants of Doubt: How a Handful of Scientists Obscured the Truth on Issues from Tobacco Smoke to Global Warming* (New York: Bloomsbury, 2010).

24 E.W. Kenworthy, "DDT. In the End the Risks Were Not Acceptable", *New York Times*, June 18, 1972.

25 Thomas H. Jukes, "To the Editor", *New York Times*, September 4, 1971.

26 See for example "Benefits and Costs of Pesticide Use. FAO/IAEA International Symposium on Agrochemicals: Fate in Food and the Environment Using Isotope Techniques, Held in FAO Headquarters in Rome, Italy, 7 to 11 June 1982", *IAEA Bulletin* 24, no. 3 (1982), 39–41.

27 Richard D. Lyons, "Pesticide: Boon and Possible Bane", *New York Times*, December 11, 1977.

28 Kinkela, *DDT and the American Century*, 136–160; Conis, "Debating the Health Effects of DDT".

29 Macekura, *Of Limits and Growth*; Matthias Schmelzer, *The Hegemony of Growth: The OECD and the Making of the Economic Growth Paradigm* (Cambridge: Cambridge Univ. Press, 2016).

30 Sean F. Johnston, "The Technological Fix as Social Cure-All: Origins and Implications", *IEEE Technology and Society Magazine* 37, no. 1 (2018), 47–54.

31 Soraya Boudia, "Managing Scientific and Political Uncertainty. Environmental Risk Assessment in a Historical Perspective", in *Powerless Science? Science and Politics in a Toxic World*, eds. Soraya Boudia and Nathalie Jas, Environment in History 2 (New York: s.n., 2014); Samuel P. Hays, *A History of Environmental Politics since 1945* (Pittsburgh: Univ. of Pittsburgh Press, 2000).

32 The *Programme of Work for the OECD 1971* was presented by the Secretary General to the Council on October 1970, quoted in Pestre, "A Weaker Form of Knowledge? The Case of Environmental Knowledge and Regulation", 306.

33 Pesticides Trust, *The Pesticide Hazard: A Global Health and Environmental Audit* (London: Zed Books, 1993), 11–37.

34 Weir and Schapiro, *Circle of Poison*. In the wake of its publication, a "circle of poison bill" was introduced repeatedly during the 1990s, but failed every time. Suzie Larsen, "Pesticide Dumping Continues; Leahy to Reintroduce Circle of Poison Bill", *Mother Jones*, July 21, 1998.

35 Backcover of Weir and Schapiro, *Circle of Poison*.

36 Richter, "For Export Only"; Leslie Ware, "The Pesticide Boomerang", *Audubon* 81, no. 5 (1979): 150.

37   Larsen, "Pesticide Dumping Continues; Leahy to Reintroduce Circle of Poison Bill".

38   Weir and Schapiro, *Circle of Poison*.

39   Quoted in Bull, *A Growing Problem*, 148. Kiano's statement has been widely quoted, see Bull, *A Growing Problem*, 19, 181.

40   Quoted in Francine Schulberg, "United States Export of Products Banned for Domestic Use", *Harvard International Law Journal* 20, no. 2 (1979): 366; Bull, *A Growing Problem*, 148.

41   Anne Schonfield, Wendy Anderson, and Monica Moore, "PAN's Dirty Dozen Campaign – the View at Ten Years", *Global Pesticide Campaigner* 5, no. 3 (1995), 8–10.

42   See also Paul Adler, *No Globalization Without Representation: U.S. Activists and World Inequality* (Philadelphia: Univ. of Pennsylvania Press, 2021), 61–71.

43   "Consumers Take the Offensive Against Multinationals. An Interview with Anwar Fazal", *The Multinational Monitor* 3, no. 7 (1982), https://multinationalmonitor.org/hyper/issues/1982/07/interview-fazal.html.

44   "Consumers Take the Offensive Against Multinationals. An Interview with Anwar Fazal".

45   On counter science during the German campaign against nuclear energy, see the contribution by Stefan Esselborn and Karin Zachmann to this volume. On "radical science" see Peter J. Taylor and Karin Patzke, "From Radical Science to STS", *Science as Culture* 30, no. 1 (2021), 1–10; Sigrid Schmalzer, Alyssa Botelho, and Daniel S. Chard, eds., *Science for the People: Documents from America's Movement of Radical Scientists* (Amherst: Univ. of Massachusetts Press, 2018); David King and Les Levidow, "Introduction: Contesting Science and Technology, from the 1970s to the Present", *Science as Culture* 25, no. 3 (2016), 367– 372.

46   For example Oxfam Project File, IND 55; Oxfam SL 12 "Planthopper Controlo Research".

47   See Stephen Bocking, *Nature's Experts: Science, Politics, and the Environment* (New Brunswick: Rutgers Univ. Press, 2004), 199–223. See also the discussion on citizen science Bruno J. Strasser, Jérôme Baudry, Dana Mahr, Gabriela Sanchez, and Elise Tancoigne, "'Citizen Science'? Rethinking Science and Public Participation", *Science & Technology Studies* 32, no. 2, (2019), 52–76.

48   Carina Weber and Peter Becker, "Die oder wir? Zur Rolle der Pestizide bei der Bekämpfung von Ernte- und Nachernteverlusten", in *Globale Ernährungssicherung und Pestizide; Besteht ein Zusammenhang?*, ed. Pestizid Aktions-Netzwerk (Hamburg: Confront, 1991), 27.

49   Industrieverband Pflanzenschutz 1983: Pflanzenschutz in der Dritten Welt, IPS Forum 5, Industrieverband Pflanzenschutz e.V., Frankfurt am Main, quoted in: Jürgen Knirsch, "Fragen, Fakten und Fiktionen. Eine Auseinandersetzung mit den Begründungszusammenhängen der chemischen Industrie zum Thema „Welternährung und Pestizide", in *Globale Ernährungssicherung und Pestizide; Besteht ein Zusammenhang?*, ed. Pestizid Aktions-Netzwerk (Hamburg: Confront, 1991), 15.

50   See for example Robert van den Bosch, *The Pesticide Conspiracy* (Garden City: Doubleday and Company, 1978); Pestizid Aktions-Netzwerk e.V., "Pestizid-Brief" (Hamburg: PAN Germany, 1988); Anna-Maria Hagerfors, *Giftexport: Pharmaka und Pestizide für die dritte Welt*, Rororo aktuell 5436 (Reinbek bei Hamburg: Rowohlt, 1984).

51   Knirsch, "Fragen, Fakten und Fiktionen", 11–12.

52   Bull, *A Growing Problem*, 96.

53 See for example Robert Richter, "For Export Only".

54 John F. Copplestone, "A Global View of Pesticide Safety", in *Pesticide Management and Insecticide Resistance*, eds. David Watson and A.W.A. Brown (Oxford: Elsevier Science, 1977); Meeting. WHO Expert Committee on Vector Biology and Control, *Safe Use of Pesticides: Third Report of the WHO Expert Committee on Vector Biology and Control; [Geneva, 3–9 October 1978]*, Technical Report Series/World Health Organization 634 (Geneva, 1979); Weltgesundheitsorganisation, *Safe Use of Pesticides: 9. Report of the WHO Expert Committee on Vector Biology and Control*, Technical Report Series/World Health Organization 720 (Geneva, 1985).

55 Weir and Schapiro, *Circle of Poison*, 7.

56 WHO Meeting. WHO Expert Committee on Vector Biology and Control, *Safe Use of Pesticides*.

57 Arnold Schwab, *Pestizideinsatz in Entwicklungsländern: Gefahren und Alternativen*, Tropical Agroecology 3 (Weikersheim: Margraf, 1989), 13, 21.

58 David Michaels, Clara Barrera, and Manuel G. Gachara, "Occupational Health and the Economic Development of Latin America", *The Multinational Monitor 5*, no. 9 (1984), 94–114; Bull, *A Growing Problem*, 37–54.

59 Quoted in Pesticides Trust, *The Pesticide Hazard: A Global Health and Environmental Audit*, 53.

60 Barry I. Castleman, "The Double Standard in Industrial Hazards", *The Multinational Monitor 5*, no. 9 (1984), https://multinationalmonitor.org/hyper/issues/1984/09/castleman.html. Iris Borowy, "Hazardous Waste: The Beginning of International Organizations Addressing a Growing Global Challenge in the 1970s", *Worldwide Waste: Journal of Interdisciplinary Studies 2*, no. 1 (2019), 1–10; Simone M. Müller, "Hidden Externalities: The Globalization of Hazardous Waste", *Business History Review 93*, no. 1 (2019), 51–74; Jennifer Clapp, *Toxic Exports. The Transfer of Hazardous Wastes from Rich to Poor Countries* (Ithaca: Cornell Univ. Press, 2010); Susanna Rankin Bohme, *Toxic Injustice: A Transnational History of Exposure and Struggle* (Oakland: Univ. of California Press, 2015).

61 Rachel E. Sherman, Steven A. Anderson, Gerald J. Dal Pan, Gerry W. Gray, Thomas Gross, Nina L. Hunter, Lisa LaVange et al., "Real-World Evidence – What Is It and What Can It Tell Us?", *The New England Journal of Medicine 375*, no. 23 (2016), 2293–2297.

62 Bull, *A Growing Problem*.

63 Moritz Epple, "The Theaetetus Problem. Some Remarks Concerning the History of Weak Knowledge", in *Weak Knowledge: Forms, Functions, and Dynamics*, Discourses of Weakness and Resource Regimes vol. 4, eds. Epple, Imhausen, and Müller (Frankfurt am Main: Campus Verlag, 2020), 36.

64 See the introduction to this volume.

65 On the dimensions of weak knowledge see Epple, "The Theaetetus Problem. Some Remarks Concerning the History of Weak Knowledge", 32.

References

Adler, Paul. *No Globalization without Representation: U.S. Activists and World Inequality*. Philadelphia: Univ. of Pennsylvania Press, 2021.

Arnold, David. *Toxic Histories: Poison and Pollution in Modern India*. Science in History. Cambridge: Cambridge Univ. Press, 2016.

"Benefits and Costs of Pesticide Use. FAO/IAEA International Symposium on Agrochemicals: Fate in Food and the Environment Using Isotope Techniques, Held in FAO Headquarters in Rome, Italy, 7 to 11 June 1982". *IAEA Bulletin* 24, no. 3 (1982): 39–41.

Bocking, Stephen. *Nature's Experts: Science, Politics, and the Environment*. New Brunswick: Rutgers Univ. Press, 2004.

Bohme, Susanna Rankin. *Toxic Injustice: A Transnational History of Exposure and Struggle*. Oakland: Univ. of California Press, 2015.

Borowy, Iris. *Defining Sustainable Development for Our Common Future: A History of the World Commission on Environment and Development (Brundtland Commission)*. London: Routledge, 2014.

Borowy, Iris. "Hazardous Waste: The Beginning of International Organizations Addressing a Growing Global Challenge in the 1970s". *Worldwide Waste: Journal of Interdisciplinary Studies* 2, no. 1 (2019): 1–10.

Boudia, Soraya. "Managing Scientific and Political Uncertainty. Environmental Risk Assessment in a Historical Perspective". In *Powerless Science? Science and Politics in a Toxic World*, edited by Soraya Boudia, and Nathalie Jas, 1335–1352. Environment in History 2. New York: Berghahn Books, 2014.

Bull, David. *A Growing Problem: Pesticides and the Third World Poor*. Oxford: OXFAM; Birmingham U.K. Supplier Third World Publications, 1982.

Carson, Rachel. *Silent Spring*. Boston: Houghton Mifflin, 1962.

Cartwright, Nancy. "Well-Ordered Science: Evidence for Use". *Philosophy of Science* 73, no. 5 (2006): 981–990. https://philpapers.org/rec/CARWSE.

Castleman, Barry I., and Vincente Navarro. "International Mobility of Hazardous Products, Industries, and Wastes". *International Journal of Health Services Planning, Administration, Evaluation* 17, no. 4 (1987): 617–633.

Castleman, Barry I. "The Double Standard in Industrial Hazards". *The Multinational Monitor* 5, no. 9 (1984).

Clapp, Jennifer. *Toxic Exports. The Transfer of Hazardous Wastes from Rich to Poor Countries*. Ithaca: Cornell Univ. Press, 2010.

Clarke, Sabine. "Rethinking the Post-War Hegemony of DDT: Insecticides Research and the British Colonial Empire". In *Environment, Health and History*, edited by Virginia Berridge, and Martin Gorsky, 133–153. Houndmills: Basingstoke; Palgrave Macmillan, 2012.

Conis, Elena. "Debating the Health Effects of DDT: Thomas Jukes, Charles Wurster, and the Fate of an Environmental Pollutant". *Public Health Reports* 125, no. 2 (2010): 337–342.

"Consumers Take the Offensive Against Multinationals. An Interview with Anwar Fazal". *The Multinational Monitor* 3, no. 7 (1982).

Copplestone, John F. "A Global View of Pesticide Safety". In *Pesticide Management and Insecticide Resistance*, edited by David Watson, and Anthony W. A. Brown, 147–155. Oxford: Elsevier Science, 1977.

Davis, Frederick Rowe. *Banned: A History of Pesticides and the Science of Toxicology*. New Haven, 2014.

Dinham, Barbara, ed. *The Pesticide Hazard: A Global Health and Environmental Audit*. London: Zed Books, 1993.

Dunlap, Thomas R. *DDT: Scientists, Citizens, and Public Policy*. Princeton: Princeton Univ. Press, 1981.

Dunlap, Thomas R., ed. *DDT, Silent Spring, and the Rise of Environmentalism: Classic Texts*. Weyerhaeuser Environmental Classics. Seattle: Univ. of Washington Press, 2008.

Ehlers, Sarah, and Stefan Esselborn. "Introduction: Evidence in Action". In *Evidence in Action between Science and Society: Constructing, Validating and Contesting Knowledge*, edited by Sarah Ehlers, and Stefan Esselborn, 1–26. Routledge Studies in the History of Science, Technology and Medicine. London, New York: Routledge, 2022.

Epple, Moritz. "The Theaetetus Problem. Some Remarks Concerning the History of Weak Knowledge". In *Weak Knowledge: Forms, Functions, and Dynamics*, edited by Epple, Imhausen, and Müller, 19–41. New York: Campus Verlag, 2020.

Epple, Moritz, Annette Imhausen, and Falk Müller, eds. *Weak Knowledge: Forms, Functions, and Dynamics*. Discourses of Weakness and Resource Regimes vol. 4. Frankfurt am Main: Campus Verlag, 2020.

Hagerfors, Anna-Maria. *Giftexport: Pharmaka und Pestizide für die Dritte Welt*. Rororo aktuell 5436 = rororo aktuell. Reinbek bei Hamburg: Rowohlt, 1984.

Hays, Samuel P. *A History of Environmental Politics Since 1945*. Pittsburgh: Univ. of Pittsburgh Press, 2000.

Hiltzik, Michael. "Rachel Carson, 'Mass Murderer'? A Right-Wing Myth About 'Silent Spring' Is Poised for a Revival". *Los Angeles Times*, February 6, 2017. https://www.latimes.com/business/hiltzik/la-fi-hiltzik-carson-myth-20170206-story.html.

Hough, Peter. *The Global Politics of Pesticides: Forging Concensus from Conflicting Interests*. Hoboken: Taylor and Francis, 2014.

Jas, Nathalie, and Soraya Boudia, eds. *Toxicants, Health and Regulation Since 1945*. Studies for the Society for the Social History Medicine 9. London: Pickering & Chatto, 2013.

Jeyaratnam, Jerry, Kwok C. Lun, and Wai-On Phoon. "Survey of Acute Pesticide Poisoning Among Agricultural Workers in Four Asian Countries". *Bulletin of the World Health Organization* 65, no. 4 (1987): 521–527.

Johnston, Sean F. "The Technological Fix as Social Cure-All: Origins and Implications". *IEEE Technology and Society Magazine* 37, no. 1 (2018): 47–54.

Jukes, Thomas H. "To the Editor". *New York Times*, September 4, 1971.

Kenworthy, E.W. "DDT. In the End the Risks Were Not Acceptable". *New York Times*, June 18, 1972.

King, David, and Les Levidow. "Introduction: Contesting Science and Technology, from the 1970s to the Present". *Science as Culture* 25, no. 3 (2016): 367–372.

Kinkela, David. *DDT and the American Century: Global Health, Environmental Politics, and the Pesticide That Changed the World*. Chapel Hill: The Univ. of North Carolina Press, 2011.

Knirsch, Jürgen, ed. *Pestizide, Ex- & Import: Folgen des Pestizidexports in Länder der Dritten Welt. Beiträge einer öffentlichen Anhörung zu den Folgen des Pestizidexports in Länder der Dritten Welt*. Köln: Kölner-Volksblatt-Verl, 1985.

Knirsch, Jürgen. "Fragen, Fakten und Fiktionen. Eine Auseinandersetzung mit den Begründungszusammenhängen der chemischen Industrie zum Thema 'Welternährung und Pestizide'". In *Globale Ernährungssicherung und Pestizide; Besteht ein Zusammenhang?*, edited by Pestizid Aktions-Netzwerk. Hamburg: Confront, 1991, 5–20.

Kueffer, Christoph, Flurina Schneider, and Urs Wiesmann. "Addressing Sustainability Challenges with a Broader Concept of Systems, Target, and Transformation Knowledge". *GAIA. Ecological Perspectives for Science and Society* 28, no. 4 (2019): 386–388.

Larsen, Suzie. "Pesticide Dumping Continues; Leahy to Reintroduce Circle of Poison Bill". *Mother Jones*, July 21, 1998.

Linear, Marcus. *Zapping the Third World: The Disaster of Development Aid.* London: Pluto Press, 1987.

Lyons, Richard D. "Pesticide: Boon and Possible Bane". *New York Times*, December 11, 1977.

Lytle, Mark H. *The Gentle Subversive: Rachel Carson, Silent Spring, and the Rise of the Environmental Movement. New Narratives in American History.* New York: Oxford Univ. Press, 2007.

Macekura, Stephen J. *Of Limits and Growth: The Rise of Global Sustainable Development in the Twentieth Century. Global and International History.* Cambridge: Cambridge Univ. Press, 2015.

Magnuson, Ed. "The Poisoning of America". *Time* 116, no. 12 (1980), 58–66.

McCormick, John. *The Global Environmental Movement: Reclaiming Paradise.* London: Belhaven Press, 1989.

Meeting. WHO Expert Committee on Vector Biology and Control. *Safe Use of Pesticides: Third Report of the WHO Expert Committee on Vector Biology and Control; [Geneva, 3–9 October 1978].* Technical Report Series/World Health Organization 634. Geneva, 1979.

Michaels, David, Clara Barrera, and Manuel G. Gachara. "Occupational Health and the Economic Development of Latin America". *The Multinational Monitor* 5, no. 9 (1984).

Milius, Peter, and Dan Morgan, "Hazardous Pesticides Sent as Aid", *Washington Post*, December 8, 1976.

Müller, Simone M. "Hidden Externalities: The Globalization of Hazardous Waste". *Business History Review* 93, no. 1 (2019): 51–74.

Nixon, Rob. *Slow Violence and the Environmentalism of the Poor.* Cambridge, MA: Harvard Univ. Press, 2011.

Norris, Ruth, ed. *Pills, Pesticides & Profits: The International Trade in Toxic Substances.* Croton-on-Hudson: North River Press, 1982.

Oppenheimer, Michael, Dale Jamieson, Jessica O'Reilly, Matthew Shindell, Milena Wazeck, and Naomi Oreskes. *Discerning Experts: The Practices of Scientific Assessment for Environmental Policy.* Chicago: Univ. of Chicago Press, 2019.

Oreskes, Naomi, and Erik M. Conway. *Merchants of Doubt: How a Handful of Scientists Obscured the Truth on Issues from Tobacco Smoke to Global Warming.* New York: Bloomsbury, 2010.

Pesticides Trust. *The Pesticide Hazard: A Global Health and Environmental Audit.* London: Zed Books, 1993.

Pestizid Aktions-Netzwerk, ed. *Globale Ernährungssicherung und Pestizide: Besteht ein Zusammenhang?* Hamburg: Confront, 1991.

Pestizid Aktions-Netzwerk e.V. *Pestizid-Brief.* Hamburg, 1988.

Pestre, Dominique. "A Weaker Form of Knowledge? The Case of Environmental Knowledge and Regulation". In *Weak Knowledge: Forms, Functions, and Dynamics*, Discourses of Weakness and Resource Regimes vol. 4, edited by Epple, Imhausen, and Müller, 295–320. New York: Campus Verlag, 2020.

Postel, Sandra L. *Defusing the Toxics Threat: Controlling Pesticides and Industrial Waste.* Worldwatch Paper 79. Washington: Worldwatch Inst., 1987.

Richter, Robert. "For Export Only – Pesticides". Aired 1981, on PBS.

Riding, Alan. "Free Use of Pesticides in Guatemala Takes a Deadly Toll". *New York Times*, November 9, 1977.

Schmalzer, Sigrid, Alyssa Botelho, and Daniel S. Chard, eds. *Science for the People: Documents from America's Movement of Radical Scientists.* Amherst: Univ. of Massachusetts Press, 2018.

Schmelzer, Matthias. *The Hegemony of Growth: The OECD and the Making of the Economic Growth Paradigm.* Cambridge: Cambridge Univ. Press, 2016.

Schonfield, Anne, Wendy Anderson, and Monica Moore. "PAN's Dirty Dozen Campaign – the View at Ten Years". *Global Pesticide Campaigner* 5, no. 3 (1995): 8.

Schulberg, Francine. "United States Export of Products Banned for Domestic Use". *Harvard International Law Journal* 20, no. 2 (1979): 331–381.

Schwab, Arnold. *Pestizideinsatz in Entwicklungsländern: Gefahren und Alternativen.* Tropical Agroecology 3. Weikersheim: Margraf, 1989.

Sherman, Rachel E., Steven A. Anderson, Gerald J. Dal Pan, Gerry W. Gray, Thomas Gross, Nina L. Hunter, Lisa LaVange et al., "Real-World Evidence – What Is It and What Can It Tell Us?". *The New England Journal of Medicine* 375, no. 23 (2016): 2293–2297.

Strasser, Bruno J., Jérôme Baudry, Dana Mahr, Gabriela Sanchez, and Elise Tancoigne, "'Citizen Science'? Rethinking Science and Public Participation". *Science & Technology Studies*, 2019, 52–76.

Taylor, Peter J., and Karin Patzke. "From Radical Science to STS". *Science as Culture* 30, no. 1 (2021): 1–10.

Torny, Didier. "Managing an Everlastingly Polluted World. Food Policies and Community Health Actions in the French West Indies". In *Toxicants, Health and Regulation since 1945*, edited by Nathalie Jas, and Soraya Boudia, 117–134. Studies for the Society for the Social History Medicine 9. London: Pickering & Chatto, 2013.

Van den Bosch, Robert. *The Pesticide Conspiracy.* Garden City: Doubleday and Company, 1978.

Ware, Leslie. "The Pesticide Boomerang". *Audubon* 81, no. 5 (1979): 150.

Weber, Carina, and Peter Becker. "Die oder wir? Zur Rolle der Pestizide bei der Bekämpfung von Ernte- und Nachernteverlusten". In *Globale Ernährungssicherung und Pestizide; Besteht ein Zusammenhang?*, edited by Pestizid Aktions-Netzwerk, 27–32. Hamburg: Confront, 1991.

Weir, David, and Mark Schapiro. *Circle of Poison: Pesticides and People in a Hungry World.* Oakland: Institute for Food and Development Policy, 1981.

Weltgesundheitsorganisation. *Safe Use of Pesticides: 9. Report of the WHO Expert Committee on Vector Biology and Control.* Technical Report Series. Geneva: WHO, 1985.

Zachmann, Karin, and Sarah Ehlers, eds. *Wissen und Begründen: Evidenz als umkämpfte Ressource in der Wissensgesellschaft.* Baden-Baden: Nomos, 2019.

Zimmer, Thomas. "In the Name of World Health and Development: The World Health Organization and Malaria Eradication in India, 1949–1970". In *International Organizations and Development, 1945–1990*, edited by Marc Frey, Sönke Kunkel, and Corinna R. Unger, 126–149. London: Palgrave Macmillan UK, 2014.

9 Participation as Evidence Contestation

The Ambiguous Balance of Social and Epistemic Involvement through Citizen Science[1]

Kevin Altmann and Andreas Wenninger

Thanks to digital technologies and the ongoing transformation of science communication, the boundaries between science and society have become progressively more porous. In the context of digital platforms and media, and in the light of science policy programs such as Public Understanding of Science, Public Engagement with Science and Technology or Open Science, the closure of science is increasingly perceived as problematic in public debates. One of the frequently mentioned problems in respect of the supposedly widening gap between science and society is that the actual human and societal needs of society are not (or no longer) perceived by science and that science therefore does not provide adequate solutions to the urgent problems of humanity. Participation and citizen science are often seen as a necessary corrective to this tendency, not inevitably by dissolving or shifting boundaries, but through the emergence of corridors where laypeople can gain access to scientific knowledge production. However, due to the fact that a corridor also has limitations of its own, the ways of access are not all-encompassing but defined in specific ways. The issue of ecological sustainability is just one example.[2] One aim of science communication is to disseminate scientific rationality from academia throughout society using more engaging forms of communication. One can see citizen science, conversely, as a movement to integrate people from outside academia as representatives of (civil) society within the social dimension of science, meaning that people are included in research as *addressable actors* with a more or less concretely defined opportunity to participate. Participation can be limited to a few mouse clicks or constitute an (equal) right to speak in scholarly discussion.[3] This is one way in which corridors enable exchange between science and society. From the perspective of science, however, such opening processes can easily be perceived as de-stabilization if lay participation is accompanied by a contestation of the evidence put forward by professional scientists. The assumption that this involvement in the social dimension also offers opportunities to influence academia in its factual

DOI: 10.4324/9781003273509-14

dimension – i.e. the ability for participants to seriously co-determine the topics, research methods and evaluation in the research process, which represents another corridor to enter scientific knowledge production – is usually accepted without reflection.[4] The following article offers a first step toward more nuanced considerations in this regard by differentiating between different corridors of exchange through citizen science, especially considering the open question of whether certain boundaries between science and society will be maintained or will shift due to lay participation. We will describe these corridors in more detail later as *modes* of citizen science.

Citizen science refers in a broad sense to research projects and activities in which professional experts and laypeople work together on scientific knowledge production. The label citizen science first appeared in the 1990s and the activities subsumed under it have been steadily increasing since then – strongly driven by digitization since 2010.[5] The spectrum of forms of participatory research ranges from the retention of a classical understanding of science, which assigns the participants comparatively marginal tasks in the production of knowledge, to projects in which laypeople organize the entire research approach autonomously. Overall, the diversity of citizen science "makes it difficult, and probably counterproductive, to speak of citizen science as a single or coherent practice, because it includes practices with different actors, missions, and values".[6] Besides the role of citizen science in present scientific funding policy, the heterogeneity of citizen science becomes apparent through various institutional formations[7] in academia.[8] The establishment of citizen science platforms like Zooniverse and Bürger schaffen Wissen (Citizens Create Knowledge) or international umbrella organizations as the European Citizen Science Association (ECSA) are examples for bundling citizen science activities and actors.[9] Publication formats have been established over time, in which actors from different backgrounds discuss and elaborate the various citizen science activities as part of the orientation of the citizen science field.[10]

Science today seems to be organized in such a way that it guarantees its thematic openness (anything can become the subject of scientific research). Nevertheless, there are various measures designed to ensure the independence of science from non-scientific interests. According to some, the ways in which research findings are obtained and the evaluation of scientific results is ideally left exclusively to certified experts:

> To produce useful results for society, such as knowledge with practical applications and policy implications, scientists should be allowed to make decisions within their domain of expertise, free from outside interference and control.[11]

We term this *professional control*. Direct participation in the epistemic core of science thus seems difficult or even impossible for laypeople (i.e. persons without appropriate credentials). Despite all the current changes regarding the relationship between science and society, the epistemic core of institutionalized science remains relatively stable:

> While a transformation is taking place in the social dimension and new actors are being admitted, authorities are being questioned and old structures are being broken up, comparatively little movement can be observed in the factual dimension [...] The opening of the social dimension is taking place with the simultaneous closing of the factual dimension.[12]

Given the rise and establishment of citizen science, this appears paradoxical. Our previous investigations of citizen science projects have shown the varying dynamics between opening and closing scientific participation:

> the inclusion of people (social dimension) is subject to multiple socio-technical restrictions and channelings. Within the project contexts, these have the function of counteracting the increased participation as a result of opening up in one dimension (social, factual or time dimension) by restricting participation in other dimensions, not only to make the projects practically manageable, but also to maintain professional control in the project contexts despite increased participation.[13]

Participation in the social dimension has tended to lead to closures in the factual dimension. Against this background, we investigate the following guiding question in this chapter: What citizen science activities enable not only participation within the social dimension of science, but also in the factual dimension? Which is to ask whether and to what extent laypeople can participate in the *creation and conception* of research projects. Such citizen science activities would potentially allow for the contestation of evidence offered by certified scientists through extra-scientific participation, e.g. by introducing new and challenging content, views or practices. Our question also involves the extent to which citizen science makes the realization of a contestation of evidence adjustable in order to mediate between the opening and closing processes of scientific knowledge production.

In the first section, we outline the different expectations associated with the emergence of citizen science and to what extent they carry narratives of evidence-critical participation. In the second section, we introduce the diverse *modes* of citizen science and examine their respective potential for evidence contestation and how these modes organize lay participation in various degrees. For this purpose, we situate the modes in a schematic relation to academia and the prevailing notion of epistemic participation. In the third section, we consider the mechanisms of structurally enabling evidence contestation for selected modes based on specific cases. Finally, we examine the interplay of the different modes, which allows for moderate evidence contestation in citizen science. We find that evidence contestation is not a phenomenon concerning citizen science in general. The heterogeneity of citizen science allows contestation only in very specific ways, and then only very moderately. This serves the safeguarding of the common production of scientific knowledge from potentially dangerous influences through lay participation but still permits the feeding in of new perspectives. Therefore, we are not focusing on the actual contestation of evidence through lay participation, but reflect on the structural prerequisites for an influence on the factual dimension of science through various modes of citizen science.

Divergent Expectations on Participatory Research

In the increasingly established and sophisticated field of citizen science, one can identify various demands and expectations of participatory research, both within the field and from outside. One can "see citizen science as essential to closing the gap between society, science, and politics. Citizen science is seen as producing relevant knowledge, but also as building connections between disparate institutions".[14]

Common self-portrayals of citizen science activities frame it as a democratizing research approach to the production of socially robust knowledge with the emphasis on research that is thus more relevant to society.[15] At the same time, the concept of citizen science is gaining political weight.[16] Now, in addition to the claim to social relevance, the development of general requirements and standards for citizen science activities points to a progressive professionalization of the field. These standards aim to secure, evaluate and control scientific evidence under the premise of lay participation, as, for example, the ECSA's ten principles of citizen science show: "Citizen science projects have a genuine science outcome. For example, answering a research question or informing conservation action, management decisions or environmental policy".[17]

We understand the cross-project formulation of guidelines as a means of ensuring the scientificity of participatory research and thus as a form of boundary work between science and non-science.[18] The elaboration of scientific principles to regulate access to participation in science is intended to filter out content and actors (including political activism, economic interests and anti-scientific efforts) that may endanger the generation of (scientific) evidence.

These different expectations give rise to a tension between the preservation of scientific premises and a possible challenge to classical evidence production by citizen science. The establishment of participatory research structures now provides the conditions for professionalized science to engage in self-reflective and yet scientifically controlled contestation of evidence. But the contestation of scientific evidence, previously reserved for professional researchers, now also encompasses civic participants who do not necessarily see themselves as bound by the standards of scientific methodology. In this regard, it is crucial to distinguish between two historical lines of citizen science approaches, which have different ways of dealing with the tension between epistemically effective and scientifically controlled lay participation.

In this respect, Caren B. Cooper and Bruce W. Lewenstein refer, on the one hand, to the US-American line of citizen science, which is mainly oriented toward the organization of cooperation between laypeople and experts. On the other hand, there is a second, British-European line of citizen science,[19] which has a much stronger science-critical and social-theoretical reflexive approach.[20] Gwen Ottinger also distinguishes between two traditional lines of public participation in scientific research: She contrasts a "social movement-based citizen science with a scientific authority-driven citizen science".[21]

From our perspective, the most globally established US citizen science approach does not aim at a fundamental critique or transformation of science, but rather at science-conforming (or scientific authority-driven) participation. However, this does not mean that this form of citizen science is less important for social transformation processes. A professionally guided and thus "solid" scientific implementation of citizen science can enhance its social relevance. Concerns of laypeople (e.g. environmental protection, rare diseases) or certain social groups (e.g. indigenous peoples) may thus have a greater chance to be taken into account within political decision-making processes. However, even within less fundamentally critical approaches of lay participation, there are different varieties of citizen science, which offer latent forms of a reconfiguration of classical knowledge production. These *modes of citizen science* – as representations of participation primarily conforming to science – can still contain evidence-critical components that open up the factual dimension to input from laypeople.

In what follows, we outline these different modes of citizen science and their underlying concepts of civic participation. We ask to what extent these respective modes show potential for not only making people and their labor power available for scientific research in social terms, but also for feeding factual-content elements (knowledge) into it. This will serve to identify the substantive positioning vis-à-vis academia and to show the opportunities for a contestation of evidence by citizen science.

The Spectrum of Participatory Research and the Positioning of Citizen Science in Relation to Academia

The Modes of Citizen Science

In order to develop answers to our guiding question, a number of conceptual considerations are required. Since it is difficult – as we mentioned in our introduction – to speak of a homogenous concept of citizen science in a useful way, it is methodologically necessary to introduce some differentiations at this point. This is essential to reduce the exuberant variety of forms of the label *citizen science* to types that are methodologically manageable and do some justice to practice. Here we use classifications of participatory activities in the social dimension of science that have become common in the citizen science literature – the so-called *modes of citizen science*.[22] Furthermore, the diversity of actor types and disciplinary fields is part of contemporary research on citizen science.[23] In contrast, scientific investigations into the spectrum of opportunities for contesting evidence through citizen science are still an open field.

> 'Citizen science' has mainly been viewed as a way of assisting scientists in reaching their research goals, ignoring the possibility that participatory research could also expand what counts as the scientific worldview.[24]

To examine the possibilities for evidence contestation in specific modes, we combine different methods of data generation and evaluation. Primarily this happens in the form of ethnographic observation of group and network meetings in the field of citizen science, supplemented by the examination of central documents and online platforms. For the present contribution, we make a heuristic reduction of this methodological approach. We enrich our theoretical-conceptual considerations with empirical case studies. The focus of our contribution is oriented toward qualitative-reconstructive social research[25] in order to open up and theoretically reflect on the fundamental structures of meaning, patterns of discourse and ethno-methods of the citizen science field.

After these conceptual and methodological considerations, we first introduce the different modes of citizen science. We note that while we do not make any empirical statements, i.e. from a sociological perspective, the question remains open whether and to what extent the different citizen science actors use the labels of the modes in the context of other interests. This may be the case, for example, when citizen science projects merely describe themselves as co-creative in order to achieve self-interested goals such as research funding or reputation but are nevertheless hierarchically structured in practice. Such inconsistencies between claims and realizations of citizen science can be critically addressed in terms of "astroturfing"[26] or in the context of a "neoliberal transformation of science".[27] Of course, one should not omit these critical reflections when dealing with the phenomenon of citizen science. By adopting the modes of citizen science from field terminology, we supplement it within the framework of our research interest with regard to the potential for contesting evidence through citizen science. We thus differentiate four modes of citizen science, which are key to our approach (see Figure 9.1).

The characterization of modes illustrates the diverse conceptions of citizen science and the various possible forms of participation. A broad

Modes of Citizen Science	
Contributory	• A larger number of laypeople contribute to a research project designed by professional academics (mostly in digital settings) • Participation is limited in time as well as to single or few project phases (e.g. classification of research objects)
Collaborative	• Professional scientists and laypeople work together in a research project over several project phases (e.g. in the design and evaluation of questionnaires) • The design and management of such projects is usually in the hands of professional scientists
Co-Creation	• Conception from the beginning in cooperation of professional scientists and citizens • Collaboration takes place jointly across all project phases
Collegiate	• Laypeople organize a project independently and without the involvement of professional scientists • They independently design their research and conduct it themselves throughout all phases of the project

Figure 9.1 Modes of Citizen Science. Figure by Kevin Altmann and Andreas Wenninger.

spectrum of participatory research has emerged, in which laypeople have both the opportunity to participate in science at a low threshold (primarily in the contributory and collaborative modes) and to participate in citizen science at various degrees of self-direction and conceptual work (primarily in the collegiate and co-creation modes). Our impression is that citizen science is not simply an addendum to science but is establishing itself as a science-related network that is largely engaged in balancing divergent expectations from different sectors of society through its heterogeneity. The importance of the analytical focus on the ideal types of participatory research is particularly evident in the "increasingly institutionalized citizen science movement".[28] In order to understand the ways in which these modes of citizen science – in addition to serving various functions, e.g. educational – can also generate opportunities to contest scientific evidence, it will help to locate them vis-à-vis academia.

The Spectrum of Participatory Research

The proximity or distance to academia generates insights into the extent to which, for example, professional control is evident in the modes of citizen science. This in turn determines the degrees of freedom of

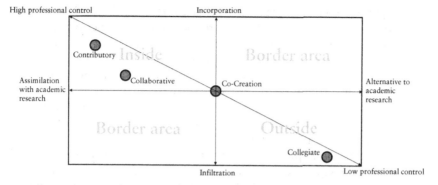

Figure 9.2 Structure of the coordinate system and location of the modes of citizen science. Graphic by Kevin Altmann and Andreas Wenninger.

participation, which we understand as structural conditions for a contestation of evidence. In particular, we are interested in the question of what we term the *capacity for influence*. This means the opportunity for laypeople to feed their specific content into science and their potential to question established procedures of scientific research. With the term capacity for influence, we do not primarily mean the scientific outcome of projects, measured in terms of peer-reviewed publications, as in the study by Dick Kasperowski and Thomas Hillman.[29] Rather, we see this in qualitative terms, i.e. the potential to exert any influence at all on the factual dimension of scientific research for laypeople.[30] Alternatively, to take up the question posed by Bruno J. Strasser et al.: Does citizen science produce a "New Science"?[31] For us, these capabilities are the analytical access to explore the possibilities of evidence contestation in citizen science. In this regard, we want to clarify: Our categorization of citizen science by modes does not mean that a mode represents "good" or "bad" science, or that modes should be normatively compared to each other in terms of "better" and "worse". A mode with a high professional control is thus not automatically better science than others. Our following scheme is likewise not designed to define the boundaries of science and non-science. We understand the modes of citizen science themselves as those definitions where the field itself locates its outer boundaries. In order to illustrate and specify our previous considerations, we outline the following schematic system of coordinates.

We localize the modes based on three scales, each consisting of two poles facing each other (see Figure 9.2):

1 *Assimilation with* and *alternative to* academic research: We understand the pole Assimilation as the (in its maximum form) quasi-silent integration of lay participation into the structures of academic research.[32] In contrast, alternative means the establishment of a

separate, civically organized research identity beyond the infrastructures and professional control of science. In other words, those modes of citizen science that are relatively close to the pole of assimilation, in our view, primarily follow the organizational structures of scientific evidence production (project design, time planning etc.). Modes in proximity to the pole alternative conversely offer more potential for feeding in factual lay content. Alternative in this case does not mean non-science, but a more self-initiated way of practicing scientific participation.

2 *High* and *low* professional control:[33] High professional control means the scientifically guided evaluation of cooperative work with laypeople. We see a high degree of professional control primarily in science-initiated activities for preserving and implementing scientific guidelines in participatory research. Low professional control indicates an opening to the possibility of developing alternative perspectives and bodies of knowledge. This potentially offers a greater influence in the factual dimension. However, (smooth) connectivity to academic research is not possible here, or only with difficulty.

3 *Incorporation* and *infiltration*: Incorporation represents (in its maximum form) the adoption of the scientific understanding of professionalized research through the complete embedding of foreign components into science (e.g. when laypeople contribute to a professionally prefabricated research conception). Infiltration also means the influx of foreign entities into science; however, they do not lose their property of otherness and challenge science epistemically to a greater or lesser extent, albeit with comparative difficulty.

The contributory and collaborative modes locate themselves in the left half of the scheme, which we frame as the *inside* area of the participatory spectrum. Therefore, they are in proximity to the poles assimilation, high professional control and incorporation. These modes are characterized above all by the fact that they offer selective access to individual research phases and that both the project conceptions and the joint research activities are determined by the interpretive authority of professional researchers.[34] So *inside* does not mean being inside science as a closed area, but being part of established processes of scientific knowledge production. "Scientists generally design projects to which members of the public contribute data but also help to refine project design, analyze data and/or disseminate findings".[35]

In our view, these modes fulfill two things: On the one hand, the reproduction of classical scientific knowledge production and thus also the safeguarding of evidence; on the other hand, they generate a socially broad anchoring of science by enabling low-threshold participation for laypeople. This can be, for example, access to different research phases without structural hurdles such as the need for an academic degree.

The relatively high degree of professional conceptualization and guidance limits the discursive space for participatory research, which offers little scope for evidence contestation. Because of the connectivity to academic research structures, we refer to them collectively below as assimilative modes.

The collegiate mode locates instead on the right half of our scheme near to the poles alternative, low professional control and infiltration and thus beyond or respectively *outside* invited participation[36] of scientific research, but still in connection with the scientific system. Collegiate opens up greater scope for a potential contestation of evidence through greater capacity of influence on the factual dimension. These capacities arise in collegiate mode from the changing constellations of expertise. Here, laypeople can have a pronounced or even the sole interpretive authority over the design of participation.

The mode co-creation is located exactly in the middle and therefore at the border areas of our scheme. Border areas indicate the hybridity of citizen science activities, which often show mixed forms of assimilative and alternative elements of civic participation. Citizen science cannot always be clearly assigned to one pole and does not always have to remain in the same relation to academia, e.g. due to changes in project conceptions. We speak of border areas to reflect that citizen science projects do not necessarily represent a particular mode and their orientation may change over time. However, by positioning itself between the assimilative and alternative pole, co-creation represents neither a complete appropriation by, nor a clear demarcation from academic research structures. The common research identity is fed by the equivalent distribution of participation between researchers and laypeople. Even without the exclusive orientation toward the formally established procedures of professional research, the connectivity to academic institutions and their control mechanisms remains. Instead, professional control can take place cooperatively, e.g. when both scientists and laypeople evaluate their collaboration. The positioning of co-creative citizen science "between Science and Civil Society"[37] enables structural opportunities for a potential contestation of scientific evidence.

We understand citizen science projects not only as partial forms of civic participation, but also as part of more general (cross-project) conceptions (*modes*). The schematic location of the modes allows us to grasp the diverse positions that citizen science can take in relation to academic research structures. The preceding mapping is thus essential, since from this localization the potential for contesting evidence becomes visible and understandable. In the following, we focus specifically on the collegiate and co-creation modes, as these two offer the greatest potential for contesting evidence through lay participation. We will elaborate the specific mechanisms for creating contesting opportunities for laypeople. At the same time, however, we illustrate which epistemic regulations go along with these opportunity structures for contesting evidence. We will illustrate this by means of reference to empirical case studies.

Collegiate: On the Topothek Project and the (De-)Realization of Epistemic Potential

As the spectrum of participatory research shows, an alternative, not (primarily) academia-initiated conception of citizen science occurs in the field of citizen science, which represents a structural response to non-scientific expectations of participation in science. The citizen science platforms themselves address these civic formats of research, which we have already identified as participation in the collegiate mode:[38]

> In this context, **cooperation between research institutions and insti-tutionally unaffiliated individuals** can take very different forms, ranging from completely self-initiated 'free' projects [...] to guidance by scientific institutions lay people [emphasis in original].[39]

Such *free* projects in the collegiate mode and their position outside pro-fessionalized research structures evade the control mechanisms found in the collaborative/contributory modes. The question arises of how to make this potential loss of professional control manageable in the collegiate mode but still recognize these alternative forms of citizen science as a stream of public participation in science. The case of the collegiate project Topothek and its discursive reappraisal in the course of the platformization of citizen science serves as a useful example here.

Topothek[40] is an online platform by and for people interested in local history, who collect historical material in their respective local topo-theques, digitize it and make it available online. The vast majority of topotheques can be viewed globally by anyone with internet access. The goal of the Topothek is to build an online archive that makes it possi-ble to store local historical material permanently and thus make local historical knowledge accessible (for historical research and the general public). The project is run exclusively by interested laypeople; profes-sional scientists are neither actively involved nor, to our knowledge, has the Topothek been used in the context of academic-historical research so far. The Topothek occasionally networks with museums/historical archives and schools, but there is little contact with traditional research institutions. Partly to explicitly demarcate themselves from such institu-tions, they describe the platform as an alternative, pre-scientific knowl-edge producer. In this regard, contributions from citizens are

> detached from the contexts of classical institutions, de-contextualised and re-contextualised within the framework of the Topothek as data close to real life. The de-contextualisation becomes [...] visible

as an evidence strategy in the sense of 'real-world evidence': local, individual, personal, contemporary and real-life 'data' and 'information' are the raw material for alternative form of 'knowledge'.[41]

Its connectivity to academia is thus not very pronounced. It is difficult to discern whether this is a deliberate strategy on the part of the platform operators, or whether it is a consequence of non-consideration by academic historians. Their self-positioning fluctuates accordingly between a supplement *to* and preliminary work *for* science on the one hand and a (more detailed, more lively) alternative to academic science, at least when it comes to local historical research. It is difficult to evaluate the scientific output, which we have not been able to determine so far. Topothek runs outside of academia but still makes some claim to scientific relevance. The capacity for influence here is great, as laypeople can potentially contribute their perspectives and bodies of knowledge without hindrance. At the same time, however, the connectivity to academia is severely limited, so that this potential cannot unfold there, at least for the time being.

The intriguing aspect, however, is not the project itself and how specifically evidence is produced there. Rather, the framing of the project by the institutional actors becomes significant. We do not intend to show whether or not citizen science in mode collegiate constitutes a concrete evidence contestation. Rather, we want to use the case to investigate which structural possibilities citizen science in the collegiate mode enables at all in order to be able to perform evidence contestation.

In the context of an international citizen science conference, the participants (mainly professional researchers, platform operators and organizers from the field of citizen science) of a particular workshop discussed the example of the Topothek. The session aimed to optimize the self-presentation of projects on citizen science platforms in order to best address interested citizens and possibly encourage them to participate. It is worth mentioning that projects listed on these citizen science platforms have to meet specific criteria in order to be visible on the platform. Without fulfilling these criteria, getting listed on the platform is not possible due to the platforms own policy. These criteria include various aspects, the most significant being both the scientific orientation of the projects, i.e. formulating a concrete research question, as well as the special emphasis on the inclusion of citizens and relevant societal goals.

Starting from these premises, a discussion arose in the course of the workshop about the extent to which the Topothek is a scientific project and thus citizen science, since no concrete research question has been formulated in the self-description of this project. Participant "A" (professional researcher and also member of a national citizen science platform) asked whether the Topothek had a scientific orientation (as required, among other things, by the criteria of the platform). Participant "B", who is involved in the listing processes of another citizen science platform, answered this question and explained the background of the listing of the Topothek on the platform. We present the excerpt from this discussion in the following:

A: What I find interesting, if I might now nitpick a bit, that's kind of the question here... we actually also say, often I think, that it's important to say that we posed a research question, and that it's a research project, and so where in fact is the research question here? But I actually understand it more like this: This is actually a large repository and starting from it, you can then ask a wide range of research questions. But if we now take a super critical view, we could say, well, there's actually no research question being asked here, is there?

B: Right, I'd certainly agree with that. Nevertheless, it is/it corresponds to our quality criteria, because we've written into the quality criteria that a project must either have a research question or a goal, or it must build up an infrastructure, so to speak [...] the goal [in another project, which is similar to the Topothek, authors' note] was simply to sequence the human genome and to be able to start new research based on that. And in principle, a topotheque is the same thing on a community-historical level.

[...]

A: That's a really major question, because I, because I have the feeling from some discussions that in citizen science this term is somehow just expanded and that you can also initiate projects that don't have a super clear research question defined down to the last detail. But I think for me it's not always very clear in my head how strict or how very close it has to be to research or what it's supposed to be. I do totally get what you're saying [participant B], yeah, and the, I think for some people, there are some projects that are kind of based on this "we'll create something first and then we'll take a look at it". (...) But I sometimes still urge people to say, what exactly is your research about, yeah?

[...]

B: More than happy to oblige!

It is apparent from this excerpt that "A" connects the question with the distinct claim to citizen science to raise a scientific question and to make it explicit in external communication. In contrast, "B" sees the criteria of the platform fulfilled in that the project formulates a clear goal by addressing a topic relevant to science and building up and providing a potentially scientifically relevant infrastructure due to the ambivalence of project alignments of the citizen science platform.

We want to point out that at the time of the listing of the Topothek, the criteria of the platform were not yet as strongly formulated as they are at present. Nevertheless, at the time of the presentation of the project at the mentioned conference, the project had been running for several years without a clear scientific question,[42] but remained a relatively free collaboration between laypeople, while the outcome of the project was not taken up by professionalized research or getting evaluated by scientific researchers during the project itself. So far the results primarily serve a socially relevant goal of documenting historical details of local communities. Especially in light of the fact that the citizen science platform, where the Topothek is listed, applies clear scientifically focused criteria to new projects, this illustrates the status that the Topothek project has on the platform.

In the case of the Topothek, "B" emphasizes (rather than stating a crystal-clear scientific question) the alternative possibilities offered by the project, e.g. to make knowledge gaps visible that academic research has not yet (sufficiently) perceived. "B" ascribes to this collegiate project a different role than to the professionally controlled and assimilative projects, which have dedicated scientific questions.

Thus, the collegiate mode serves as a kind of antithesis to traditional research, on which professional scientists (in our case actor "A") can explain their claims as to what a decidedly scientifically oriented citizen science should look like in contrast to such an alternative mode like collegiate. Instead of these scientifically driven projects, citizen science in the mode collegiate ensures the visibility of alternative perspectives in science in order to meet the expectations of extra-institutional actors, e.g. local communities. This enables the occurrence of lay perspectives and unattended topics in the scientific field, which represents a structural opening of the factual dimension due to the enhanced opportunities of doing self-determined lay research beyond scientific research structures. This legitimates the listing of the Topothek on the platform.

This is not only evident in the style of self-representation of projects. Citizen science platforms normalize the possible forms of alternative participation by explicitly not presenting them as activist action or even branding them as anti-scientific endeavors. Due to Thomas F. Gieryn, those brandings can be functional to draw a line between scientific and non-scientific interests, as he states in his findings about scientific boundary work and the establishment of science against former gatekeepers of knowledge like the church.[43] Against such a branding, the emphasis on the societal relevance of the Topothek intends that the project not only remains acceptable but can also be promoted as desirable and innovative, especially to those who might criticize the lack of a decidedly scientific question (in this case actor "A"). The literature in the citizen science field also specifically addresses alternative modes such as collegiate. In this regard, Jennifer L. Shirk et. al explicitly refer to the programmatic innovation potential of the collegiate mode. A specific scientific role is thereby attributed to the collegiate mode, even if it lacks practical integration in the processes of professionalized project activities:

In this model, professional and amateur researchers may collaborate only when an amateur writes and submits findings for peer review and publication. Although often overlooked or highly critiqued, committed amateurs can make critical contributions that may not otherwise transpire owing to a lack of resources, time, skills or inclinations in the professional scientific community.[44]

In addition to the social and factual dimension of participation, we take the temporal dimension also into account, in which the collegiate activities can also become functionally effective in practice. As explained in the contribution by Karin Zachmann and Stefan Esselborn in the present volume, social movements can call for alternative forms of research, which historically have led to alternative scientific initiatives. According to Zachmann and Esselborn, these initiatives have themselves taken on a variety of forms. For example, different efforts to establish parallel research that aim to examine and correct research results from established science. Other actors have sought to establish decidedly alternative forms of knowledge production. The Öko-Institut (Institute for Applied Ecology), which emerged from the environmental movement of the 1970s/1980s as an alternative research institution, should be mentioned here as an example of the institutionalization of scientific research. The Öko-Institut originated outside academia but achieved acceptance by established science and the scientific policy advice sector in the long run. The institute integrated itself into academia and thus changed its alternative status. However, this in turn also achieved a certain transformation of research through the institutionalization of the Öko-Institut. The significance of citizen science in the mode collegiate as a representative of alternative perspectives offers the possibility of future permanence. Whether a development similar to that of the Öko-Institut is possible, is a task for further research.

In the context of citizen science, the collegiate mode offers potentionally the greatest opportunities for evidence contestation, despite, or perhaps because of, the largely missing connectivity to the professionalized research structures of academia. But this structural opening does not mean that these greater potentials of influencing knowledge production lead to an actual contestation of scientific evidence. Because at the same time, the collegiate mode also assumes a semi-epistemic role, in which this mode is potentially not part of concrete elaborations of participatory frameworks, publications and practical applications inside the academic field. Although the literature in the field of citizen science does identify the collegiate mode as part of participatory streams, it shows that the collegiate mode is less relevant when it comes to analyzing the influence or concrete content orientation of citizen science.[45] In research practice, too, the collegiate mode is the least implemented form of participatory engagement among the four modes.[46] The shown excerpts illustrate the more representational relevance than the actual questioning of scientific evidence that is assigned to the Topothek project by the discussing scientists. The possibility of contesting evidence through citizen science in the collegiate mode is thus double-edged: On the one hand, the structural prerequisites are created to challenge scientific evidence beyond professional control mechanisms, due to the intense and more autonomous involvement of laypeople in the collegiate mode. Nevertheless, in the practice of scientific usage, this potential often remains unused because the precarious connectivity to the formats of academia is missing. Therefore, citizen science in the mode collegiate remains in a position outside the structures of professionalized criticism of scientific knowledge. The example of the collegiate project Topothek illustrates that mass participation by laypeople is possible in modes other than the assimilative. However, the lack of connectivity to academia does limit an epistemically effective integration of alternative citizen science in scientific institutions, their infrastructures and their respective academic reputations. In the special case of the Topothek, the given framings of the project by professional scientists (as a socially relevant form of self-initiated lay participation) and the missing connectivity to the criteria and control structures of professional science imply a limitation of realizing the potentials for an actual contestation of evidence.

Given the lack of actual contestation of evidence in the modes of citizen science presented thus far, the question arises: Which forms of citizen science now enable a contestation of evidence? In the following, we describe the mode of co-creation, especially with regard to the aspect of connecting citizen science to various interests. From our perspective, this mode is most suitable for balancing the asymmetries between the assimilation and alternative modes. It serves the different expectations of science and non-science for citizen science *simultaneously and equally,* without creating too one-sided connectivity neither with regard to the scientific premises of professionalized research nor the non-scientific demands for participation. We will also illustrate this with reference to a specific case to consider how evidence contestation within the co-creation mode can be realized and managed.

Co-Creation between Exchange and Persistence: The Approaches of Questioning Scientific Claims

In our previous case studies, we investigated one co-creation project named Patient Science.[47] Certified scientific experts from academic and non-academic fields recruited citizen scientists as so-called patient researchers. Professional and patient researchers together designed and organized a research project. A general goal was predefined, which was concretized with questions prepared (by the professional researchers) in small groups, intending to address problems in the everyday life of cystic fibrosis patients and to identify a specific need for research. We illustrate with the following summary field notes taken during ethnographic observations of this project, to which extent evidence contestation and influence for patient researchers on the factual dimension can take place within this mode of citizen science. The quotes are anonymous literal statements of the observed actors in the field from one of the authors.

When comparing several possible research topics, the involved actors compared various dimensions with each other. These included the relevance for science, the benefit for the citizen scientists and the science literacy (here it was determined whether the topic could be implemented more as traditional academic research or whether it required more application-oriented research). Despite the intensive involvement of the patient researchers, the professional scientists dominated the regular events for long periods. This was mainly because they usually took over the moderation and organization of the meetings. Presumably, this type of organization by the professional scientists results unintentionally in the patient researchers remaining in relatively passive roles. This generates only limited opportunities for patient researchers to significantly influence the discussions. Overall, one can say that the professional control is very prominent since there is also training on scientific research by professionals for the patient researchers. In addition, topics that the professional scientists judged to be not scientifically feasible or scientific enough were dismissed without much contestation. Nevertheless, moments occurred repeatedly that could be regarded as contestation of the authority of the certified academics and thus also of the evidence provided by them. Once, for example, a professional scientist reacted with the statement "I feel discriminated against as a professional researcher, this is not a level playing field", because their positive evaluation of a topic was not taken up in full. Here the group, through the strong influence of the patient researchers, had agreed on another topic. This triggered for a moment heated, confusing discussion in the group. The topic, which many patient researchers rated as a priority, provoked fierce counter-reactions and devaluations by individual professional scientists, who said that this topic did not entail a "scientific question" and thus did not allow for "operationalization in the scientific sense". If this topic was elaborated, it would take the form of a recommendation, similar to those found in lifestyle magazines, for example, such as answers to the question: "Is this the right diet for me"? As a result, the professional researchers could only partially prevail by merging the two topics into one. At the same time, the actors involved mutually agreed that the result does not have to be/cannot be a scientific result in a traditional sense. Ultimately, an innovative result emerges within the framework of co-creative cooperation with the clear signature of citizen scientists, but at the same time, boundary work takes place in order to secure the scientific authority over the area of academic research.

The Patient Science project illustrates to what extent non-academic actors can question the authority of professional science and its evidence claims. Based on this, we now look at the mode of co-creation as an overarching concept of epistemic participation.

The co-creation mode offers large capacities for influence since not only can laypeople be involved in data collection or analysis in large numbers (social dimension) but are also represented in the research conception (factual dimension) and thus in all research phases (temporal dimension). However, the increased capacity for influence remains moderate, as the researchers themselves remain an equivalent part of the research process and professional control continues to exist. Where the contributory, collaborative and collegiate modes limit the possibilities for critically assessing evidence in various ways, co-creation offers the option of being able to become effective as a layperson in the structures of evidence production without creating non-productive tensions, such as major negotiations over the conception of a research project. We will illustrate the possibilities for evidence contestation through co-creation using the following case.

For this, we focus on the td-net toolbox of the Swiss Academy of Sciences as a co-creative approach. This toolbox is dedicated to the further development of co-creative research as a scientific method.[48] According to its self-description, the toolbox's overarching purpose is to enable heterogeneous cooperation:

> The methods and tools offered by the td-net toolbox specifically focus on jointly developing projects, conducting research and exploring ways to impact in heterogeneous groups. They are intended to help shape collaboration between experts and stakeholders from science and practice in systematic and traceable ways.[49]

The toolbox's idea of cooperation between laypeople and experts builds on the concept of thought styles developed by Ludwik Fleck. According to Fleck's concept of thought collectives, mutual influences arise between the "esoteric circles" of specialists, the "exoteric circle" of laypeople and the gradations in between. This can further lead to a democratization of the esoteric circles, if a majority (e.g. the public) takes a stronger position and thus forces an exchange between the thought collectives of laypeople and experts.[50]

The intercollective exchange of thought styles as the overall approach and specific criterion of the td-net toolbox[51] can be exemplified by one of the concrete tools. The tool *emancipatory boundary critique* focuses on the empowerment of laypeople and their epistemic capacity. This tool serves as an example of the possibilities of evidence contestation in the mode co-creation:

> Emancipatory boundary critique consists of a set of questions to empower non-experts to uncover normative assumptions underlying an expert's solution to a problem along with the solution's social

and ecological implications [...] All solutions to a problem include underlying assumptions [...] In the critical discussion of these underlying assumptions and their consequences the expert is as much a lay-person as the non-expert.[52]

The aim of the tool is to enable laypeople to deconstruct the normative assumptions of presented scientific solutions in order to not only accept solutions, but also to reflect on their backgrounds as a form of emancipated participation. However, in our view, those questions are not only tools for laypeople to contribute their perspectives to a greater extent. The formulation of these questions ensures that the exchange of thought styles does not lead the scientist to lose their role as the provider of solutions to a lay audience: "How are thought styles bridged? Enabling a dialogue on equal footing between the experts who propose solutions and the affected non-experts bridges thought styles".[53]

Some examples of the set of questions will illustrate the ambiguous balance between the distinct interests of experts and laypeople. The questions divide into the following categories: Sources of (1) motivation (2) power (3) knowledge and (4) legitimation. We give an example for each category:[54]

1 What is (ought to be) the purpose? That is, what are (should be) the consequences?
2 Who is (ought to be) the decision-maker? That is, who is (should be) in a position to change the measure of improvement?
3 What kind of expertise is (ought to be) consulted? That is, what counts (should count) as relevant knowledge?
4 What secures (ought to secure) the emancipation of those affected from the premises and promises of those involved? That is, where does (should) legitimacy lie?

The explicit focus of the questions on the backgrounds of the scientific solution proposals enables the laypeople to become part of the content discussions and also to demand justifications from professional researchers about scientific knowledge. On the one hand, this concerns both the level of knowledge itself and what can become the object of knowledge here in the first place (3), as well as the consequences of scientific knowledge and its social impact (4). On the other hand, laypeople are enabled to (critically) question the social conditionality of research; especially not only in the questions of power relations (2) in research, but also the purposes (1) of scientific research. This opens up advanced possibilities to contest scientific knowledge that are potentially much stronger here than in other modes of citizen science.

However, it is precisely the content-related focus of the questions that shifts the structural component of empowering citizens out of focus, namely who decides which solutions are worth discussing in the context of this tool on the one hand and who designs the solutions on the other. The agreement of the involved actors still refers to the fact that the questions do not intend to challenge the monopoly role of the researcher but primarily to stake out the distribution of roles between laypeople and experts. The tool is thus based on the agreement of not contesting the roles of experts and laypeople due to the fact that the latter are still not providing the solutions. They become equal partners with the experts in critically discussing solutions, but not as providers, as becomes apparent through the output definition of the tool:

> The output is a dialogue about a proposed solution, that openly deliberates underlying normative assumptions as well as social and ecological implications. The overall outcome is a broader (and emancipated) understanding of the proposed solution amongst the people involved in the dialog.[55]

In addition to the (project internal) interpretation by the td-net toolbox of Fleck's concept of thought styles, we would add the following: Fleck emphasizes the persistent tendencies of thought styles. Among other things, Fleck points out that contradictory evidence (as it can occur, for example, through another thought collective) remains, for example, largely ignored or is very strongly attempted to be integrated into one's own thought style.[56] This insight coincides with Gieryn's work on scientific boundary work, in which one's own boundaries (and so the thought style) are defended against others. Despite intercollective exchange, the actual influences thus remain rather marginal. Even under the assumption of a fundamental change of the primary thought style of a collective, it would not disappear completely. It will persist as a foundation for new ways of thinking.[57]

Thus, the approach of this co-creative tool is precisely not about symmetrizing thought styles unilaterally. Rather, it is about maintaining the *different* thought styles of various actors by formulating the questions in such a way that they protect the interests of the respective other. This balancing enables the scientists *and* laypeople to continuously enforce their own interests. Therefore, we observe a moderate contestation of evidence through lay participation in this case. The conception of the emancipatory boundary critique tool offers possibilities for questioning scientific evidence, but it does not touch the hierarchical structures of evidence production and the dominance of established science in it.

In the following, we open a network-theoretical perspective on our case from the co-creation mode using the sociological concept of boundary objects. We do this not only to explore how evidence-critical participation occurs, but also to ask ourselves what functional meaning the modes and their connectivity to academia have for handling a possible contestation of evidence.

Co-Creation as a Point of Convergence? A Network-theoretical Reflection on Realizations of Evidence Contestation

The persistence of different interests and thought styles leads us to the assumption that an actual contestation of evidence through co-creation does not appear merely through identifying consensus and dissent. Moreover, epistemic participation is not realized through the mere recognition of other thought styles. Instead we hold that, in terms of the sociology of science, co-creation approaches such as the td-net toolbox do fulfill the conditions of "heterogeneous cooperation",[58] but more in the sense of "boundary objects".[59] In the classic example by Susan Leigh Star and James R. Griesemer, a museum constitutes boundary objects. It offers different artifacts that can have various and locally reconfigurable meanings for specific actors, depending on their respective uses and disciplinary backgrounds.[60] In this context, Star also speaks of "cooperation without consensus".[61] Cooperation between researchers and laypeople, however, requires agreements that not only retain their validity within science, but also "across time, space and local contingencies"[62] to acknowledge the heterogeneity of actors and their respective point of view.[63] This would be the case if, for example, participatory research on air pollution in cities is not only used for scientific data gathering, but also leads to concrete measures of air improvement for the citizens who participate in the data collection. According to Star and Griesemer, the possibility of such compatibility is not bound to a consensus on the content of the collaboration.[64]

What becomes necessary instead are *translations* between the actor's interests. This means that the different interests of the actors unite under *one* programmatic goal. Star and Griesemer describe this in reference to Michel Callon as obligatory passage points.[65] In the case of the assimilative modes, this passage point is the scientific evidence orientation, while for the collegiate mode it is the distinctiveness of lay participation. With the orientation to a specific passage point, the assertion of scientific authority takes place.[66] However, by focusing on the translation to the passage point of *one* actor and its defense, there would be a danger that cooperation would break down if there was no safeguarding of the interests of other participants.[67]

In terms of evidence contestation, this means that both a clear restriction of opportunity structures for contesting evidence and too much openness to evidence-critical influences are problematic in contexts of collaboration among heterogeneous actors. From their perspective, Star and Griesemer refer to boundary objects that enable the translation of such interests into different points of passage. These translations then replace consensus on content and maintain the relevance of the overall endeavor (in our case, participation in science) to different social worlds.[68] Boundary objects create cooperation by translating the respective expectations of all involved actors.

If we relate the perspective of boundary objects to the mode of co-creation, it can be understood as that form of citizen science which (in its ideal-typical conception) shows itself as a frame of reference to which different actors with different interests can appeal without serving the partial expectations of science or non-science asymmetrically. Under these conditions, evidence contestation becomes realizable. However, this does not mean that professional control becomes obsolete. The interrelationship between epistemic safeguarding and contestation of evidence characterizes epistemic participation through citizen science. The concept of boundary objects acknowledges the persistence of the actors' interests and thought styles (as Fleck mentions) through the act of translation.

In the case of the emancipatory boundary critique tool, it is not primarily a matter of emancipation, but of serving the interests of laypeople in co-creation, and yet the structural conditions and structural limits mean that the researcher still has control over who offers solutions and puts them up for discussion. Cooperation always occurs here in the context of professionally organized research and the existing limits of roles between laypeople and experts. Co-creation seems to serve as the *primary mode for epistemic collaboration* in order to serve the expectations of non-science on the one hand, but without giving up the passage point of science on the other.

To what extent this can be observed in practice and how co-creation processes thus enable lay participation in the social, factual and temporal dimensions in concrete terms is something that needs to be discussed beyond the scope of this chapter in the form of further comparative case studies. However, co-creation approaches maintain the structural conditions of professional control to secure evidence and, at the same time, enforce the capacity for influence through lay participation. Co-creation thus generates, in our view, currently the greatest possibilities for lay inclusion not only in the social and temporal dimension, but also and particularly in the factual dimension of science, as an equilibrium of participatory and yet scholarly research. A possible contestation of evidence would thus not be a fundamental questioning of science as the primary producer of knowledge. It would take place in the form of the inclusion of laypeople in the structures of science and its premises, in which potentially critical participation remains adjustable. The possibility of challenging evidence through extra-scientific perspectives is realized here in a factual and not system-related way. We would describe this as a moderate contestation of evidence in the co-creation mode.

In the following outlook, we discuss the reciprocity of all the referred modes. We illustrate that the heterogeneity of modes does not simply represent constitutional processes of diverse participatory conceptions that are only powerful in their own right. We understand the contestation of evidence in the field of citizen science precisely in view of the arrangement of modes as a network of heterogeneous participation. It is *only* through their interconnectedness that the modes fulfill the functional requirements for constituting citizen science as a participatory yet evidence-oriented approach.

Outlook: Modes of Contestation? Citizen Science as Regulated Critique

If we look at the modes of citizen science as a whole, we can see that they each serve different tasks in shaping participation. The assimilative contributory and collaborative modes characterize themselves by a high connectivity to academia. Through this, we see the reproduction of common procedures of professionalized research by the incorporation of citizen science activities into established structures of scientific research. The structural possibilities of evidence contestation thus remain largely limited, which, however, enables low-threshold and broad participation.

The other modes, collegiate and co-creation, instead lead to certain opportunities of contesting scientific evidence. The alternative collegiate mode operates primarily outside of professionalized research, but cannot realize its high potential for evidence contestation because it lacks the necessary connectivity to academia. The collegiate mode serves here rather as a representative of non-scientific perspectives. Nevertheless, it offers the opportunity to try out alternative content and methods within their framework that can become scientifically relevant at a later point in time (see on this the contribution by Zachmann and Esselborn in this volume).

Co-creation, on the other hand, translates the perspectives of various actors to produce epistemic cooperation and a controlled opening of the factual dimension. Co-creation enables controlled yet epistemically effective participation that assimilative and alternative modes can hardly provide, as they increasingly regulate or restrict participation in science. However, the concession of epistemic impact does not mean that co-creation implies a departure from traditional research structures. In our view, a possible epistemic participation through co-creation refers rather to the involvement of laypeople on the conceptual-content level than to the structural reconfiguration of scientific knowledge production and its producers. The contestation of scientific evidence thus remains moderate in the co-creation mode, without developing into a general systemic critique of scientific evidence. Here we need to distinguish between challenging scientific evidence and challenging the structures of scientific evidence production.

Even though we have distinguished various forms of participation, we see a possible answer to the question of how citizen science as a field generates evidence-critical participation precisely in the interplay of the modes through mechanisms of opening and closing of scientific knowledge production. This reciprocity enables a moderate contestation of evidence for more than a single mode. The interplay of opening and closing means, in simplified terms: A specific mode functions as the advocate of the one perspective but is at the same time the antagonist of the other, as a kind of trade-off. If the expectation of scientificity is fulfilled by an assimilative mode, epistemic participation cannot be fulfilled without restrictions, and vice versa.

A broad spectrum of modes alone does not ensure adequate possibility of actual epistemic participation. This oscillates between an opening of the social dimension for factually compliant participation (assimilative modes) and a closing of the factual dimension through the restriction of overly large potential for evidence criticism (alternative modes). In these modes, evidence contestation is thus either structurally limited or remains practically unrealized. The mode co-creation complements and increases the enormous diversity of citizen science and, at the same time, directs the epistemic effectiveness of laypeople into controllable channels (thus limiting it at the same time).

Regarding our guiding question (Which citizen science activities offer the potential for a contestation of evidence?), we state that the opportunities for contesting scientific evidence differ depending on the mode of citizen science. However, we suggest that the interplay of heterogeneous modes and their multiple connectivities to diverse participatory expectations is functionally meaningful for the field to enable epistemic participation, but to keep it at a tolerable level for science. We thus formulate our thesis of a moderate contestation of evidence not only for specific modes, but also as a mechanism of structural arrangement of heterogeneous concepts of civic participation. For a further (network-)analysis of the functionality of the different positioning between citizen science and academia (which we have hinted at in this chapter), the inclusion of a further system- and differentiation-theoretical vocabulary would be required. We also see the need for the extension of a sociologically informed mapping approach to conceptions of citizen science.[69]

The crux of the diversity of modes of citizen science is that the opening of such a network of heterogeneous participation is at the same time accompanied by new forms of epistemic restriction for lay participation. This is the case if different modes allow partial forms of participation (intra-scientific, extra-scientific, conceptual) but do not generate entire participation. Although the diverse participation structure initially meets the various expectations of different actors, this risks a sustained disappointment of expectations in practice. This applies both not only for the public but also for traditional science that has an interest in a more participatory research. In the sense of truly epistemically effective participation of laypeople, it is therefore important to raise awareness of the extent to which a pluralistic program of citizen science contains not only inclusion, but also exclusion potential, and, moreover, whether this corresponds to or runs counter to the intentions of various citizen science actors. In this respect, we hope that further empirical studies will be conducted within the framework of such conceptual considerations. In particular, to investigate the realization of capacity for influence through citizen science for a productive contestation of evidence.

Notes

1 We would like to thank the reviewers for the good suggestions for improvement, as well as our project partners Michael Kitzing, Claudia Göbel and Sascha Dickel for important hints.
2 See Henry Sauermann, Katrin Vohland, Vyron Antoniou, Bálint Balázs, Claudia Göbel, Kostas Karatzas and Peter Mooney et al., "Citizen Science and Sustainability Transitions", *Research Policy* 49, no. 5 (2020): 1–16.
3 For more detailed reflections on this point: Andreas Wenninger and Sascha Dickel, "Paradoxien digital-partizipativer Wissenschaft", *Österreichische Zeitschrift für Soziologie* 44, no. S1 (2019), 257–286.
4 On sociological systems theory and the further differentiation between the factual and social dimension, see Niklas Luhmann and Rhodes Barrett, *Theory of Society*, Vol. 1 (Stanford: Stanford Univ. Press, 2012); Niklas Luhmann and Rhodes Barrett, *Theory of Society*, Vol. 2 (Stanford: Stanford Univ. Press, 2013); Dirk Baecker and Niklas Luhmann, *Introduction to Systems Theory* (Cambridge: Polity, 2013).
5 See for example Christopher Kullenberg and Dick Kasperowski, "What Is Citizen Science? – A Scientometric Meta-Analysis", *PLoS One*, 11, no. 1 (2016): 6. See also Sascha Dickel, "Öffnung für Alle? Einlösung oder Erosion des Projekts moderner Wissenschaft?", *Technikfolgenabschätzung – Theorie und Praxis* 26, no. 1–2 (2017): 57.
6 Katrin Vohland, Maike Weißpflug, and Lisa Pettibone, "Citizen Science and the Neoliberal Transformation of Science – an Ambivalent Relationship", *Citizen Science: Theory and Practice* 4, no. 1 (2019): 3.
7 For the process of anchoring citizen science in the institutional contexts of science, see "First Citizen Science Professorship in Germany", Bürger schaffen Wissen, accessed September 2, 2021, https://www.buergerschaffen wissen.de/blog/erste-citizen-science-professur-deutschland.
8 We understand the term academia as the academic world, which encompasses any disciplines, institutions, practices and actors from natural sciences, social sciences and humanities.
9 See a selection of these on the platform Österreich forscht, accessed September 2, 2021, https://www.citizen-science.at/eintauchen/weltweit.
10 See Susanne Hecker, Lisa Garbe, and Aletta Bonn, "The European Citizen Science Landscape – A Snapshot", in *Innovation in Open Science, Society and Policy*, eds. Susanne Hecker, Muki Haklay, Anne Bowser, Zen Makuch, Johannes Vogel, and Aletta Bonn. (London: UCL Press, 2018).
11 David B. Resnik, "The Autonomy of Science", in *Playing Politics with Science. Balancing Scientific Independence and Government Oversight* (Oxford: Oxford Univ. Press, 2009), 1. Regarding the certification of

professional scientists, Collins and Evans also state that "The standard way to try to measure expertise externally is by reference to credentials such as certificates attesting to past achievement of proficiency", see Harry Collins and Robert Evans, *Rethinking Expertise* (Chicago: Univ. of Chicago Press, 2007), 67.

12 Own translation of the quote by the authors. Sarah Ehlers and Karin Zachmann, "Wissen und Begründen: Evidenz als umkämpfte Ressource in der Wissensgesellschaft. Einleitung", in *Wissen und Begründen: Evidenz als umkämpfte Ressource in der Wissensgesellschaft*, eds. Karin Zachmann and Sarah Ehlers (Baden-Baden: Nomos, 2019): 27.

13 Own translation of the quote by the authors. Andreas Wenninger, Fabienne Will, Sascha Dickel, Sabine Maasen, and Helmuth Trischler, "Ein- und Ausschließen: Evidenzpraktiken in der Anthropozändebatte und der Citizen Science", in *Wissen und Begründen: Evidenz als umkämpfte Ressource in der Wissensgesellschaft*, eds. Karin Zachmann and Sarah Ehlers (Baden-Baden: Nomos, 2019): 54.

14 Vohland, Weißpflug and Pettibone, "Citizen Science and the Neoliberal Transformation of Science", 3.

15 On the notion of socially robust knowledge, Helga Nowotny, Peter Scott, and Michael Gibbons, *Wissenschaft neu denken: Wissen und Öffentlichkeit in einem Zeitalter der Ungewißheit* (Weilerswist: Wellbrück, 2004), 149.

16 Federal Ministry of Education and Research's Communication on the Sustainable Integration of Citizen Science into the Science System, accessed February 4, 2022, https://www.bmbf.de/bmbf/shareddocs/pressemitteilungen/de/karliczek-wir-wollen-die-buerg–wissenschaftssystem-verankern.html.

17 ECSA (European Citizen Science Association), "Ten Principles of Citizen Science", Berlin, 2015, 1. Accessed November 22, 2022, http://doi.org/10.17605/OSF.IO/XPR2N.

18 See Thomas F. Gieryn, "Boundary-Work and the Demarcation of Science from Non-Science: Strains and Interests in Professional Ideologies of Scientists", *American Sociological Review* 48, no. 6 (1983), 789. Moreover, see Thomas F. Gieryn, Cultural *Boundaries of Science. Credibility on the Line* (Chicago: Univ. of Chicago Press, 1999).

19 One of the significant representatives of the British line of citizen science is the sociologist Alan Irwin; see Alan Irwin, *Citizen Science. A Study of People, Expertise, and Sustainable Development* (London: Routledge, 1995).

20 See Caren B. Cooper and Bruce V. Lewenstein, "Two Meanings of Citizen Science", in *The Rightful Place of Science: Citizen Science, a series by the Consortium for Science*, Policy & Outcomes (Arizona State: Univ. Press, 2016), 51–62.

21 See Gwen Ottinger, "Reconstructing or Reproducing? Scientific Authority and Models of Change in Two Traditions of Citizen Science", in *The Routledge Handbook of the Political Economy of Science* (London: Taylor and Francis, 2017), 351–363, here 351.

22 Among other typification approaches, we refer to the typology of project models of participation in scientific research by Jennifer L. Shirk, Heidi L. Ballard, Candie C. Wilderman, Tina Phillips, Andrea Wiggins, Rebecca Jordan, and Ellen McCallie, "Public Participation in Scientific Research: A Framework for Deliberative Design", *Ecology and Society* 17, no. 2 (2012): Art. 29. Shirk et al., "Public Participation" also list a fifth model of participation, contractual projects, which are designed for laypeople to commission professional research to investigate specific topics for them. Although we do not list this model here as part of potentially epistemic participation, this model should

not be unmentioned. An early typology of modes comes from Bonney et al., see Rick Bonney, Heidi Ballard, Rebecca Jordan, Ellen McCallie, Tina Phillips, Jennifer Shirk, and Candie C. Wilderman, *Public Participation in Scientific Research: Defining the Field and Assessing its Potential for Informal Science Education. A CAISE Inquiry Group Report* (Washington: Center for Advancement of Informal Science Education (CAISE), 2009).

23   Lisa Pettibone, Katrin Vohland, and David Ziegler, "Understanding the (Inter)Disciplinary and Institutional Diversity of Citizen Science: A Survey of Current Practice in Germany and Austria", *PLoS One* 12, no. 6 (2017): 1–16.

24   Bruno J. Strasser, Jérome Baudry, Dana Mahr, Gabriela Sanchez, and Elise Tancoigne, "'Citizen Science'? Rethinking Science and Public Participation", *Science & Technology Studies* 32, no. 2 (2019): 64.

25   David Silverman, *Interpreting Qualitative Data* (London: SAGE, 2014).

26   "Against Citizen Science", accessed 11/23/2021, https://aeon.co/essays/is-grassroots-citizen-science-a-front-for-big-business.

27   Vohland, Weißpflug, and Pettibone, "Citizen Science and the Neoliberal Transformation of Science", 1.

28   Vohland, Weißpflug, and Pettibone, "Citizen Science and the Neoliberal Transformation of Science", 3.

29   Dick Kasperowski and Thomas Hillman, "The Epistemic Culture in an Online Citizen Science Project: Programs, Antiprograms and Epistemic Subjects", *Social Studies of Science*, Vol. 48, no. 4 (2018): 564–588.

30   Christopher Kullenberg and Dick Kasperowski, "What Is Citizen Science? – A Scientometric Meta-Analysis", *PLoS One* 11, no. 1 (2016): 1–16.

31   See Strasser et al., "'Citizen Science'? Rethinking Science and Public Participation", *Science & Technology Studies* 32, no. 2 (2019), 52–76.

32   Following Park and Burgess, we understand assimilation as the latent process of structural subordination of foreign actors into the dominant community of the field, whereby internal conflicts are regulated and finally accommodated as the creation of a common basis of interests. For the case of assimilative citizen science, we observe similar processes in which participatory efforts are (smoothly) absorbed into the scientific field structures, so that citizen science can no longer be clearly distinguished from professionalized research, but itself becomes an integral part of the field and its practices. See Robert E. Park and Ernest W. Burgess, *Introduction to the Science of Sociology: Including an Index to Basic Sociological Concepts* (Chicago: Univ. of Chicago Press, 1969). On the migration sociological concept of assimilation, see further Milton M. Gordon, *Assimilation in American Life. The Role of Race, Religion, and National Origins* (New York: Oxford Univ. Press, 1964).

33   It should be noted that various control mechanisms are also found in the citizen initiatives, but these controls are not guided by professional researchers to evaluate scientificity.

34   See for example Bonney et al., *Public Participation in Scientific Research*, 11.

35   Hecker, Garbe, and Bonn, "The European Citizen Science Landscape", 194.

36   On the differentiation of invited and uninvited participation in science, see Brian Wynne, "Public Participation in Science and Technology: Performing and Obscuring a Political-Conceptual Category Mistake", *East Asian Science, Technology and Society* 1 (2007): 99–110.

37   Vohland, Weißpflug, and Pettibone, "Citizen Science and the Neoliberal Transformation of Science", 2.

38   In Shirk et al., "Public Participation", collegiate essentially refers to amateur researchers working individually who have little or no contact with certified experts in the process of their research. However, in a few rare

cases, it is precisely within this mode that particularly highly acknowledged achievements within scientific communities may be found: "[T]he degree of amateur participation in the research process is so extensive and independent that expert amateurs arguably adopt the traditional role of scientist-as-knowledge-producer"; see Shirk et al., "Public Participation in Scientific Research", 5. Our example illustrates that within this mode, however, larger project contexts can also take place.

39 Own translation of the quote by the authors. "Definitions, Standards, Quality Criteria", Bürger schaffen Wissen, accessed September 2, 2021, https://www.buergerschaffenwissen.de/citizen-science/definitionen-standards-qualitaetskriterien.

40 Topothek, website, accessed September 9, 2021, https://www.topothek.at/de/.

41 Own translation of the quote by the authors. Mariacarla Gadebusch Bondio, Emilia Lehmann, Andreas Wenninger, and Tommaso Bruni, "De- und Re-Kontextualisieren: Die 'reale Evidenz' von Patient_innen und Citizen Scientists", in *Wissen und Begründen: Evidenz als umkämpfte Ressource in der Wissensgesellschaft*, eds. Karin Zachmann and Sarah Ehlers (Baden-Baden: Nomos, 2019): 74.

42 On the Topothek project and its thematic focus, see "Topotheque. Our Memories", accessed November 26, 2021, https://www.topothek.at/en/.

43 See Thomas F. Gieryn, "Boundary-Work and the Demarcation of Science from Non-Science: Strains and Interests in Professional Ideologies of Scientists", *American Sociological Review* 48, no. 6 (1983), 789. Moreover, see Thomas F. Gieryn, *Cultural Boundaries of Science. Credibility on the Line* (Chicago: Univ. of Chicago Press, 1999).

44 Shirk et al., "Public Participation", 5.

45 Hecker et al. also mention collegiate, but not as part of the elaboration of a citizen science based approach of scientific collaboration. See Susanne Hecker, Muki Haklay, Anne Bowser, Zen Makuch, Johannes Vogel, and Aletta Bonn, *Innovation in Open Science, Society and Policy* (London: UCL Press, 2018). The elaboration of the different modes is done within this publication, see Hecker, Garbe and Bonn, "The European Citizen Science Landscape", 194.

46 According to the research by Hecker et al., the collegiate mode accounts for only 4.6 percent of the citizen science projects studied. See Hecker, Garbe and Bonn, "The European Citizen Science Landscape", 194.

47 Project Patient Science, accessed February 11, 2022, https://www.isi.fraunhofer.de/de/competence-center/neue-technologien/projekte/patient-science.html. In the meantime, the organizing team of the 'Patient Science' project has published a journal article on chronically ill people as co-researchers, accessed November 22, 2022, https://doi.org/10.35844/001c.35634.

48 "td-net toolbox", td-net-toolbox, accessed February 1, 2022, https://naturalsciences.ch/co-producing-knowledge-explained/methods/td-net_toolbox.

49 "About the td-net toolbox", td-net-toolbox, accessed January 31, 2022, https://naturalsciences.ch/co-producing-knowledge-explained/methods/td-net_toolbox/about_td_net_toolbox.

50 See Ludwik Fleck, *Entstehung und Entwicklung einer wissenschaftlichen Tatsache. Einführung in die Lehre vom Denkstil und Denkkollektiv* (Frankfurt am Main: Suhrkamp, 1993), 148–150.

51 "What is a Method or Tool for Co-producing Knowledge?", td-net-toolbox, accessed February 4, 2022, https://naturalsciences.ch/co-producing-knowledge-explained/methods/td-net_toolbox/about_td_net_toolbox.

52 "Emancipatory boundary critique", td-net-toolbox, accessed February 7, 2022, https://naturwissenschaften.ch/co-producing-knowledge-explained/methods/td-net_toolbox/emancipatory_boundary_critique_final_.
53 "Emancipatory Boundary Critique".
54 "Emancipatory Boundary Critique". The boundary questions for this tool are taken from Ulrich's work on critical systems heuristics, see Walter Ulrich, *A Brief Introduction to Critical Systems Heuristics* (CSH) (Milton Keynes: The Open Univ., 2005).
55 "Emancipatory boundary critique".
56 See Fleck, *Wissenschaftliche Tatsache,* 43–50.
57 See Fleck, *Wissenschaftliche Tatsache,* 131.
58 Jörg Strübing, Ingo Schulz-Schaeffer, Martin Meister, and Jochen Gläser, *Kooperation im Niemandsland. Neue Perspektiven auf Zusammenarbeit in Wissenschaft und Technik* (Wiesbaden: VS Verlag für Sozialwissenschaften, 2004).
59 See Susan Leigh Star and James R. Griesemer, "Institutional Ecology, 'Translations' and Boundary Objects: Amateurs and Professionals in Berkeley's Museum of Vertebrate Zoology, 1907–1939", *Social Studies of Science* 19 (1989): 387–420.
60 Star and Griesemer, "Boundary Objects", 404 f.
61 Susan Leigh Star. "Cooperation Without Consensus in Scientific Problem Solving: Dynamics of Closure in Open Systems", in *CSCW: Cooperation or Conflict?,* ed. Steve Easterbrook (London: Springer, 1993), 93–105.
62 Star and Griesemer, "Boundary Objects", 387.
63 Star and Griesemer, "Boundary Objects", 387.
64 Star and Griesemer, "Boundary Objects", 388.
65 See Michel Callon, "Some Elements of a Sociology of Translation. Domestication of the Scallops and the Fishermen of St Brieuc Bay", *The Sociological Review* 32, 1_suppl (1984): 196–233.
66 Star and Griesemer borrow their concept here from the work of Callon and Law, among others, on the problem of translation, see John Law, "Technology, Closure and Heterogeneous Engineering: The Case of Portuguese Expansion", in *The Social Construction of Technological Systems. New Directions in the Sociology and History of Technology,* eds. Wiebe Bijker, Thomas P. Hughes, and Trevor Pinch. (Cambridge, MA: MIT Press, 12th pr., 1993), 111–134. See also Michel Callon and John Law, "On Interests and Their Transformation: Enrolment and Counter-Enrolment", *Social Studies of Science* 12, no. 4 (1982): 615–625.
67 Star and Griesemer, "Boundary Objects", 390f.
68 Star and Griesemer, "Boundary Objects", 389f.
69 For example, Chilvers et al. characterize the reciprocal formations of participation formats in the context of the low-carbon energy transition as participatory ecology, see Jason Chilvers, Rob Bellamy, Helen Pallett, and Tom Hargreaves, "A Systemic Approach to Mapping Participation with Low-Carbon Energy Transitions", 251.

References

Bonney, Rick, Heidi Ballard, Rebecca Jordan, Ellen McCallie, Tina Phillips, Jennifer Shirk, and Candie C. Wilderman. *Public Participation in Scientific Research: Defining the Field and Assessing Its Potential for Informal Science Education. A CAISE Inquiry Group Report.* Washington: Center for Advancement of Informal Science Education (CAISE), 2009.

Callon, Michel, "Some Elements of a Sociology of Translation. Domestication of the Scallops and the Fishermen of St Brieuc Bay". *The Sociological Review* 32, no. 1_suppl (1984): 196–233.

Callon, Michel, and John Law. "On Interests and Their Transformation: Enrolment and Counter-Enrolment". *Social Studies of Science* 12, no. 4 (1982): 615–625.

Chilvers, Jason, Rob Bellamy, Helen Pallett, and Tom Hargreaves. "A Systemic Approach to Mapping Participation With Low-Carbon Energy Transitions". *Nature Energy*, Vol. 6 (2021): 250–259.

Collins, Harry, and Robert Evans. *Rethinking Expertise.* Chicago: Univ. of Chicago Press, 2007.

Cooper, Caren B., and Bruce V. Lewenstein. "Two Meanings of Citizen Science". In *The Rightful Place of Science: Citizen Science, a series by the Consortium for Science.* Arizona State: Univ. Press, 2016.

Dickel, Sascha. "Öffnung für Alle? Einlösung oder Erosion des Projekts moderner Wissenschaft?. *Technikfolgenabschätzung – Theorie und Praxis* 26, no. 1–2 (2017): 55–59.

Ehlers, Sarah, and Karin, Zachmann. "Wissen und Begründen: Evidenz als umkämpfte Ressource in der Wissensgesellschaft. Einleitung". In *Wissen und Begründen: Evidenz als umkämpfte Ressource in der Wissensgesellschaft,* edited by Karin Zachmann, and Sarah Ehlers, 9–29. Baden-Baden: Nomos, 2019.

Gadebusch Bondio, Mariacarla, Emilia Lehmann, Andreas Wenninger, and Tommaso Bruni. "De- und Re-Kontextualisieren: Die 'reale Evidenz' von Patient_innen und Citizen Scientists". In *Wissen und Begründen: Evidenz als umkämpfte Ressource in der Wissensgesellschaft,* edited by Karin Zachmann, and Sarah Ehlers, 59–82. Baden-Baden: Nomos, 2019.

Gieryn, Thomas F. "Boundary-Work and the Demarcation of Science from Non-Science: Strains and Interests in Professional Ideologies of Scientists". *American Sociological Review* 48, no. 6 (1983): 781–795.

Gieryn, Thomas F. *Cultural Boundaries of Science. Credibility on the Line.* Chicago: Univ. of Chicago Press, 1999.

Gordon, Milton M. *Assimilation in American Life. The Role of Race, Religion, and National Origins.* New York: Oxford Univ. Press, 1964.

Hecker, Susanne, Muki Haklay, Anne Bowser, Zen Makuch, Johannes Vogel, and Aletta Bonn. *Citizen Science: Innovation in Open Science, Society and Policy.* London: UCL Press, 2018.

Hecker, Susanne, Lisa Garbe, and Aletta Bonn. "The European Citizen Science Landscape – a Snapshot". In *Innovation in Open Science, Society and Policy,* edited by Susanne Hecker, Muki Haklay, Anne Bowser, Zen Makuch, Johannes Vogel, and Aletta Bonn, 190–200. London: UCL Press, 2018.

Irwin, Alan. *Citizen Science. A Study of People, Expertise, and Sustainable Development.* London: Routledge, 1995.

Kasperowski, Dick, and Thomas Hillman. "The Epistemic Culture in an Online Citizen Science Project: Programs, Antiprograms and Epistemic Subjects". *Social Studies of Science,* Vol. 48 (4) (2018): 564–588.

Kullenberg, Christopher, and Dick Kasperowski. "What Is Citizen Science? – A Scientometric Meta-Analysis". *PLoS One* 11, no. 1 (2016): 1–16.

Law, John. "Technology, Closure and Heterogeneous Engineering: The Case of Portuguese Expansion". In *The Social Construction of Technological Systems.*

New Directions in the Sociology and History of Technology, edited by Wiebe Bijker, Thomas P. Hughes, Trevor Pinch, 111–134. Cambridge, MA: MIT Press, 1993.

Luhmann, Niklas, and Rhodes Barrett. *Theory of Society*, Vol. 1. Stanford: Stanford Univ. Press, 2012.

Luhmann, Niklas, and Rhodes Barrett. *Theory of Society*, Vol. 2. Stanford: Stanford Univ. Press, 2013.

Nowotny, Helga, Peter Scott, and Michael Gibbons. *Wissenschaft neu denken. Wissen und Öffentlichkeit in einem Zeitalter der Ungewißheit*. Weilerswist: Wellbrück, 2004.

Ottinger, Gwen. "Reconstructing or Reproducing? Scientific Authority and Models of Change in Two Traditions of Citizen Science". In *The Routledge Handbook of the Political Economy of Science*. First edition, edited by David Tyfield, Charles Thorpe, Rebecca Lave, Samuel Randalls, 351–363. London: Taylor and Francis, 2017.

Park, Robert E., and Ernest W. Burgess. *Introduction to the Science of Sociology. Including an Index to Basic Sociological Concepts*. Chicago: Univ. of Chicago Press, 1969.

Pettibone, Lisa, Katrin Vohland, and David Ziegler. "Understanding the (Inter) Disciplinary and Institutional Diversity of Citizen Science: A Survey of Current Practice in Germany and Austria". *PLoS One*, 12, no. 6 (2017): 1–16.

Resnik, David B. "The Autonomy of Science". In *Playing Politics with Science. Balancing Scientific Independence and Government Oversight*, 51–88. Oxford: Oxford Univ. Press, 2009.

Sauermann, Henry, Katrin Vohland, Vyron Antoniou, Bálint Balázs, Claudia Göbel, Kostas Karatzas, Peter Mooney et al. "Citizen Science and Sustainability Transitions". *Research Policy* 49, no. 5 (2020): 1–16.

Shirk, Jennifer L., Heidi L. Ballard, Candie C. Wilderman, Tina Phillips, Andrea Wiggins, Rebecca Jordan, and Ellen McCallie. "Public Participation in Scientific Research: A Framework for Deliberative Design". *Ecology and Society* 17, no. 2 (2012): Art. 29.

Silverman, David. *Interpreting Qualitative Data*. London: SAGE, 2014.

Star, Susan Leigh. "Cooperation Without Consensus in Scientific Problem Solving: Dynamics of Closure in Open Systems". In *CSCW: Cooperation or Conflict?* edited by Steve Easterbrook, 93–105. London: Springer, 1993.

Star, Susan Leigh, and James R. Griesemer. "Institutional Ecology, 'Translations' and Boundary Objects: Amateurs and Professionals in Berkeley's Museum of Vertebrate Zoology, 1907–1939". *Social Studies of Science* 19 (1989): 387–420.

Strasser, Bruno J., Jérome Baudry, Dana Mahr, Gabriela Sanchez, and Elise Tancoigne. "'Citizen Science'? Rethinking Science and Public Participation". *Science & Technology Studies* 32, no. 2 (2019): 52–76.

Strübing, Jörg, Ingo Schulz-Schaeffer, Martin Meister, and Jochen Gläser. *Kooperation im Niemandsland. Neue Perspektiven auf Zusammenarbeit in Wissenschaft und Technik*. Wiesbaden: VS Verlag für Sozialwissenschaften, 2004.

Ulrich, Walter. *A Brief Introduction to Critical Systems Heuristics (CSH)*. Milton Keynes: The Open Univ, 2005.

Vohland, Kathrin, Maike Weißpflug, and Lisa Pettibone. "Citizen Science and the Neoliberal Transformation of Science – An Ambivalent Relationship". *Citizen Science: Theory and Practice* 4, no. 1: 25 (2019): 1–9.

Wenninger, Andreas, and Sascha Dickel. "Paradoxien Digital-Partizipativer Wissenschaft". *Österreichische Zeitschrift für Soziologie 44*, no. *S1* (2019): 257–286.

Wenninger, Andreas, Fabienne Will, Sascha Dickel, Sabine Maasen, and Helmuth Trischler. "Ein- und Ausschließen: Evidenzpraktiken in der Anthropozändebatte und der Citizen Science". In *Wissen und Begründen: Evidenz als umkämpfte Ressource in der Wissensgesellschaft*, edited by Karin Zachmann, and Sarah Ehlers, 31–58. Baden-Baden: Nomos, 2019.

Wynne, Brian. "Public Participation in Science and Technology: Performing and Obscuring a Political–Conceptual Category Mistake". *East Asian Science, Technology and Society 1* (2007): 99–110.

Part V

Interpreting and Communicating Academic Evidence

Consumer Behavior and the Media

10 Exploring Consumers' Interpretation of Contested Nutritional Evidence

The Relevance of the Moral Foundations Theory

Edoardo Maria Pelli and Jutta Roosen

In 1977, the Select Committee on Nutrition and Human Needs to the United States Senate published the *Dietary Goals for the United States*.[1] In its recommendation to limit the consumption of red meat and dairy products, it was met with controversy by the food industry. Changes in those goals reduced the dietary recommendation to the level of nutrient composition of energy intake, that then recommended how much energy was to be taken from carbohydrates, sugars, fat and protein. *Dietary Goals* was followed by *The Surgeon General's Report on Nutrition and Health*,[2] which recognized the relationship between nutritional problems and overconsumption as well as imbalances of dietary intake. What had been a problem of scarcity before World War II had become an issue of food affluence by the 1970s.[3] This presented a major shift from addressing deficiencies in single nutrients.[4] The resulting changes in nutrition recommendations illustrate well why exploring consumers' interpretation of nutritional evidence is relevant. It still holds true that variations in nutritional needs between different people make it impossible to define simple nutritional standards.[5] Recommended daily allowances can be calculated for different population groups but have limited effectiveness in terms of guiding consumers' food choices.

Nutritional sciences are a relatively young scientific discipline. While dietetic nutrition knowledge existed in ancient Greece in the form of nutrition guidelines for a good life, a natural science based approach to nutrition only emerged in the 19th century.[6] At that time, food security and safety were the focus of concern. The primary preoccupation was with the provision of sufficient energy, considered not only the foundation of life but also a means of avoiding social unrest.[7] Over time, scientific advances made it possible to describe nutrition in more refined terms, focusing not only on energy but also on the supply of various nutrients. For example, the first vitamin was scientifically described less than a century ago in 1926.[8] Until approximately the 1990s, nutritional science research and recommendations were

DOI: 10.4324/9781003273509-16

devoted to identifying key nutrients (such as vitamins, carbohydrates, fats) and the specific effect of single nutrients on human health.[9] The consequence of this was that governments developed the recommended dietary allowances for each nutrient, advising a minimum intake of "good" nutrients (such as vitamins, minerals and proteins) and reduction of "bad" nutrients (mostly fats and sugars).[10] However, the scientific evidence and efficacy of this approach has been contested since the 1980s, when it was discovered that the effects on human health may be determined less by single nutrients than by the complexity of dietary patterns. This led to the development of nutrition pyramids, as developed in the United States and Germany.[11] This gradual shift in scientific knowledge shows that the evidence in the field of nutritional sciences is still controversial, as evident in the ongoing debates among scientists on the possible relationships among foods, nutrients and human health.[12]

In fact, results from studies supporting these new approaches have become available only recently, and many scientists criticize the relatively weak evidence existing between nutrition and health as it refers to single foods or nutrients.[13] Part of the debate among the scientific community originates from the fact that randomized control trials (RCTs) are often considered the highest source of scientific evidence in the context of evidence-based medicine.[14] Therefore, the quality of nutritional studies is often evaluated with tools that confine the concept of quality within the realm of RCTs. However, RCTs are not always suitable or possible. While RCTs are easier to conduct when assessing the short-term effects of nutrients on human health,[15] they are not appropriate when establishing the long-term and lifestyle effects of nutrients, for practical, ethical and methodological reasons.[16] This is why evidence for the latter is based on observational and cohort studies.[17] In order to study the issue of nutritional evidence, it is important to recognize both the complexity and controversy involved. As is often the case at the scientific frontier, established scientific knowledge is thrown into question by new research results. Furthermore, nutritional evidence is a contested issue because the food industries provide substantial funding for nutritional research to universities and to nutritional and health institutions, leading to conflicts of interest and the so-called funding effect.[18] This effect relates to the risk that scientific studies funded by the food industry tend to lead to results that support the sponsors' interests.

The complexity and ambiguity in the field of nutritional sciences creates confusion among consumers, also because the simple provision of evidence on what constitutes healthy eating seems insufficient to convince them to change their current behaviors.[19] The behavioral impact of nutrition advice remains limited although findings from the nutritional sciences receive a lot of public attention and media coverage, where the contested nature of nutritional evidence is also a factor propagating the confusion among consumers. Nutritional information is often sourced from incomplete and contradictory scientific studies.[20] In fact, the media often report results from contradictory studies that link several types of foods to specific health effects, such as the controversial debate regarding the possible relationship between consuming (red) meat and developing cancer in recent years. Furthermore, newspapers are under pressure to maximize their readership while journalists often do not have the time and the scientific skills to evaluate scientific studies critically.[21]

Given that nutritional evidence is in trouble, the objective of this essay is to develop a conceptual framework to study how consumers deal with contested nutritional evidence. We refer to consumers as independent individuals that build their own evidence practices regarding food and nutrition.[22] As such, they are users but not passive recipients of scientific evidence. The maintained hypothesis of our theoretical reasoning is that evidence in the field of nutrition is controversial and contested for many reasons, and it may therefore be difficult for consumers to base their views regarding their eating behavior on scientific evidence alone. Our hypothesis is that consumers form their attitudes toward nutritional evidence not fully rationally on scientific evidence and knowledge, but that this process is influenced by non-epistemic values through heuristic information processes. In this context, psychological approaches can be useful in developing the conceptual framework and deriving testable hypotheses. Our approach identifies one specific category of values following the moral foundations theory (MFT). The theory assumes that intuitive judgments guide the evaluation of objects and behavior, which we hold to be applicable in the context of consumers' food decision-making. These decisions and evaluations often occur in low involvement situations so that an intuitive approach seems particularly useful. This chapter develops a moral foundation approach to study consumers' interpretation of contested nutritional evidence. We present a theoretical exploration of consumers' interpretation of contested nutritional evidence, highlighting the importance of MFT as part of this process. We introduce an example where nutritional evidence is contested: Namely the phenomenon of the so-called superfoods. Here we show that from a theoretical perspective, considering the difficulty of basing their views on nutritional evidence, consumers may not judge the nutritional aspects of foods completely rationally, as suggested by neoclassical economic models. We therefore propose that individuals base their perceptions of nutritional aspects of foods on more intuitive and heuristic processes, largely driven by moral values. In this context, we explore the MFT and demonstrate in-depth how this theory fits into our theoretical framework. We do so via a series of four theoretical examples. Finally, we summarize our theoretical framework and explain how the issue of contested nutritional evidence relates to the wider discussion on contested evidence in this volume.

The Contested Nutritional Evidence for Superfoods

The current marketing trend for the so-called superfoods is especially characterized by contested nutritional evidence. Superfood is a marketing term used to promote several foods for their presumed exceptional nutritional characteristics and benefits for health, despite lacking and/or controversial scientific evidence.[23] Superfoods are mostly exotic fresh produce coming from the Global South and also consumed in the Global North, where they represent a current important marketing trend,[24] although there is legally no regulated definition of the term superfoods worldwide.[25] Prime examples of these superfoods include, among others, avocado, quinoa, chia seeds, acai berries, goji berries and maca root. These superfoods, which for centuries were consumed only by the local communities in the Global South, have recently been in increasing demand among consumers in the Global North.[26] Studies that have measured the health benefits of superfoods (considered by Joseph Bassaganya-Riera et al. as belonging to the broad category of functional foods) often have limited evidence from human trials.[27] Bassaganya-Riera et al. remark that it is important to have complementary studies on humans, that is both interventional RCTs and observational nutritional epidemiological studies. This need for complementarity is confirmed also by David L. Katz et al.,[28] who showed that the "relative primacy in adjudicating medical evidence" often attributed to RCTs does not always represent the case. As mentioned in the introduction, while for studies on short-term effects of nutrients on human health RCTs are easier to conduct, for studies on long-term and lifestyle effects of nutrients on human health RCTs are not applicable. In these latter cases, observational nutritional epidemiology is necessary. Taking this general argument back to our specific case of superfoods, we can infer that the complementarity of evidence is urgently needed. To measure a short-term effect of superfoods on health (e.g. the reduction in cholesterol level by the healthy fatty acids contained in avocado), a RCT would be applicable. However, to establish a long-term effect (e.g. the relationship between avocado consumption and the prevention on the development of several non-communicable diseases), a RCT would not be appropriate and observational nutritional epidemiology would be needed. However, evidence coming from complementary approaches to human studies is lacking, and this underlines the contested and controversial nature of nutritional evidence with regard to superfoods and health.

Avocado is held to possess numerous nutritional and health benefits, such as antioxidant capacities.[29] Acai[30] and goji berries[31] are also said to possess high levels of antioxidants. Regarding chia seeds and quinoa, their presumed health benefits are due to the high level of proteins. Last but not least, maca root is considered beneficial for health due to its high levels of vitamins and minerals.[32] However, as mentioned previously, the scientific evidence for the nutritional and health benefits of these so-called superfoods is generally still limited and contested.

Furthermore, despite the presumed nutritional benefits, the consumption of superfoods is linked to several environmental and socio-economic issues. A prime example is the case of avocado, which has become a very trendy superfood worldwide. In Mexico, the main producing country, avocado production has led to the problem of severe water depletion.[33] Even beyond Mexico, a study considering global avocado production in 2018 indicated that: "Globally, around 6.96 km³ of water is used or the equivalent of around 2.82 million Olympic size swimming pools (assuming a volume of 2500 m³ each) for avocado production in 2018".[34] In addition, avocado production has several socio-economic consequences. For example, as more land is devoted to avocado production for export, food insecurity has risen in some regions of Mexico. Stimulated by the increasing international demand, avocado production often replaces subsistence crops. These socio-economic and environmental aspects of superfoods are not only related to the primary production and the environment but also have ethical implications across the agri-food supply chain: consumers at the end of the chain may have strong concerns about the harm done to the environment and producers in the Global South.

Another example related to the possible negative environmental impact is the case of quinoa produced in the coastal area of Peru. Here the use of pesticides has intensified together with the expansion of quinoa production. The rise of pesticide use may hamper Peruvian quinoa's export to the Global North, due to possible lacks in meeting the requirements of international pesticide residue limits.[35] The situation exemplifies the fact that the controversy around excess/inappropriate pesticide use in the Global South is still present today, even if the major controversy on this issue took place in the 1970s–1980s, as described in Chapter 8 of this volume. The concern over the excessive use of pesticides in the Global South among environmentalists was a critique of previous established scientific evidence, which justified the use of pesticides for economic and agricultural productivity reasons. These concerns were expressed in scientific arguments together with their moral implications. In the case of quinoa, the excessive use of pesticides in the Global South demonstrates that even nowadays the environmental evidence regarding some superfoods may be contested and may provoke moral concerns. In fact, nowadays consumers in the Global North may feel ethical concerns about the negative conditions of the environment related to their consumption choices and the negative implications for the populations in the Global South living in areas where these products are cultivated.

Another feature that makes superfoods a very interesting case study for contested nutritional evidence relates to the fact that superfoods are heavily promoted in the media for their presumed nutritional and health benefits. One of the extreme examples of this phenomenon is the promotion of the nutritional and health benefits of some superfoods (such as acai berries) by the famous TV-host Oprah Winfrey.[36] This advancement of superfoods by influencers and the media not only has marketing but also ethical implications for promoters and consumers. With influencers and media providing incomplete and inaccurate information, consumers may perceive the misinformation as unfair and misleading.

Given all these multiple facets, we explore consumers' interpretation of contested nutritional evidence using the example of superfoods. Nutritional evidence is contested, and neither actors in public policy making nor individual consumers can rely on a strong evidence base for their decision-making.

The Need for Psychological Approaches to Study Consumers' Interpretation of Contested Nutritional Evidence

To explore theoretical approaches for studying consumers' interpretation of contested nutritional evidence, it is important to define the concepts of consumer beliefs and consumer attitude. A consumer belief about an object can be defined as the perceived probability that an object is associated with another concept. For example, a possible belief is that superfoods (the object) are healthy (the other concept). Furthermore, the cognitive, affective and behavioral consumer interpretation of an object is considered in relation to a consumer's attitude. According to the multi-attribute attitude model, the consumer attitude about an object can be defined as a function of all consumer beliefs about the object and the evaluative aspects of those beliefs. Mathematically it is possible to define a consumer attitude toward an object as:

$$\sum_{i=1}^{N} B_i a_i$$

where

B_i = belief "i" about the object,
a_i = evaluative aspects of B_i,
N = the number of beliefs.[37]

Further research on the concept of beliefs was developed in the context of the means-end-chain theory. This theory assumes that beliefs are part of a hierarchical evaluation and relate the object of evaluation to a chain of other concepts, i.e. attributes, consequences and values.[38] In their empirical study on beliefs, Klaus G. Grunert and Tino Bech-Larsen presented a framework linking the concepts of beliefs, means-end-chain theory and the choice option attractiveness, where the latter can be considered a synonym of attitude.[39] The research aim of their study was to understand empirically whether the attractiveness of or the attitude toward a choice option can be explained by attributes only, or whether beliefs linking the choice attributes to consequences and values improve their explanatory power. From the results of this study, obtained through a specific methodological procedure (the laddering method), the authors write: "We have concluded that beliefs linking the product to constructs of higher levels of abstraction – consequences and values – improve the explanation of choice option attractiveness beyond the explanation achieved by beliefs linking the product to attributes only".[40] These results are in line with the results of a previous empirical study, obtained through the same methodological procedure.[41] It indicated that values had significant explanatory power in explaining product preference beyond the explanatory power provided only by attributes and consequences. The term used in this study, namely product preference, could be considered a synonym of product acceptance. In conclusion, both empirical studies suggest that values linked to a product may be very important because they explain how the concrete attributes of a product influence consumers' beliefs and consequently attitudes and finally determine consumer acceptance or non-acceptance of the product. This insight is especially relevant in our research context because the nutritional evidence of superfoods is scarce and consequently it is difficult for consumers to judge the scientific nutritional attributes of superfoods. Consumers may interpret the superfoods and form their beliefs not fully rationally based on scientific attributes and their consequences but founded on abstract values linked to these attributes through heuristic information processes.

The strictly economic rational model grounded in neoclassical economics, where the consumer tries to maximize utility through a strictly rational judgment of the quality aspects of food products, is not sufficient to study the topic of consumers' attitudes toward nutritional evidence for superfoods. Instead, a psychological model is needed that takes into account consumer strategies for dealing with incomplete knowledge such as heuristic information processing and values.[42] This entails the following questions: Which type of psychological model would be appropriate? Which kinds of values may be more appropriate to explain, through heuristic information processes, consumers' beliefs and consequently attitudes and acceptance with regard to the nutritional evidence of superfoods? We provide answers to these questions in the next section.

The Contribution of the Moral Foundations Theory to Investigating Consumers' Interpretation of Contested Nutritional Evidence

Moral values can be important moderators in explaining consumer beliefs, and consequently attitudes and acceptance or non-acceptance with regard to the nutritional evidence for foods, through heuristic information processes. In fact, in this context of uncertainty around the nutritional aspects of foods, several types of foods are often considered "good foods for health" or "bad foods for health", and these definitions may have a strong moral connotation. Furthermore, superfoods are often perceived as "good foods for health" even if consumers lack the specific nutritional knowledge of superfoods to make such judgments. Moreover, previous literature suggests that the attribute "healthy" (which, according to us, is the main characteristic attributed to superfoods) implicitly hints at specific discourses of the "submerged iceberg of moral values".[43]

We conjecture that moral values can act as moderators for interpreting scientific knowledge and we consider MFT as a promising and comprehensive psychological theory that is appropriate to studying consumers' interpretation of nutritional evidence. The first instantiations of MFT go back to Jonathan Haidt[44] and Haidt and Jesse Graham.[45] We introduce the theory here as described by Graham et al.[46] MFT was developed to give an answer to the question of the origins of morality. The authors ask: "Where does morality come from? Why are moral judgments often so similar across cultures, yet sometimes so variable? Is morality one thing, or many? MFT was created to answer these questions".[47] Of course, MFT was born in the context of an earlier extensive and significant literature on moral development as summarized in Graham et al. The development of modern moral psychology started from the work of Lawrence Kohlberg, who assumed that there was only one moral foundation, namely the concept of justice. Kohlberg has thus been considered a monist moral psychologist. Carol Gilligan later criticized Kohlberg on the grounds that women's morality presents two moral values, i.e. not only the concept of justice but also the concept of care. Kohlberg accepted Gilligan's view and this dualistic approach (justice and care) has gained general consent among moral psychologists, e.g. Elliot Turiel. The assumption that morality relates only to individuals and how individuals establish relationships between one another was challenged by Richard A. Shweder, who supported the idea of a broader pluralism, using the example of a non-western culture, i.e. India, where it was evident that morality relates not only to individuals but also to collective phenomena such as groups, organizations, rules and cultural inheritances and religious beliefs. Shweder introduced the idea that across cultures human beings are constituted by three moral aspects: Autonomy (which refers to moral characteristics such as care and justice), community (which refers to moral characteristics such as loyalty, obligation and respect) and divinity (which refers to spiritual and moral characteristics such as purity and sanctity). In relation to this categorization of explicit moral discourse by Shweder, Alan P. Fiske claimed that moral evaluations were based on four relational models: "Communal Sharing, Authority Ranking, Equality Matching, and Market Pricing". Jonathan Haidt (one of the authors of MFT) tried to combine the theories of Shweder and Fiske but it was difficult to merge two different perspectives (the manifestly moral discourse of Shweder and the concept of interpersonal relationships in Fiske). Jonathan Haidt (along with his colleague Craig Joseph) set out to construct a more comprehensive theory.

Taking a pluralistic approach, Haidt and the other authors of MFT asked themselves: How many basic elements of morality can be identified and what are these basic elements of morality?[48] To elaborate the answer to this question, the authors interestingly used the metaphor of food taste:

> The human tongue has five discrete taste receptors (for sweet, sour, salt, bitter, and umami). Cultures vary enormously in their cuisines, which are cultural constructions shaped by historical events, yet the world's many cuisines must ultimately please tongues equipped with just five innate and universal taste receptors. What are the best candidates for being the innate and universal 'moral taste receptors' upon which the world's many cultures construct their moral cuisines? What are the concerns, perceptions, and emotional reactions that consistently turn up in moral codes around the world, and for which there are already-existing evolutionary explanations?[49]

As an answer to this metaphorical question, the authors identified five basic elements of morality or five moral values, which they defined as the five moral foundations: Care/harm; fairness/cheating; loyalty/betrayal; authority/subversion; purity/sanctity/degradation.

Care/harm refers to the preoccupation with taking care of others or for feeling compassion for people who have been caused harm. The authors give the example of mothers (not only humans but also other mammals) who are very concerned to take care of and nurture their offspring. Another example given by the authors is the compassion for victims. Fairness/cheating relates to the feeling of being honest or dishonest with other people. An example would be the perceived accuracy of media in reporting news. Readers may feel that the media are fair and report the news accurately, or that the media are cheating and they report false news, perhaps because of a conflict of interests. Loyalty/betrayal refers to the level of loyalty between individuals. An example would be a loyal friend who helps you when needed, while a disloyal friend does not. Authority/subversion refers to the obedience or disobedience to authority. An example could be citizens who trust and obey legislators who make the laws and citizens who do not obey the laws. Purity/sanctity/degradation refers to feelings of delight or disgust. An example could be a delight with foods that are perceived as nutritious and safe, and disgust for foods that are perceived as unhealthy and unsafe (e.g. contaminated by pathogens).

Let us now briefly examine the assumptions that sustain this theory and the development of the moral foundations. Firstly, these foundations are considered to be innate in the "first draft of the moral mind",[50] conceived and organized in advance of experience. However, despite this nativist approach, the authors believe that they are shaped differently for each individual according to their different experiences and cultural learnings, and through the process of cultural development.

Another important aspect that characterizes the moral foundations, as conceived by the authors, relates to the concept of intuitionism. Moral judgments (motivated through the moral foundations) happen quickly and intuitively, while the moral reasoning comes after the moral judgment, as a support and a justification for the moral judgment. This concept was developed through the social intuitionism model (SIM).[51] We will come back to this concept later when we show how these concepts relate to our conceptual framework explaining consumers' judgment of superfoods.

Furthermore, Graham, Haidt and their collaborators have always welcomed the possibility that the moral values could exceed five, being open to any update and testing of the theory. In fact, many methods have been developed to test MFT. The concepts have been developed in a method-theory coevolution: The theory can inspire new methods to test the theory, while at the same time the results from the application of methods can inspire the further development of the theory itself. Furthermore, this theory has been applied within the field of social psychology. The authors claim that it should also be applied beyond this field and there have indeed been such applications. In fact, the authors believe that MFT is a practical theory that may prove useful in many fields. This theory is thus in perpetual evolution, to be updated and in development, and thus particularly suitable for research that is highly cross-disciplinary.

We believe that our conceptual model, which crosses different disciplines, i.e. economics, psychology as well as food and nutritional sciences, can contribute on updating and developing the theory. In fact, our application of MFT to food consumption and particularly nutritional evidence covers a new area. Moreover, although it is a relatively new field, there have been already very interesting empirical applications of MFT to food consumption. Given the importance of these empirical applications of MFT to food consumption, we briefly review some of those studies here.

One study found that respondents with stronger agreements on care and fairness statements were more likely to purchase environmentally sustainable dairy products and pork from swine raised with limited antibiotic usage. They were also more inclined to vote for stricter livestock environmental standards and disease protocols.[52] Another study found that the agreement with care was positively correlated with the likelihood of being vegetarian rather than flexitarian, while the agreement with authority was positively correlated with the likelihood of being a full-time meat eater rather than a flexitarian.[53] Furthermore, another study revealed that the importance of the moral value of purity mediated a positive relationship between religiosity and diet-minded food consumption, which can be considered a diet based on foods free from fats, sugars or allergens.[54]

Although these studies apply MFT to the study of consumers' attitudes toward food consumption, the relationship between MFT and nutritional evidence has not yet been fully investigated with the explicit acknowledgment that nutritional evidence is a contested issue. We therefore develop a conceptual framework to explain how the MFT can contribute to understanding consumers' attitudes toward contested nutritional evidence using the example of the superfoods. Our conceptual framework is displayed in Figure 10.1.

As we can see from Figure 10.1, our theoretical assumptions consist in the following process: Consumers have scarce knowledge about the scientific nutritional attributes of superfoods because the nutritional evidence for superfoods is limited and controversial and consumers lack the expertise to judge the nutritional properties. Given this scarce knowledge, consumers form their beliefs and consequently their attitudes and acceptance of the nutritional evidence for superfoods through a heuristic information process moderated by the moral values that consumers hold. This means that consumers base their beliefs and consequently attitudes and acceptance or non-acceptance of superfoods by relying on

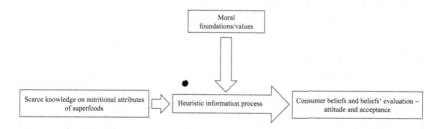

Figure 10.1 Conceptual framework. Figure by Edoardo Maria Pelli and Jutta Roosen.

moral values linked to superfoods rather than on scientific nutritional attributes and information.

MFT is particularly suited to studying consumers' attitudes toward nutritional evidence of superfoods. In fact, foods promoted as healthy are generally perceived to have strong moral connotation, e.g. "good foods" which are free from "bad nutrients or additives". In addition, previous literature suggests that the attribute "healthy" implies specific discourses about moral values.[55]

These moral values somehow substitute the needs of consumers for nutritional evidence in order to form their attitudes. Therefore, consumers bypass the issue of nutritional evidence by referring to moral values. As such, consumers are not actively engaging in evidence critique by questioning specific methods and results in the nutritional sciences. Rather, they are developing their own intuitive judgments about the benefits of foods. By doing so, they participate in the de-stabilization of the evidence at the interface of science and public. At the same time, consumers use these intuitive judgments to re-stabilize their everyday practices in the face of contested nutritional evidence.

We now present two more possible examples of how consumers' beliefs (and consequently attitudes and acceptance) with regard to the nutritional evidence of superfoods may be moderated by specific moral values as defined by MFT. The first example, displayed in Figure 10.2, relates to how consumers may form their beliefs (and consequently attitude and acceptance) with regard to the presumed healthiness of superfoods.

We can interpret Figure 10.2 as follows: Superfoods are presented as healthy to consumers, who cannot judge this attribute "healthy" because

Superfoods promoted as healthy may be perceived by consumers as pure, uncontaminated by "bad" nutrients

+ correlated with the importance of the moral foundation/value of Purity

+ consumer acceptance of superfoods

Figure 10.2 Possible relationship between the attribute: "healthy" and the moral foundations/values. Figure by Edoardo Maria Pelli and Jutta Roosen.

of the lack of sufficient nutritional knowledge and expertise. Therefore, consumers may accept this attribute "healthy" not based on its scientific meaning but based on the association to a concept of a "good" and "pure" food free from "bad nutrients", which may be positively correlated to the importance of the moral foundation of purity/sanctity/degradation. Therefore, through a heuristic information perception process moderated by the moral foundation of purity/sanctity, consumers may form their beliefs and consequently attitudes toward and acceptance of superfoods. Namely, the more consumers are attached to the moral value of purity/sanctity, the more consumers may perceive and accept superfoods as healthy foods.

The second example relates to the consumers' trust on influencers who promote superfoods through the media and this example is displayed in Figure 10.3. The conceptual framework illustrated can be interpreted as follows. "Pseudo-scientific" authorities,[56] such as influencers, may promote superfoods as healthy through the media and consumers may trust them. This trust may lead consumers to think that influencers are role-model authorities and there may be a positive correlation between trust and the moral foundation of authority. Therefore, through a heuristic information process moderated by the moral foundation of authority, consumers may form their beliefs about superfoods promoted as healthy by the influencers. The more consumers trust influencers and the more consumers are attached to the moral foundation of authority, the more consumers may perceive and accept superfoods as healthy foods. Furthermore, the correlation between authority and trust can be

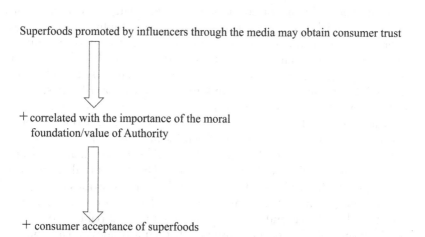

Superfoods promoted by influencers through the media may obtain consumer trust

+ correlated with the importance of the moral foundation/value of Authority

+ consumer acceptance of superfoods

Figure 10.3 Possible relationship between the specific role of communication of some influencers and moral foundations/values. Figure by Edoardo Maria Pelli and Jutta Roosen.

observed in the previous literature, where authority, even though not addressed specifically within the MFT, is linked to the recommendations of role models (influencers).[57] The meaning of the term influencer as part of the role of authority may itself imply that influencers are able to sway consumer decisions because consumers trust them.

After these examples, we want to discover in more detail why the MFT is very suited for our conceptual framework. Our conceptual framework is based on the idea that, given that the nutritional evidence for superfoods is scarce, consumers form their beliefs and consequently their attitudes and acceptance or non-acceptance of superfoods not completely rationally based on scientific attributes, but by the moderation of moral values linked to superfoods, through a heuristic information process. The authors of MFT explicitly state that moral values are activated intuitively and based on heuristic information processes, rather than evoked rationally. The theory explains this through the development of the SIM:

> the sudden appearance in consciousness, or at the fringe of consciousness, of an evaluative feeling (like–dislike, good–bad) about the character or actions of a person, without any conscious awareness of having gone through steps of search, weighing evidence, or inferring a conclusion.[58]

From this quote, we see that our conceptual framework perfectly fits with MFT, also because both MFT and our conceptual framework assume that moral judgments are intuitive and do not weigh the evidence in a computational manner. In our case, the nutritional evidence for superfoods cannot be weighted because it is contested and scarce and therefore difficult for consumers to assess. The moral foundations serve as a guideline to assessing the value of acceptability of superfoods. Moreover, in this quote, we can see that the explicit moral judgments are expressed through words such as "good" or "bad", which are the ones also pertinent to our conceptual framework, such as "good food for health". For these reasons, we think that the MFT contributes substantially to our conceptual framework.

Lastly, MFT acknowledges the fact that cultural development plays a role in shaping the innate moral values of consumers, therefore we can extend our conceptual framework to include the concept of cultural development, as displayed in Figure 10.4. The conceptual framework displayed can be interpreted as follows. As in the previous conceptual framework (Figure 10.1), consumers, who have scarce knowledge of nutritional evidence and the scientific nutritional attributes of superfoods, form their beliefs and consequently their attitudes and acceptance or non-acceptance with regard to superfoods via the moderation of the moral values, through a heuristic information process. The difference in our new conceptual framework (Figure 10.4) is that the moderating role

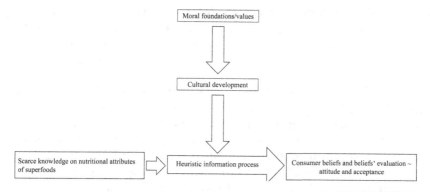

Figure 10.4 Extension of the conceptual framework through the role of cultural development. Figure by Edoardo Maria Pelli and Jutta Roosen.

of moral values is shaped and affected by cultural development. Hence, cultural development also has a role in forming consumers' beliefs and consequently attitudes and acceptance with regard to superfoods. For example, in a culture where environmental impact has a great importance for consumers, this cultural attitude would reinforce the concept of care for the environment and therefore it could decrease consumers' acceptance of superfoods, the production of which is associated with negative environmental impact. On the other hand, in a culture where environmental impact has not gained yet the attention of consumers, this cultural attitude would not reinforce the concept of care for the environment and therefore it would not decrease consumers' acceptance of superfoods whose production has a negative environmental impact. With this last conceptual framework, we conclude our development of the theoretical approach. This theoretical development has shown that MFT can contribute substantially to the exploration of consumers' attitudes toward contested nutritional evidence.

Conclusion

In this chapter, we have explored the role that MFT can play in attempts to explain consumers' strategies with respect to "re-stabilizing" the contested nature of evidence coming from the nutritional sciences. By applying MFT to the study of consumers' attitudes toward nutritional evidence about superfoods, it seems that we have served our theoretical hypothesis that consumers face scarce nutritional evidence. Therefore, their attitudes will not be based on scientific claims (of which consumers have scarce knowledge), but on heuristic information processes linked to moral values. Furthermore, we base our theoretical framework on the SIM developed by Haidt,[59] which posits that initial moral judgments are rather intuitive and driven more by heuristic than fully rational ones.

This volume is concerned with different situations in which scientific evidence is contested and therefore subjected to processes of destabilization and re-stabilization. Our case of nutritional evidence represents an interesting although slightly different aspect of these processes. Evidence in the field of nutrition is contested because corporate funding may bias research[60] and methodological challenges lead to sometimes contradictory and conflicting results, thus increasing the confusion among the general public and consumers. Particularly for the case of superfoods, the health benefits conferred on these products are built on contradictory studies and often lack the evidence coming from human clinical trials.[61] This phenomenon of controversial evidence is further amplified through the widespread and often inaccurate coverage and promotion of the presumed health benefits by the media, advertising and influencers.[62,63] Therefore, consumers bridge the lack of nutritional evidence by using intuitive judgments based on moral values in order to form their beliefs, attitudes and acceptance of superfoods.

Notes

1 US Senate, *Dietary Goals for the United States* (Washington: U.S. Government Printing Office, 1977).
2 Surgeon General, *Healthy People. The Surgeon's General Report on Health Promotion and Disease Prevention* (Washington: United States Public Health Service, 1979).
3 Fabrice Etilé, "Food Consumption and Health", in *The Oxford Handbook of the Economics of Food Consumption and Policy*, eds. Jayson L. Lusk, Jutta Roosen, and Jason F. Shogren (Oxford: Oxford Univ. Press, 2011), 716–746.
4 J. Michael McGinnis and Marion Nestle, "The Surgeon General's Report on Nutrition and Health: Policy Implications and Implementation Strategies", *American Journal of Clinical Nutrition* 49, no. 1 (1989): 23–28.
5 Surgeon General, *Healthy People.*
6 Eva Barlösius, *Soziologie des Essens: Eine sozial- und kulturwissenschaftliche Einführung in die Ernährungsforschung* (Weinheim: Beltz Juventa, 2016), 58–69.
7 The World Food Summit defined food security in 1996 as follows: "Food security exists when all people, at all times, have physical and economic access to sufficient, safe and nutritious food that meets their dietary needs and food preferences for an active and healthy life", World Food Summit 1996, *Rome Declaration on World Food Security* (Rome: 1996). Food security is commonly described as resting on the three pillars of availability, access and utilization. Sufficient availability and utilization was for long concerned with sufficiency of energy provision (calories). This was also what governments were concerned with up to the mid-20th century. Nutrition policies related to assuring sufficient access to energy from food and were deemed important to maintain the work forces, and the armed forces, see also Eva Barlösius, *Soziologie des Essens*, 2016. In recent times, other aspects of accessibility and healthfulness have been added to the conceptualization and measurement of food security. A prime example is the use of dietary diversity indicators, see C.B. Barret, "Measuring Food

Insecurity", *Science* 327 (February 12, 2010): 825–828. Therefore, organizations of the United Nations now mostly speak of "malnutrition in all its forms" when referring to food insecurity, see FAO, IFAD, UNICEF, WFP, and WHO, *The State of Food Security and Nutrition in the World 2021. Transforming Food Systems for Food Security, Improved Nutrition and Affordable Healthy Diets for All* (Rome: FAO, 2021).

  8 Dariush Mozaffarian, Irwing Rosenberg, and Ricardo Uauy, "History of Modern Nutrition Science: Implications for Current Research, Dietary Guidelines, and Food Policy", *BMJ* 361, no. k2392 (2018): 1–6.

  9 Mozaffarian et al., "History", 1–6.

10 Mozaffarian et al., "History", 1–6.

11 Barlösius, *Soziologie des Essens*, 58–69.

12 Mozaffarian et al., "History", 1–6.

13 Mozaffarian et al., "History", 1–6.

14 David L. Katz, Micaela C. Karlsen, Mei Chung, Marissa M. Shams-White, Lawrence W. Green, Jonathan Fielding, Ayumi Saito, and Walter Willett, "Hierarchies of Evidence applied to Lifestyle Medicine (HEALM): Introduction of a Strength-of-Evidence Approach Based on a Methodological Systematic Review", *BMC Medical Research Methodology* 19, no. 178 (2019): 1–16.

15 Juan B. Bengoetxea and Oliver Todt, "Decision-Making in the Nutrition Sciences: A Critical Analysis of Scientific Evidence for Assessing Health Claims", *Manuscrito* 44, no. 3 (2021): 42–69.

16 Saana Jukola, "On the Evidentiary Standards for Nutrition Advice", *Studies in History and Philosophy of Science* 73 (2019): 1–9.

17 Saana Jukola, "On the Evidentiary Standards for Nutrition Advice", 1–9.

18 Marion Nestle, "Conflicts of Interest in Food and Nutrition Research", in *Feeding the World Well: A Framework for Ethical Food Systems*, ed. Alan M. Goldberg (Baltimore: Johns Hopkins Press, 2020), 89–97.

19 Mozaffarian et al., "History", 1–6.

20 Naomi Oreskes, *Why Trust Science?* (Princeton: Princeton Univ. Press, 2019), 73.

21 Navjoyt Ladher, "Nutrition Science in the Media: You Are What You Read", *BMJ* 353, no. i1879 (2016): 1–2.

22 Christine Hassauer and Jutta Roosen, "Toward a Conceptual Framework for Food Safety Criteria: Analyzing Evidence Practices Using the Case of Plant Protection Products", *Safety Science* 127, no. 104683 (2020): 1–17.

23 Joseph Bassaganya-Riera, Elliot M. Berry, Ellen E. Blaak, Barbara Burlingame, Johannes le Coutre, Willem van Eden, and Ahmed El-Sohemy et al., "Goals in Nutrition Science 2020–2025", *Frontiers in Nutrition* 7, no. 606378 (2021): 1–3.

24 Mintel Press Office, "Super growth for 'super' foods: New product development shoots up 202% globally over the past five years", accessed January 6, 2022, https://www.mintel.com/press-centre/food-and-drink/super-growth-for-super-foods-new-product-development-shoots-up-202-globally-over-the-past-five-years.

25 "Superfoods or Superhype?", Harvard T.H. Chan School of Public Health, accessed January 6, 2022, https://www.hsph.harvard.edu/nutritionsource/superfoods/.

26 Graciela Andrango and Trent Blare, "Theme Overview: Functional Foods: Fad or Path to Prosperity?", *Choices* 4 (2020): 1–2.

27 Bassaganya-Riera et al., "Goals", 1–3.

28 Katz et al., "Hierarchies", 1–16.

29 Deep J. Bhuyan, Muhammad A. Alsherbiny, Saumya Perera, Mitchell Low, Amrita Basu, Okram A. Devi, Mridula S. Barooah, Chun G. Li, and Konstantinos. Papoutsis, "The Odyssey of Bioactive Compounds in Avocado (*Persea americana*) and Their Health Benefits", *Antioxidants* 8, 10, 426 (2019): 1–53.

30 Graciela Andrango, Trent Blare, and Guy Hareau, "Functional Foods: Fad or Path to Prosperity? Data Visualization", *Choices* 4 (2020): 1–2.

31 Zheng F. Ma, Hongxia Zhang, Sue S. Teh, Chee W. Wang, Yutong Zhang, Frank Hayford, Liuyi Wang, Tong Ma, Zihan Dong, Yan Zhang, and Yifan Zhu, "Goji Berries as a Potential Natural Antioxidant Medicine: An Insight into their Molecular Mechanisms of Action", *Oxidative Medicine and Cellular Longevity*, no. 2437397 (2019): 1–9.

32 Graciela Andrango, Amy Johnson, and Marc F. Bellemare, "Quinoa Production and Growth Potential in Bolivia, Ecuador, and Peru", *Choices* 4 (2020): 1–10.

33 Ruben Sommaruga and Honor May Eldridge, "Avocado Production: Water Footprint and Socio-Economic Implications", *Euro Choices* 20, no. 2 (2020): 48–53.

34 Sommaruga and Eldridge, "Avocado Production", 50.

35 Andrango et al., "Quinoa", 1–10.

36 Casimir MacGregor, Alan Petersen, and Christine Parker, "Promoting a Healthier, Younger You: The Media Marketing of Anti-Aging Superfoods", *Journal of Consumer Culture* 21, no. 2 (2021): 164–179.

37 Martin Fishbein, "An Investigation of the Relationships between Beliefs about an Object and the Attitude toward that Object", *Human Relations* 16 (1963): 233–239.

38 Jonathan Gutman, "A Means-End Chain Model Based on Consumer Categorization Processes", *Journal of Marketing* 46, no. 2 (1982): 60–72.

39 Klaus G. Grunert and Tino Bech-Larsen, "Explaining Choice Option Attractiveness by Beliefs Elicited by the Laddering Method", *Journal of Economic Psychology* 26, no. 2 (2005): 223–241.

40 Grunert and Bech-Larsen, "Explaining Choice Option Attractiveness", 237.

41 W. Steven Perkins and Thomas J. Reynolds, "The Explanatory Power of Values in Preference Judgements: Validation of the Means-End Perspective", in *Advances in Consumer Research*, Vol. 15, ed. Micheal J. Houston (Provo: Association for Consumer Research, 1988), 122–126.

42 The relevance of rational choice to decision-making where a consumer maximizes utility subject to a budget constraint has been questioned in consumer behavior research. The idea of *homo economicus* dates back to John Stuart Mill (1806–1873) who describes a hypothetical, self-interested individual seeking to maximize individual utility, see Michael S. Aßländer and Hans G. Nutzinger, "John Stuart Mill", in *Klassiker des ökonomischen Denkens*, ed. Heinz D. Kurz (München: Verlag C.H. Beck, 2008). Rational utility maximization requires complete knowledge of all options available and the consequences of these options. In 1957, Herbert A. Simon published his book *Models of Man* (New York: John Wiley) where he criticized that this *homo economicus* ignores insights from psychology and conceived a cognitively limited agent. This idea of bounded rationality translates this idea of *homo economicus* to one that considers effort in information access in its rational choice, see Gregory Wheeler, "Bounded Rationality", in *The Stanford Encyclopedia of Philosophy, Fall 2020*, ed. Edward N. Zalta (Stanford, CA: Metaphysics Research Lab, Stanford Univ., 2020), https://plato.stanford.edu/archives/fall2020/entries/bounded-rationality/. The roles of imperfect information, biases and heuristics in

information processing and decision-making have led to a rich literature in economic psychology and behavioral economics. A history on the development of behavioral economics is provided by Floris Heukelom, *Behavioral Economics: A History* (New York: Cambridge Univ. Press, 2014).

43 Theo Van Leeuwen, "Legitimation in Discourse and Communication", *Discourse & Communication* 1, no. 1 (2007): 97.

44 Jonathan Haidt, "The Emotional Dog and its Rational Tail: A Social Intuitionist Approach to Moral Judgment", *Psychological Review* 108, no. 4 (2001): 814–834.

45 Jonathan Haidt and Jesse Graham, "When Morality Opposes Justice: Conservatives Have Moral Intuitions that Liberals May not Recognize", *Social Justice Research* 20 (2007): 98–116.

46 Jesse Graham, Jonathan Haidt, Sena Koleva, Matt Motyl, Ravi Iyer, Sean P. Wojcik, and Peter H. Ditto, "Chapter Two – Moral Foundations Theory: The Pragmatic Validity of Moral Pluralism", *Advances in Experimental Social Psychology* 47 (2013): 55–130.

47 Graham et al., "Moral Foundations Theory", 56.

48 Graham et al., "Moral Foundations Theory", 56.

49 Graham et al., "Moral Foundations Theory", 60.

50 Graham et al., "Moral Foundations Theory", 61.

51 Haidt, "The Emotional Dog", 814–834.

52 Ellen Goddard, Violet Muringai, and Albert Boaitey, "Moral Foundations and Credence Attributes in Livestock Production: Canada", *Journal of Consumer Marketing* 36, no. 3 (2019): 418–428.

53 Charlotte J.S. De Backer and Liselot Hudders, "Meat Morals: Relationship between Meat Consumption Consumer Attitudes towards Human and Animal Welfare and Moral Behavior", *Meat Science* 99 (2015): 68–74.

54 Elisabeth A. Minton, Kathryn A. Johnson, and Richie L. Liu, "Religiosity and Special Food Consumption: The Explanatory Effects of Moral Priorities", *Journal of Business Research* 95 (2019): 442–454.

55 Van Leeuwen, "Legitimation", 97.

56 MacGregor et al., "Promoting a Healthier, Younger You", 164–179.

57 Joop De Boer and Harry Aiking, "Favoring Plant Instead of Animal Protein Sources: Legitimation by Authority, Morality, Rationality and Story Logic", *Food Quality and Preference* 88, no. 104098 (2021): 1–10.

58 Jonathan Haidt and Fredrik Bjorklund, "Social Intuitionists Answer Six Questions about Moral Psychology", in *Moral Psychology, Vol. 2: The Cognitive Science of Morality: Intuition and Diversity*, ed. Walter Sinnott-Armstrong (Cambridge: MIT Press, 2008), 188. Modified from: Jonathan Haidt, "The Emotional Dog and its Rational Tail: A Social Intuitionist Approach to Moral Judgment", *Psychological Review* 108 (2001): 814–834.

59 Haidt and Bjorklund, "Social Intuitionists Answer Six Questions", 814–834.

60 Nestle, "Conflicts", 89–97.

61 Bassaganya-Riera, "Goals", 1–3.

62 Ladher, "Nutrition", 1–2.

63 MacGregor et al., "Promoting a Healthier, Younger You", 164–179.

References

Andrango, Graciela, Amy Johnson, and Marc F. Bellemare. "Quinoa Production and Growth Potential in Bolivia, Ecuador, and Peru". *Choices* 4 (2020): 1–10.

Andrango, Graciela, and Trent Blare. "Theme Overview: Functional Foods: Fad or Path to Prosperity?". *Choices* 4 (2020): 1–2.

Andrango, Graciela, Trent Blare, and Guy Hareau. "Functional Foods: Fad or Path to Prosperity? Data Visualization". *Choices* 4 (2020): 1–2.

Aßländer, Michael S., and Hans G. Nutzinger. "John Stuart Mill". In *Klassiker des ökonomischen Denkens*, edited by Heinz D. Kurz, 176–195. München: Verlag C.H. Beck, 2008.

Barlösius, Eva. *Soziologie des Essens: Eine sozial- und kulturwissenschaftliche Einführung in die Ernährungsforschung.* Weinheim: Beltz Juventa, 2016, 58–69.

Barret, Christopher B. "Measuring Food Insecurity". *Science* 327, no. 5967 (2010): 825–828.

Bassaganya-Riera, Joseph, Elliot M. Berry, Ellen E. Blaak, Barbara Burlingame, Johannes le Coutre, Willem van Eden, and Ahmed El-Sohemy et al. "Goals in Nutrition Science 2020–2025". *Frontiers in Nutrition* 7, no. 606378 (2021): 1–3.

Bengoetxea, Juan B., and Oliver Todt. "Decision-Making in the Nutrition Sciences: A Critical Analysis of Scientific Evidence for Assessing Health Claims". *Manuscrito* 44, no. 3 (2021): 42–69.

Bhuyan, Deep J., Muhammad A. Alsherbiny, Saumya Perera, Mitchell Low, Amrita Basu, Okram A. Devi, Mridula S. Barooah, Chun G. Li, and Konstantinos Papoutsis. "The Odyssey of Bioactive Compounds in Avocado (*Persea americana*) and their Health Benefits". *Antioxidants* 8, no. 10, 426 (2019): 1–53.

De Backer, Charlotte J.S., and Liselot Hudders. "Meat Morals: Relationship between Meat Consumption Consumer Attitudes towards Human and Animal Welfare and Moral Behavior". *Meat Science* 99 (2015): 68–74.

De Boer, Joop, and Harry Aiking. "Favoring Plant Instead of Animal Protein Sources: Legitimation by Authority, Morality, Rationality and Story Logic". *Food Quality and Preference* 88, no. 104098 (2021): 1–10.

Etilé, Fabrice. "Food Consumption and Health". In *The Oxford Handbook of the Economics of Food Consumption and Policy*, edited by Jayson L. Lusk, Jutta Roosen, and Jason F. Shogren, 716–746. Oxford: Oxford Univ. Press, 2011.

FAO, IFAD, UNICEF, WFP, and WHO. *The State of Food Security and Nutrition in the World 2021. Transforming Food Systems for Food Security, Improved Nutrition and Affordable Healthy Diets for All.* Rome: FAO, 2021.

Fishbein, Martin. "An Investigation of the Relationships between Beliefs about an Object and the Attitude toward that Object". *Human Relations* 16 (1963): 233–239.

Fiske, Alan P. *Structures of Social Life: The Four Elementary Forms of Human Relations: Communal Sharing, Authority Ranking, Equality Matching, Market Pricing.* New York: Free Press, 1991.

Gilligan, Carol. *In a Different Voice: Psychological Theory and Women's Development.* Cambridge, MA: Harvard Univ. Press, 1982.

Goddard, Ellen, Violet Muringai, and Albert Boaitey. "Moral Foundations and Credence Attributes in Livestock Production: Canada". *Journal of Consumer Marketing* 36, no. 3 (2019): 418–428.

Graham, Jesse, Jonathan Haidt, Sena Koleva, Matt Motyl, Ravi Iyer, Sean P. Wojcik, and Peter H. Ditto. "Chapter Two – Moral Foundations Theory: The Pragmatic Validity of Moral Pluralism". *Advances in Experimental Social Psychology* 47 (2013): 55–130.

Grunert, Klaus G., and Tino Bech-Larsen. "Explaining Choice Option Attractiveness by Beliefs Elicited by the Laddering Method". *Journal of Economic Psychology* 26, no. 2, (2005): 223–241.

Gutman, Jonathan. "A Means-End Chain Model Based on Consumer Categorization Processes". *Journal of Marketing* 46, no. 2 (1982): 60–72.

Haidt, Jonathan, and Fredrik Bjorklund. "Social Intuitionists Answer Six Questions about Moral Psychology". In *Moral Psychology, Vol. 2: The Cognitive Science of Morality: Intuition and Diversity*, Vol. 188, edited by Walter Sinnott-Armstrong. Cambridge: MIT Press, 2008. Modified from: Haidt, Jonathan, "The Emotional Dog and its Rational Tail: A Social Intuitionist Approach to Moral Judgment", *Psychological Review* 108, no. 4 (2001): 814–834.

Haidt, Jonathan, and Jesse Graham. "When Morality Opposes Justice: Conservatives Have Moral Intuitions that Liberals May not Recognize". *Social Justice Research* 20 (2007): 98–116.

Haidt, Jonathan. "The Emotional Dog and its Rational Tail: A Social Intuitionist Approach to Moral Judgment". *Psychological Review* 108, no. 4 (2001): 814–834.

Harvard T.H. Chan School of Public Health. "Superfoods or Superhype?". Accessed January 6, 2022. https://www.hsph.harvard.edu/nutritionsource/superfoods/.

Hassauer, Christine, and Jutta Roosen. "Toward a Conceptual Framework for Food Safety Criteria: Analyzing Evidence Practices Using the Case of Plant Protection Products". *Safety Science* 127, no. 104683 (2020): 1–17.

Heukelom, Floris. *Behavioral Economics. A History*. New York: Cambridge Univ. Press, 2014.

Jukola, Saana. "On the Evidentiary Standards for Nutrition Advice". *Studies in History and Philosophy of Science* 73 (2019): 1–9.

Katz, David L., Micaela C. Karlsen, Mei Chung, Marissa M. Shams-White, Lawrence W. Green, Jonathan Fielding, Ayumi Saito, and Walter Willett. "Hierarchies of Evidence applied to Lifestyle Medicine (HEALM): Introduction of a Strength-of-Evidence Approach Based on a Methodological Systematic Review". *BMC Medical Research Methodology* 19, no. 178, (2019): 1–16.

Kohlberg, Lawrence. "Stage and Sequence: The Cognitive-Developmental Approach to Socialization", In *Handbook of Socialization Theory and Research*, edited by David A. Goslin, 347–480. Chicago: Rand McNally, 1969.

Ladher, Navjoyt. "Nutrition Science in the Media: You Are What You Read", *BMJ* 353, no. i1879 (2016): 1–2.

MacGregor, Casimir, Alan Petersen, and Christine Parker. "Promoting a Healthier, Younger You: The Media Marketing of Anti-Aging Superfoods". *Journal of Consumer Culture* 21, no. 2 (2021): 164–179.

Ma, Zheng F., Hongxia Zhang, Sue S. Teh, Chee W. Wang, Yutong Zhang, Frank Hayford, Liuyi Wang, Tong Ma, Zihan Dong Yan Zhang, and Yifan Zhu. "Goji Berries as a Potential Natural Antioxidant Medicine: An Insight into their Molecular Mechanisms of Action". *Oxidative Medicine and Cellular Longevity* no. 2437397 (2019): 1–9.

McGinnis, J. Michael, and Marion Nestle. "The Surgeon General's Report on Nutrition and Health: Policy Implications and Implementation Strategies". *American Journal of Clinical Nutrition* 49, no. 1 (1989): 23–28.

Mintel Press Office. "Super growth for 'super' foods: New product development shoots up 202% globally over the past five years". Accessed January 6, 2022. https://www.mintel.com/press-centre/food-and-drink/super-growth-for-super-foods-new-product-development-shoots-up-202-globally-over-the-past-five-years.

Minton, Elisabeth A., Kathryn A. Johnson, and Richie L. Liu. "Religiosity and Special Food Consumption: The Explanatory Effects of Moral Priorities". *Journal of Business Research* 95 (2019): 442–454.

Mozaffarian, Dariush, Irwing Rosenberg, and Ricardo Uauy. "History of Modern Nutrition Science: Implications for Current Research, Dietary Guidelines, and Food Policy". *BMJ* 361, no. 2392 (2018): 1–6.

Nestle, Marion. "Conflicts of Interest in Food and Nutrition Research". In *Feeding the World Well: A Framework for Ethical Food Systems*, edited by Goldberg Alan M., 89–97. Baltimore: Johns Hopkins Press, 2020.

Oreskes, Naomi. *Why Trust Science?* Princeton: Princeton Univ. Press, 2019.

Perkins, W. Steven, and Thomas J. Reynolds. "The Explanatory Power of Values in Preference Judgements: Validation of the Means–End Perspective". In *Advances in Consumer Research*, Vol. 15, edited by Micheal J. Houston, 122–126. Provo: Association for Consumer Research, 1988.

Rai, Tage S., and Alan P. Fiske. "Moral Psychology is Relationship Regulation: Moral Motives for Unity, Hierarchy, Equality, and Proportionality". *Psychological Review*, 118, n. 1 (2011): 57–75.

Shweder, Richard A. "In Defense of Moral Realism: Reply to Gabennesch". *Child Development*, 61, n. 6 (1990): 2060–2067.

Shweder, Richard A., Nancy C. Much, Manamohan Mahapatra, and Lawrence Park. "The 'Big Three' of Morality (Autonomy, Community, and Divinity), and the 'Big Three' Explanations of Suffering". In *Morality and Health*, edited by Allan M. Brandt and Paul Rozin, 119–169. New York: Routledge, 1997.

Simon, Herbert A. *Models of Man*. New York: John Wiley, 1957.

Sommaruga, Ruben, and Honor May Eldridge. "Avocado Production: Water Footprint and Socio-Economic Implications". *Euro Choices* 20, no. 2 (2020): 48–53.

Surgeon General. *Healthy People. The Surgeon's General Report on Health Promotion and Disease Prevention*. Washington: United States Public Health Service, 1979.

Turiel, Elliot. *The Development of Social Knowledge: Morality and Convention*. Cambridge, England: Cambridge Univ. Press, 1983.

US Senate. *Dietary Goals for the United States*. Washington: U.S. Government Printing Office, 1977.

Van Leeuwen, Theo. "Legitimation in Discourse and Communication". *Discourse & Communication* 1, no. 1 (2007): 97.

Wheeler, Gregory. "Bounded Rationality". In *The Stanford Encyclopedia of Philosophy, Fall 2020*, edited by Edward N. Zalta. Stanford, CA: Metaphysics Research Lab, Stanford Univ., 2020. https://plato.stanford.edu/archives/fall2020/entries/bounded-rationality/.

World Food Summit. *Rome Declaration on World Food Security*. Rome, 1996.

11 Stories about Villains, Mad Scientists and Failure

Patterns of Evidence Criticism in Media Coverage of Genomic Research

Susanne Kinnebrock and Helena Bilandzic

Media coverage of science is double-edged. On the one hand, journalism is expected to cover science as objectively as possible: Science journalists should select new and important findings and present them correctly to their non-expert audiences. On the other hand, journalists are expected to transform findings and their evidence according to media logics and the common sense of a public at large.[1] As a result, complex scientific findings and evidencing practices become transformed into news stories that are supposed to be easy to understand and even entertaining.[2]

The word "story" already indicates that the transformation of scientific findings into understandable news items is mainly a process of narrativization in which archetypical protagonists (e.g. villains or mad scientists) as well as archetypical plots (e.g. hero stories or stories of failure) are employed. Stories, or to use the scientific term, narratives, can be considered a common and efficient tool for conveying meaning.[3] We grow up listening to narratives such as fairy tales; we get to know typical storylines, archetypical protagonists and antagonists. As a result, narratives are easily accessible and, not without reason, culturally and religiously formative texts are usually presented in a narrative form. Central mythological and religious records (e.g. the Iliad or the Bible) convey their moral messages in narratives.[4]

Against this background, it is reasonable that journalism also uses narratives to inform audiences about scientific findings.[5] However, an understandable and convincing story also offers a way to question the outcome of scientific research and to present problematic research. In this chapter, we analyze how news stories are used to question, criticize or even argue against scientific findings. For our analysis, we use media coverage of genomic research. We have chosen this field of research for several reasons, among them its importance for society and its rapid development. Additionally, genomic research receives considerable attention in the media. Not only is it a popular and controversial topic in news media whose coverage is full of hope and fear,[6] but it is also the subject of fictional narratives in novels and films. According to Rosalynn D. Haynes, who has analyzed recurring patterns in fiction about

DOI: 10.4324/9781003273509-17

science, it is especially narratives about genomic engineering that most often have recourse to stereotypes of scientists and criticisms of big business.[7] And, as we will demonstrate later, these features characterizing fiction about genetics can also be found in print coverage, which is to say, news tends to echo fiction when genomic research and its evidence is being criticized.

We first outline our theoretical framework and locate science journalism in the realm of science itself as well as journalism. In doing so, we focus on the strategies journalism usually uses to convey or even support scientific evidence. Since this chapter addresses the contrary – how coverage is used to weaken evidence – we focus our discussion on textual strategies employed to question and criticize evidence against the background of typical coverage of genomic research. Following this, we present a hermeneutic analysis of news stories, which reveals how journalism contests scientific evidence using its own (non-scientific) devices and creates counter-narratives. We focus on five linguistic and culturally contextualized strategies, which we explain in detail.

Theoretical Framework

Practices of (Science) Journalism

Publics seek to learn about scientific results to gain knowledge about technological, social and ecological developments that help to satisfy their curiosity about the future. Applications of scientific innovations (e.g. new medical treatments) may have a direct effect on their own lives. As laypersons, they usually do not have the access or expertise to inform themselves from genuine scientific sources. In this situation, journalism serves the function of mediating between science and the public and generating information that is understandable for the non-scientist.[8] Rather than merely dumbing down complex scientific content, journalists engage in a number of translations that gradually transform scientific content into a (re)construction for media presentation.[9]

In general, journalists' (re)construction of reality follows certain rules. For example, news values such as focus on conflicts, catastrophes, elites or the emphasis on human action (= personalization) can be found in almost every media report. In terms of content, news values ensure the newsworthiness of media coverage and can be considered a recurring pattern.[10] These event-related patterns are complemented by conventions concerning the form of news reports. Journalism relies on a set of well-established formats (editorials, reports, glosses, etc.) whose characteristics are clearly defined and are usually respected – at least by professional journalists. For the most part, all these journalistic norms, content- as well as form-related, determine media representations. And, as Sharon Dunwoody demonstrates for the field of science journalism, science journalists tend to adhere to journalistic norms to a greater extent than to scientific standards.[11]

Science Journalism and Second Level Evidencing Practices

The adherence to journalistic norms is not surprising when comparing the basic systemic logics of journalism and science. Science strives for scientific validity. The function of journalism is to transform social reality in such a way that it becomes accessible to society (as media reality).[12] Therefore, social reality is transmuted into generally understandable and preferably new topics. Science journalism deals with a specific part of social reality, i.e. science. It reports on new scientific findings, including the associated evidencing processes, and may also criticize them.[13] This happens from an observational perspective because science journalists do not conduct their own studies; they do not produce scientific evidence in the narrow sense (this is mainly reserved to the science system). However, if "evidencing practices" are conceived of as the process of presenting, embedding and using evidence,[14] they also encompass textual strategies to support a claim as evident (in the realm of science), or, true and valid (in the realm of journalism).

Science journalism can describe scientific evidencing processes and, additionally, attribute evidence to certain findings linguistically, e.g. by emphasizing the quality of a study, the accuracy of its findings or its importance for further research. In a similar vein, science journalism can question the adequacy of evidencing processes or even allege fraud, e.g. by ascribing negative attributes to certain studies and methods or by using rhetorical strategies to discredit them. As a result, evidence deriving from science is either underlined or undermined by journalistic means. Thus, on an initial level, evidencing practices are located in science itself. But on a second level, journalism steps in and affirms or disconfirms evidence – resulting in what we call "second level evidencing practices" or "second level evidencing critiques". These textual strategies should not be underestimated regarding their potential influence on public opinion and the repercussions on the science system. They can either support public knowledge and acceptance of scientific research,[15] or they can destabilize science by undermining faith in scientific research. There are quite a few historical examples where (critical) media discourse and public contestations of scientific evidence have affected public support for certain scientific fields – which can even result in restrictions on research.[16] The repertoire of textual second level evidencing or critiquing practices is large, as our analysis will show, and does not necessarily focus on scientific evidencing practices. News stories give context to scientific findings and evidencing procedures; they place research in a certain setting (which can be depicted as adequate or dubious), create characters (who can be described as moral, or untrustworthy and even evil) and, in sum, often convey a moral message.[17] As a result, journalistic critiquing practices in particular can focus on these contextual factors and simultaneously sow doubt about a scientific finding and evidencing processes.

Evidencing Practices in Media Coverage

In order to support a claim as true or valid, journalism has developed its own ways of substantiating scientific knowledge,[18] which range from science-affiliated to journalism-savvy. Susanne Kinnebrock, Helena Bilandzic and Magdalena Klingler distinguish three textual strategies to underline the evidence of a finding.[19] First, the *data and methods* of a study can be presented in news reports to justify the study's conclusions – which ultimately mirrors "evidencing practices"[20] common in the epistemic culture of science and less typical of journalism.[21] Second, experts, institutions or journals as renowned *authorities* in their specific research fields can be quoted.[22] This evidencing practice is not only applied in scientific writing; quoting authorities as sources of information is also at the heart of news reporting. Quoting authorities and, in doing so, specifying their professional roles, affiliations and positions within institutions are well-established journalistic routines to underline the quality of a source and thereby the factuality of the information. As a result, references to authorities are equally common in the epistemic cultures of science and journalism. And third, evidence claims can be supported by telling a convincing story or *narrative* (e.g. of a successful healing process).

Narratives are defined as a representation of events and characters.[23] Narratives are an everyday, natural mode of communication, widely used in science journalism because they turn scientific findings into equally understandable and tangible stories.[24] At the same time, (science) journalists can build on existing stories – on eternal stories or myths, as Jack Lule puts it[25] – to build on the audience's prior knowledge. A brief reference to mythological stories like the tragic fall of Icarus or Victor Frankenstein's incapability to control his creature can guide the reader's interpretation of a current science story and its presumed end.[26] As a result, the use of narratives can trigger different responses among the audience. Narratives can help to understand scientific evidencing procedures and thus strengthen the reader's belief in scientific evidence claims. However, narratives can also distract the audience's attention from the actual scientific point by emphasizing human interest. And additionally, a moral message is often conveyed, especially when news stories refer to myths.[27]

These three typical journalistic evidencing practices (data and methods, authorities, narratives) also provide starting points for the analysis of evidence criticism in news media. The validity of data as well as measurement reliability can be doubted, authorities can be questioned or even be unmasked as charlatans and, most of all, narratives can be used to refute the results of a study as a whole by providing counterexamples.

Coverage of Genomic Research

As with science coverage in general,[28] coverage of genomics has increased over time.[29] In the early 1990s – with Dolly, the cloned sheep, and genetically modified food – genomic research received a lot of media attention. Topics like the human genome project, human stem cells and emerging fields such as synthetic biology or xenotransplantation were frequently covered.[30] Early studies showed that coverage of the emerging field of genomics was usually more positive than negative.[31] Genetic engineering in agriculture, however, can be regarded as an exception because news reports about this topic were more critical than news reports about genomics in general.[32] In sum, the contextualization of "green" genetic engineering and "red" genetic technologies differs. Genetic engineering in agriculture is more often covered using a risk frame, genetic treatments in the field of medicine, however, tend to be reported using a progress frame.[33] Even if media debates on genomic research are quite specific to national contexts, genomics has become an internationally relevant topic in science coverage, comprising heterogeneous evaluations,[34] which makes media coverage of genomics suitable for an analysis of second level evidence critique.

Strategies of Evidence Criticism in Science Journalism

As already outlined, the daily business of science journalists is reporting, not research. Therefore, they cannot reasonably be expected to provide scientific counter-evidence. They can, however, create critical news stories. It is up to science journalists to craft their reports, i.e. they enjoy a great deal of freedom in choosing their topics and sources, which consequently affects the conclusions to be drawn from their reports.[35] To question the evidence of a specific finding, journalists can (1) rely on the description of other, alternative or "better" *data and methods*, (2) question the credibility of *authorities* and, instead, quote other, alternative or "better" experts and (3) tell compelling (human-interest) *stories* about the "victims" of science or "failed" research.

In our analysis, we show that challenging the credibility of authorities is a common pattern of evidence criticism in journalism; the strategies used for this are *personalization* and *negative stereotyping* – strategies that resonate with fictional accounts of science and genetics.[36] For lay audiences, especially criticism that focuses on scientists as people, on their actions and their morality (= personalization) is easier to understand than lengthy explanations as to why certain data and methods are problematic. As a result, our analysis pays attention to those depicted as dubious charlatans or mad scientists in conflict with groups of honorable scientists or even society as a whole.

It is not only characters that can be criticized. Critiques can also emerge from the plot of a story. *Conflict* is an essential feature of many plots.[37] Thus, a focus on various conflicts within an article (e.g. among researchers or between researchers and civil society) can serve to call scientific findings into question. Additionally, a narrative within an article can convey a moral message since evaluations also are vital features of narratives.[38] Including cues for a bad ending can be a strategy for questioning an area of research and its evidence in general. More generally, *narrativization* can be used in various ways to indirectly criticize science and question evidence.

Despite many language conventions in daily reporting for appropriate wording, journalists enjoy remarkable freedom in choosing their very own words to describe a particular situation or scientific result. Language can underline and deepen certain stereotypes and conflicts. And the particular choice of words can either create a narrative world or deconstruct it. At the same time, language is very domain-specific: Human interest topics, for example, are usually described with other, more emotive words than economic news. And, in science coverage, plain, prosaic language is predominant, which serves to convey the rationality of the field.[39] If wording typical of other domains (e.g. human interest, religion, esotericism) is used for science coverage, readers may build mental associations between the non-scientific domain with the research presented. The use of the incongruous language can also be a strategy to elicit doubts about the correctness of the presented scientific conclusions.

This brief overview of key journalistic strategies for criticizing science and questioning the evidence for scientific findings leads us to our research questions:

1 What types of personalization, negative stereotyping and conflict depictions can be identified in articles on genomic research?
2 What references to archetypical narratives are made and what kind of moral messages are suggested?
3 And what language and wording is used that is not common in the domain of science?
4 How are these textual strategies applied in the coverage of genomic research?

Evidencing Practices in Media Coverage of Genomic Research

The following hermeneutic analysis is part of a larger project that analyzes evidencing practices in reports on genomic research.[40] The sample of the content analysis consisted of 1,023 articles on genomic research published by German print media and included national quality newspapers (*Frankfurter Allgemeine Zeitung, Süddeutsche Zeitung, TAZ*), regional newspapers (*Hamburger Abendblatt, Nürnberger Nachrichten, Mitteldeutsche Zeitung*), tabloids (*Bild, Express, Berliner Kurier*) as well as weekly news magazines (*Der Spiegel, Die Zeit, Focus*). A random sample with representative layers for each medium and each year was compiled and articles from the year 2000 to 2018 were included. As the main focus of this study was to investigate evidencing practices, only articles that contained a scientific finding were included.

To briefly summarize the most prominent result, one of our main insights was that scientific findings are usually evidenced by more than one journalistic evidencing practice. Explanations of data and methods, references to authorities and finally narrative elements are often used together to underline the validity of a scientific finding. Additionally, these three practices are usually used to support, not to question, the evidence of the findings. Notably, counter-narratives are rare: Among the 1,023 articles analyzed, roughly half (n = 447) the articles used narratives to illustrate the evidence of a finding. However, among these 447 articles, only 27 used a narrative to question and criticize scientific findings, which is less than three percent of all analyzed articles (or six percent of articles applying narrative elements). Consequently, the vast majority of narratives presents findings and evidencing processes in a neutral or supportive way. Focusing on second level evidencing critique, these 27 articles containing narratives that argue against scientific evidencing practices represent the basis of the present hermeneutic analysis. Given the sampling strategy of the content analysis, the compilation of the resulting 27 articles is systematic and different from selection procedures typically used in case studies. Nevertheless, some cases (like Monsanto) show up in our material because their practices are repeatedly questioned in counter-narratives. As our hermeneutic analysis will show, textual evidencing critique might be a comparatively rare, but, when used, strong rhetorical strategy in science journalism. As soon as a counter-narrative is employed to question scientific evidence, the possible textual strategies are used intensively, as our analysis demonstrates.

In-Depth Analysis of Narrative Strategies to Question Evidence and Criticize Science

Personalization

Research on personalization has a long tradition in the field of communication. It can be defined as an editing process transforming social reality into media reality by condensing complex (research) processes into a few actions and decisions by a single person.[41] A character, mostly a scientist, and his or her experiences, motives and emotions are vividly described, which allows for a better understanding of the protagonist's point of view. Our analysis showed the dominance of two characters that are repeatedly depicted in the counter-narratives on genomic research: The mad scientist and the ruthless company. Although a company cannot be directly equated with a person, it can be depicted as a unit that acts and that has intentions and both can be judged for their morality.

In the articles analyzed, it is especially companies dealing with genetic sequencing or genetically modified plants that are criticized. These companies appear to act like human beings – and they are endowed with human attributes. According to many articles, the companies concerned seek to gain financially from the new findings – no matter the costs and consequences. Greed drives them, and Monsanto in particular is described as a ruthless, corrupt and manipulative liar. The news magazine *Der Spiegel*, for example, states with reference to Monsanto: "A global industry is trying everything to make the world dependent on genetically modified plants",[42] and the quality newspaper *Süddeutsche Zeitung* summarizes simply: "Monsanto is evil".[43]

It is no surprise that ruthless companies and mad scientists dominate counter-narratives. In their role as protagonists, they act, they fight and they have dubious intentions – which makes it easy for journalists to create a highly personalized depiction. In general, active perpetrators are more suitable characters for highly narrative news stories than passive victims. However, in some cases, the victims' stories are also told: Apart from farmers harmed by genetically modified plants, animals, especially cloned animals, are victims. They are exploited as laboratory animals, suffer and die early. One example is Dolly, the cloned sheep. Like Monsanto, Dolly is not a person. The description of her fate in the newspaper *Hamburger Abendblatt* is nevertheless touching, especially if the reader takes the perspective of a human being and makes the common idea of "a good life" the yardstick for judging Dolly's life: "Dolly lived a mere 6 years …, never knew what the sun looked like and never tasted grass. For security reasons, the cloned sheep lived in a heavily guarded concrete block, where she munched pills containing concentrated food".[44]

Stereotyping

Personalization can be regarded as a precondition for stereotyping, which we discuss with regard to the character type of the mad scientist. The importance and ambivalence of stereotypes in media coverage – as helpful organizing structures that reduce complexity as well as bundles of attributes that might be used to discriminate against particular groups of people – was already outlined a century ago by Walter Lippman in *Public Opinion*.[45] Nevertheless, content analyses dedicated to stereotypes of scientists in media coverage are rare. With reference to fiction, Matthew C. Nisbet and Anthony Dudo identify four characters: (1) The sinister, mad scientist; (2) The powerless pawn; (3) The anti-social geek; and (4) The action hero.[46] Rosalynn D. Haynes added two more characters: (5) The stupid virtuoso and (6) The scientist as idealist.[47] While negative character depictions were prevalent in the past, Haynes points out that depictions of the mad scientist are becoming less common – at least in the realm of fiction.[48] Likewise, Dudo et al. conclude that scientists are "cast in good or mixed roles, rather than as the 'evil scientist'".[49]

Regarding the 27 articles in our sample, one negative stereotype is dominant, namely that of the *mad scientist*.[50] However, media coverage on genomics characterizes the mad scientist type as less sinister but more obsessed with scientific work. The scientist appears as a maniac impervious to moderating influences. One example is the characterization of George Church, known for his work on genomic sequencing:

> George Church, molecular geneticist at Harvard University [...] is known as someone who considers very few ideas too crazy to try out himself. He and a few of his coworkers have been trying for some years to revive the mammoth [...] to what end? Is de-extinction just one of those researchers running wild ideas? Stuart Pimm is even more explicit: 'De-Extinction is nothing but a way for people who otherwise have no clue about how to solve the problems of the world to get attention'.[51]

Madness combined with craving for attention is quite often attributed to scientists. And in the quotation above from the quality newspaper *Süddeutsche Zeitung*, another pattern of criticism becomes obvious: In conveying criticism or negative stereotyping, another scientist is quoted, which allows journalists to hide their views behind quotes and to keep the appearance of journalistic objectivity.

The stereotype of the mad scientist has many facets, and it is frequently gendered.[52] In our sample, an obviously gendered sub-stereotype of the mad scientist is the image of the *grumpy old man* who is unteachable and stubborn. An example is Len Hayflick, known for criticizing anti-aging medicine. The way his looks and mode of expression are described characterizes him as a grumpy old man:

> Len Hayflick's tone becomes ominous. He turns all the energy that he would like to expend stamping his feet into a low rumble. He was, in fact, only asked whether he could help non-geneticists to prolong people's lives. 'Genes have absolutely nothing to do with aging,' barks the grand old man of aging research. And his eyebrows are so bushy that they briefly protrude from behind the thick rims of his large glasses.[53]

In stark contrast to these grumpy old men, who decorate quite a few articles critical of genomic research, is a young female scientist who extended the life of threadworms genetically.

> The young woman breeds worms. Tiny roundworms [...] that wriggle harmlessly in the test tube but look like monsters with huge gullets when you look at them under a microscope. 'I like this face', gushes the researcher, showing a close-up of one of her protégés: 'Isn't it lovely?'[54]

This example from *Süddeutsche Zeitung* (like quite a few others) indicates that enthusiasm paired with detachment from common points of view seems to be particularly typical of female researchers. While the sub-stereotype of the grumpy old man appears to be reserved for male scientists, the sub-stereotype of the *unworldly enthusiast* is mainly applied to female scientists. This is not surprising as unworldliness is also part of the traditional female stereotype[55] as well as of scientist stereotypes in general.[56]

Another sub-stereotype of the mad scientist is the *angry brawler*. The weekly newspaper *Die Zeit* presents Craig Venter, a competitor in the Human Genome Project, as an "evil and angry underdog of the gene scene".[57] Competition between different scientific projects is reduced to negative emotions and quarrelsomeness as characteristic of the scientists involved. To underline how angry these competitors are, emotive expressions are used. According to conventions of journalistic writing, quotations are usually introduced with neutral formulations such as "researcher x says" or "researcher y comments". But especially in the context of the Human Genome Project, the scientists "rant", "mock", "boast", "scoff" and "badmouth".[58] *Die Zeit* concludes:

> In short, the matadors of the gene scene are boiling – partly their soup, partly with rage. 'There's too much vanity involved,' says Friedrich von Bohlen, head of the Heidelberg-based bioinformatics company Lion Bioscience. 'Prima donnas of the worst kind' are at work there. But if you take a closer look, you'll see that there's a bit more to the wrangling than hypertrophied egos. It's about merit, but also about business and, in the end, even about science.[59]

In this quotation, another facet of the mad scientist is mentioned: The mad scientist can be quite a *peacock*. Hubris and vanity are typical of the peacock sub-stereotype of the mad scientist. These attributes are often used in counter-narratives about genomic research and are usually applied to male scientists, not female ones. It is remarkable that the peacocks are unmasked in the articles – usually by colleagues who are presented as honest scientists and reflective thinkers. The strategy behind this negative stereotyping is to depict a person in a poor light and hence to question his (rarely her) findings. Thus, it is a roundabout, but nevertheless an effective strategy for criticizing evidence. Since we do not trust braggarts or villains in our daily lives, there are few reasons to trust the findings of a peacock scientist.

The stereotype of the mad scientist deriving from fiction often includes characteristics like sinisterness and power-hunger.[60] Within our sample, however, the role of the sinister villain was not assigned to scientists. Instead, it was exclusively reserved for companies in the field of genomics. According to the counter-narratives on genomics we analyzed, the most obvious villain is Monsanto, which is described as a ruthless and greedy company:

> The peoples of the world rarely agree. But when they look at this company, everyone yells: Monsanto is evil. [...] The new group has control over what humanity eats and what penetrates the earth. It is this power that scares many people. They feel that power is in the wrong hands with Monsanto. The rise of the group was rapid. And as with almost every rapid ascent, there have been sacrificial lambs and skeletons in the closet.[61]

After this introduction, the article from the *Süddeutsche Zeitung* relates the history of Monsanto, focusing on the trail of devastation the company has left. Among other things, the article mentions pollution in the US village of Monsanto, where the company was founded, the production of glyphosate and how cancer cases were ignored, Monsanto's rise to a monopolist that blocks competitors and dictates prices, the production of the herbicide Agent Orange during the Vietnam War and finally Monsanto's sinister lobbying practices at institutions of the EU in Brussels. The article clearly suggests only one conclusion: Monsanto is wicked to the core.

Excursus Normative and Cultural Foundations of Negative Stereotyping

It is remarkable that deeply negative moral attributes are used to characterize scientists and companies in the field of genomic research. As mentioned before, less than three percent of all articles in a representative sample used a narrative to question or criticize the evidence of a scientific finding. It seems that, in general, criticizing evidence with a narrative rarely happens in media coverage of genomic research. In the few cases in which an article presents criticism incorporated in a story, however, very rich stories have been created. That means, for example, that the characters were given clear attributes, and the events were clearly evaluated. Both is surprising when considering that, thanks to the ideal of objectivity, (science) journalism usually tends to avoid too much attribution and evaluation.[62] The intensity of the attribution results from the fact that highly moralizing attributes are used in the description (and stereotyping) of scientists. Many of these negative attributes are deeply rooted in Christian culture, especially when they refer to the seven deadly sins (sloth, lust, anger, pride, envy, gluttony and greed). Stanford Lyman has analyzed these sins as moral laws, which were historically used in many cultures (not only the Christian) to describe the evil and still affect societies all over the world.[63] Similar to myths or the holy books (be it the Bible, the Koran, the Torah or similar sacred texts), the seven deadly sins can be used to quickly classify behavior, or more precisely, to brand it as evil. Given the fact that journalism has to be understood by and resonate with lay audiences, it is not surprising that a common and unambiguous reference system based on the seven deadly sins is used to mark science and scientists as evil – which is, of course, a very harsh criticism.

The seven deadly sins and some of the negative attributes conferred to scientists in the articles analyzed coincide to a remarkable extent. The stereotypical mad scientist is a glutton for (scientific) work, but at the same time indifferent to real life and social concerns. Hence, lack of moderation as well as unworldliness is associated with *gluttony*.[64] Consequently, the love for roundworms can be read as both unworldliness and indifference to real life, which are linked to *gluttony*. The stereotypical grumpy old man shows *anger*. Angry brawlers are not only angry, but they are also envious, whereby *envy* as a deadly sin also encompasses jealousy and malevolence.[65] A peacock is full of *pride and vainglory*. And, finally, the ruthless company is definitely characterized by *greed*. Without overusing the seven deadly sins, a very important pattern of criticizing scientists and thereby science and scientific evidence has become obvious: Coverage of genomic research points toward the moral deficits of some scientists and companies. Stereotyping takes place along a dimension of deeply moral attribution.

Conflict

The counter-narratives we analyzed were full of conflicts. And the pattern just described for stereotyping – the reference to deeply moral categories – also becomes visible when we consider the whole story rather than a single character. The articles often tell the story of a fight between good and evil. The lines of conflict primarily lie between the evil and money-grubbing genetic engineering companies and their opponents, who are mostly honest organic farmers, eco-activists or upright, research-oriented scientists. Additionally, conflicts arise between scientists themselves. Good and reflective scientists, who respect the limits of what is ethical or feasible, struggle with mad scientists who do not care about limits and consequences.

The conflicts are underlined by two rhetorical strategies: Scandalizing and ridiculing. Scandalizing is a well-known strategy in journalism. The roles (victim or perpetrator) are clearly assigned. The perpetrator is publicly accused of violating a norm, the event is explicitly called a scandal and indignation is articulated.[66] Since trust in a scandalized perpetrator is compromised, scandalizing scientists can be used as a strategy apparently to question the researcher as a person, but actually denigrating their research and evidencing practices. Scandalizing, therefore, can be regarded as a subtle or indirect form of evidence critique.

Another way of sowing doubts about science and evidence is ridiculing scientists, their findings or even a whole research field. Some counter-narratives in the articles analyzed were full of irony and mockery. Genetic sequencing and its results are described by the weekly newspaper *Die Zeit* in the following way:

Genome experts want only to assign about 35,000 genes to humans. And once again, this stirs controversy. Although everyone was tremendously excited about it, the outcome is somewhat unwelcome. Some bemoan the third narcissistic affront to humankind by science. First we were downgraded to a product of evolution along with monkeys and lice (Darwin did not dare claim this, but it is nevertheless the case), then we were declared to be the oppressed of the subconscious (Freud did in fact claim this, but it is nonsense), and now this: A threadworm of just 959 cells manages to have 19,098 genes, and humans with their 100 trillion cells have only a third more.[67]

Myths and Master Plots

Master plots contain an evaluative and moral dimension that frame media coverage and guide its interpretation.[68] The counter-narratives investigated in this analysis feature a master plot of failure. On a micro-level, single events of failure – a failed experiment, a big error or the death of a cloned creature – are widely reported. On a macro-level, the master plot and the evaluation of the narrative suggest the futility of the endeavor or the inevitably bad ending. For example, the stories emphasize that nature will revenge itself for human interventions, hubris will be punished and interference with the divine order will lead to disaster. To underline these messages, explicit references are made to well-known myths such as the Frankenstein myth, which stands for a monstrous creation as a consequence of a scientist's arrogance and hubris in seeking to be the creator of life.[69]

In sum, the failure master plot either emphasizes that the scientists fail to produce evidence, or it frames the whole research enterprise as extremely negative. As a result, the articles cast doubt on the evidence of the research presented. And the recurring moral message is "Keep the end in mind!" – which is both the headline from the newspaper *Frankfurter Allgemeine Zeitung* and a biblical phrase.[70] The reference to the divine order leads to the last of the five strategies of science criticism. The use of language that is typical of another domain – for example, religion or esotericism.

Language

Just as the name Frankenstein evokes ideas of the myth associated with it, words from domains other than science can be used to associate science with fields that, at first sight, do not have much in common with science and its evidencing practices. In the counter-narratives we analyzed, we were confronted with many words coming from the fields of fortune telling ("oracle", "crystal ball", "fortune teller"), magic ("wizard", "magic words"), science fiction ("chimeras", "mixed creatures") and dubious quackery ("promises of healing", "truth serum", "wishful thinking"). And it is remarkable that these words are not only used occasionally; they pervade our analyzed counter-narratives. They suggest that the respective article deals with an esoteric or fictional plot, not with science. And this strategy, placing genomic research and its evidencing practices in completely science-free domains, can be regarded as an attempt to cast doubt on the correctness of the presented scientific findings and conclusions.

In the same manner, references to religious language and biblical sayings are used and create religious allusions: "Their magic word is stem cells. Whoever can breed and train these all-healing cells like the shepherd trains his dogs, the lame and sick will make a pilgrimage to the Holy Land like the pious used to do".[71] Sentences like this, which seem strange when readers are expecting solid journalistic prose, re-emerge frequently. Biblical expressions are also used extensively. Scientists are referred to as "creator" or even "god"; "satan" or the "devil" comes into play and leads to "temptation"; scientific communities are called "parish" or "gene church"; scientific controversies "wars of faith" and scientific findings "promises of god". These are only a few examples which give the impression that religious phenomena are being described in the articles, not science and its evidencing practices.

Finally, antiquated language (*"zum gleichen Behufe"* – *for the same purpose, "Ein Tor, wer glaubt"* – *a fool who thinks that)* is occasionally used, suggesting time travel to pre-modern, even medieval times, in which theology rather than science was the dominant knowledge system and the natural sciences were associated with alchemy and the production of gold.[72] The extensive use of words and phrases untypical of journalistic writing is not coincidental. Rather, the use of a language that is not appropriate for covering science can be regarded as a strategy to achieve an image transfer from irrational, ideological domains to the field of science. And as soon as science is associated with such domains, doubts about its findings, conclusions and evidencing practices are easily sown.

Conclusion

We have presented counter-narratives that question and criticize the scientific findings and evidencing practices in the news coverage of genomic research. In all, these counter-narratives are rare; the vast majority of narratives function to support the evidence of findings in genomic research. This might not be surprising since our quantitative content analysis focused on articles containing an empirical finding from a study. When science and its core activity – carrying out studies and producing evidence – are reported, it might be easier for journalists to construct narrative descriptions of studies than argue against study results and their evidence. In the rare case that a narrative is used to undermine research, the criticism is usually indirect in the sense that it is not the finding as such that is criticized or questioned, but the moral integrity of the scientists and companies involved. The criticism is quite sharp and attuned to the readers' everyday experience and life: Counter-narratives are stories about evil villains, mad scientists and failure.

Patterns of prototypical counter-narratives emerge and they have a striking resemblance to archetypal myths – notably in a format (science journalism) that is dedicated to conveying hard scientific facts to a wider audience. A prototypical counter-narrative follows common steps that – in an evil master's handbook of counter-narratives – might read like this:

a Identify the perpetrators! (personalization)
b Describe their bad character! (negative stereotyping and its moral foundation)
c Evoke the conflict between the good and the evil! (conflict)
d Refer to archetypical narratives and suggest that the story will end badly! (master plots)
e Use language from other well-known mystic domains to make research and its evidencing practices appear in a dubious and irrational light! (language)

At first sight, the elements of these archetypical science stories on genomic research might have little to do with scientific evidencing processes because of their non-scientific focus on the (bad) character of researchers, on recurring human conflicts or stylistic devices like language. However, these narrative elements can be used as strategies of critique. They create the context in which research and evidencing practices are depicted and therefore can affect lay audience's perceptions. It has to be remembered that, outside the science system, first level evidence practices are mainly perceived through the lens of second level evidencing practices – and among these are narratives.

The emphasis on morality reveals the basic function of counternarratives – to warn against potential dangers and interests of actors that are located outside of science itself. Notably, such discourse, borrowed from well-known myths and master plots, is much more intelligible and familiar to a non-scientific audience than the scientific facts themselves. Whether this serves to make audience judgments more nuanced and critical, or to shift the focus away from public engagement with science to a generalized distrust toward science due to more or less fuzzy moral concerns, must be the subject of future research.

Notes

1   J. Scott Brennen, "Magnetologists on the Beat: The Epistemology of Science Journalism Reconsidered", *Communication Theory* 28, no. 4 (2018): 424–443.

2   Amanda Hinnant and Maria E. Len-Rios, "Tacit Understandings of Health Literacy Interview and Survey Research with Health Journalists", *Science Communication* 31, no. 1 (2009): 84–115.

3   Jerome Bruner, "The Narrative Construction of Reality", *Critical Inquiry* 18, no. 1 (1991): 1–21.

4   See Jack Lule, *Daily News, Eternal Stories: The Mythological Role of Journalism* (New York: Guilford Press, 2001); Jonathan Gottschall, *The Storytelling Animal: How Stories Make Us Human* (New York: Houghton Mifflin, 2012).

5   Michael F. Dahlstrom, "Using Narratives and Storytelling to Communicate Science with Nonexpert Audiences", *Proceedings of the National Academy of Sciences of the United States of America* 111, no. 4 (2014): 13614–13620.

6   Martin W. Bauer and Jan M. Gutteling, "Issue Salience and Media Framing over 30 Years", in *Genomics & Society: Legal, Ethical and Social Aspects*, eds. Georg Gaskell and Martin W. Bauer (London: Routledge, 2006), 113–130; Jürgen Hampel, "Die Darstellung der Gentechnik in den Medien", in *Biotechnologie-Kommunikation: Kontroversen, Analysen, Aktivitäten (Acatech DISCUSSION)*, eds. Marc-Denis Weitze, Alfred Pühler, Wolfgang M. Heckl, Bernd Müller-Röber, Ortwin Renn, Peter Weingart, and Günther Wess (Berlin: Springer, 2012), 253–285.

7   Rosalynn D. Haynes, *From Madman to Crimefighter: The Scientist in Western Culture* (Baltimore: John Hopkins Univ. Press, 2017), 337.

8   Michaela Maier, Tobias Rothmund, Andrea Retzbach, Lukas Otto, and John C. Besley, "Informal Learning Through Science Media Usage", *Educational Psychologist* 49, no. 2 (2014): 86–103.

9   Brennen, "Magnetologists".

10 Christiane Eilders, "News Factors and News Decisions: Theoretical and Methodological Advances in Germany", *Communications: The European Journal of Communication Research* 31, no. 1 (2006): 5–24; Michaela Maier, Joachim Retzback, Isabella Glogger, and Karin Stengel. *Nachrichtenwerttheorie.* 2nd edition. (Baden-Baden: Nomos, 2018).

11 Sharon Dunwoody, "Science Journalism: Prospects in the Digital Age", in *Routledge Handbook of Public Communication of Science and Technology*, eds. Massimiano Bucchi and Brian Trench (New York: Routledge, 2014), 27–39; see also Bernd Blöbaum, "Wissenschaftsjournalismus", in *Forschungsfeld Wissenschaftskommunikation*, eds. Heinz Bonfadelli, Birte Fähnrich, Corinna Lüthje, Jutta Milde, Markus Rhomberg, and Mike Schäfer (Wiesbaden: Springer VS, 2017), 221–238, 224.

12 Winfried Schulz, *Die Konstruktion von Realität in den Nachrichtenmedien: Analyse der aktuellen Berichterstattung*, 2nd. edition (Freiburg, München: Karl Alber, 1990), 27–29.

13 Blöbaum, "Wissenschaftsjournalismus", 224.

14 Sarah Ehlers and Karin Zachmann, "Wissen und Begründen: Evidenz als umkämpfte Ressource in der Wissensgesellschaft", in *Wissen und Begründen: Evidenz als umkämpfte Ressource in der Wissensgesellschaft*, eds. Karin Zachmann and Sarah Ehlers (Baden-Baden: Nomos, 2019), 17–18.

15 Science journalism, in general, tends to be rather science-friendly, see Holger Wormer, "Vom Public Understanding of Science zum Public Understanding of Journalism", in *Forschungsfeld Wissenschaftskommunikation*, eds. Heinz Bonfadelli, Birte Fähnrich, Corinna Lüthje, Jutta Milde, Markus Rhomberg, and Mike Schäfer (Wiesbaden: Springer VS, 2017), 433.

16 For example, genetics or nuclear energy in Germany, see Marc-Denis Weitze and Wolfgang M. Heckl, *Wissenschaftskommunikation: Schlüsselideen, Akteure, Fallbeispiele* (Berlin: Springer, 2016), 247–265.

17 Helena Bilandzic, "Narrativer Journalismus, narrative Wirkungen", in *Medien und Journalismus im 21. Jahrhundert*, eds. Nina Springer, Johannes Raabe, Hannes Haas and Wolfgang Eichhorn (Konstanz: UVK, 2012), 481.

18 Yigal Godler and Zvi Reich, "Journalistic Evidence: Cross-Verification as a Constituent of Mediated Knowledge", *Journalism: Theory, Practice & Criticism* 18, no. 5 (2017): 558–574.

19 Susanne Kinnebrock, Helena Bilandzic, and Magdalena Klingler, "Erzählen und Analysieren", in *Wissen und Begründen: Evidenz als umkämpfte Ressource in der Wissensgesellschaft*, eds. Karin Zachmann and Sarah Ehlers (Baden-Baden: Nomos, 2019), 137–165.

20 Ehlers and Zachmann, "Wissen und Begründen", 9.

21 Blöbaum, "Wissenschaftsjournalismus", 231.

22 Blöbaum, "Wissenschaftsjournalismus", 229.

23 Porter H. Abbott, *The Cambridge Introduction to Narrative* (Cambridge: Univ. Press); Monika Fludernik, *Towards a 'Natural' Narratology* (London: Routledge, 2010).

24 Dahlstrom, "Using Narratives"; Lucy Avraamidou and Jonathan Osborne, "The Role of Narrative in Communicating Science", *International Journal of Science Education* 31, no. 2 (2009): 1683–1707.

25 Lule, *Daily News*.

26 Kurt W. Back, "Frankenstein and Brave New World: Two Cautionary Myths on the Boundaries of Science", *History of European Ideas* 20, no. 1–3 (1995): 327–332.

27 Elizabeth S. Bird and Robert W. Dardenne, "Rethinking News as Myth and Storytelling", in *The Handbook of Journalism Studies*, eds. Karin Wahl-Jorgensen and Thomas Hanitzsch (New York: Routledge, 2009), 205–217.

28 Julia Bockelmann, "Wissenschaftsberichterstattung im SPIEGEL: Eine Inhaltsanalyse im Zeitverlauf", in *Molekulare Medizin und Medien: Zur Darstellung und Wirkung eines kontroversen Wissenschaftsthemas*, eds.

Georg Ruhrmann, Jutta Milde, and Arne Freya Zillich (Wiesbaden: VS Verlag für Sozialwissenschaften, 2011), 41–69; Bienvenido Leon, "Science Related Information in European Television: A Study of Prime-Time News" *Public Understanding of Science* 17, no. 4 (2008): 443–460.

29   Bauer and Gutteling, "Issue Salience".

30   Bauer and Gutteling, "Issue Salience"; Hampel, "Die Darstellung der Gentechnik in den Medien".

31   Matthias Kohring, Alexander Görke, and Georg Ruhrmann, "Das Bild der Gentechnik in den internationalen Medien: Eine Inhaltsanalyse meinungsführender Zeitschriften", in *Gentechnik in der Öffentlichkeit: Wahrnehmung und Bewertung einer umstrittenen Technologie*, ed. Jürgen Hampel (Frankfurt am Main: Campus, 2001), 292–316; Celine Lewis, Mahrufa Choudhury, and Lyn S. Chitty, "'Hope for Safe Prenatal Gene Tests': A Content Analysis of How the UK Press Media are Reporting Advances in Non-Invasive Prenatal Testing", *Prenatal Diagnosis* 35, no. 5 (2015): 420–427; Carolyn Michelle, "'Human Clones Talk About Their Lives': Media Representations of Assisted Reproductive and Biogenetic Technologies", *Media Culture & Society* 29, no. 4 (2007): 639–663; Matthew C. Nisbet and Bruce V. Lewenstein, "Biotechnology and the American Media: The Policy Process and the Elite Press, 1970–1999", *Science Communication* 23, no. 4 (2002): 351–391; Mary L. Nucci and Robert Kubey, "'We Begin Tonight With Fruits and Vegetables': Genetically Modified Food on the Evening News 1980–2003", *Science Communication* 29, no. 2 (2007): 147–176; Eric Racine, Isabelle Gareau, Hubert Doucet, Danielle Laudy, Guy Jobin, and Pamela Schraedley-Desmond, "Hyped Biomedical Science or Uncritical Reporting? Press Coverage of Genomics (1992–2001) in Québec", *Social Science & Medicine* 62, no. 5 (2006): 1278–1290; Mike S. Schäfer, *Wissenschaft in den Medien: Die Medialisierung naturwissenschaftlicher Themen* (Wiesbaden: Verlag für Sozialwissenschaften, 2007).

32   Bauer and Gutteling, "Issue Salience"; Toby A. Ten Eyck and Melissa Williment, "The National Media and Things Genetic – Coverage in the New York Times (1971–2001) and the Washington Post (1977–2001)", *Science Communication* 25, no. 2 (2003): 129–152.

33   Heinz Bonfadelli, "Fokus Grüne Gentechnik: Analyse des Medienvermittelten Diskurses", in *Biotechnologie-Kommunikation: Kontroversen, Analysen, Aktivitäten (Acatech DISCUSSION)*, eds. Marc-Denis Weitze, Alfred Pühler, Wolfgang M. Heckl, Bernd Müller-Röber, Ortwin Renn, Peter Weingart, and Günther Wess (Berlin: Springer, 2012), 205–252.

34   Bonfadelli, "Fokus Grüne Gentechnik", 205–252.

35   For the field of nanotechnology, see Lars Guenther and Georg Ruhrmann, "Science Journalists' Selection Criteria and Depiction of Nanotechnology in German Media", *Journal of Science Communication* 12, no 3 (2013): 1–17.

36   Haynes, *From Madman*, 5.

37   Gerald Prince, *Narratology: The Form and Functioning of Narrative* (Berlin: Walter de Gruyter).

38   William Labov and Joshua Waletzky, "Erzählanalyse: Mündliche Versionen persönlicher Erfahrung", in *Literaturwissenschaft und Linguistik: Vol. 2*, ed. Jens Ihwe (Frankfurt am Main: Fischer, 1973), 78–126.

39   Blöbaum, "Wissenschaftsjournalismus", 230.

40   Helena Bilandzic, Susanne Kinnebrock, and Lena Klingler, "Evidencing Practices of Science Journalism in the Newspaper Coverage of Genomic Research", *Manuscript Under Review*.

41   Schulz, *Konstruktion*, 45.

42 Philip Bethge, "Satt durch Designer-Pflanzen?", *Der Spiegel*, September 13, 2004.
43 Kathrin Werner, "Sähen, spritzen und vielleicht ernten", *Süddeutsche Zeitung*, September 15, 2016.
44 "Rückschlag für die Wissenschaft", Hamburger Abendblatt, February, 15, 2003.
45 Walter Lippman, *Public Opinion* (New York: Harcourt, Brace & Co, 1922).
46 Matthew C. Nisbet and Anthony Dudo, "Entertainment Media Portrayals and Their Effects on the Public Understanding of Science", in *Hollywood Chemistry: When Science Met Entertainment,* eds. Donna J. Nelson, Kevom R. Grazier, Jaime Paglia, and Sidney Perkowitz (Washington, DC: American Chemical Society, 2013), 241–249.
47 Haynes, *From Madman;* Rosalynn D. Haynes, *From Faust to Strangelove: Representations of the Scientist in Western Literature* (Baltimore: Johns Hopkins Univ. Press, 1994).
48 Haynes, *From Faust to Strangelove*, 337–339.
49 Anthony Dudo, Dominique Brossard, James Shanahan, Dietram A. Scheufele, Michael Morgan, and Nancy Signorielli, "Science on Television in the 21st Century: Recent Trends in Portrayals and Their Contributions to Public Attitudes Toward Science", *Communication Research* 38, no. 6 (2011): 763.
50 Georg Seeßlen, "Mad Scientist: Repräsentation des Wissenschaftlers im Film", *Gegenworte* 21, no. 3 (1999): 44–48.
51 Karin Blawat, "Auferstehung im Labor", *Süddeutsche Zeitung*, November 7, 2015.
52 Jenny Kitzinger, Mwenya Diana Chimba, Andy Williams, Joan Haran, and Tammy Boyce. *Gender, Stereotypes and Expertise in the Press: How Newspapers Represent Female and Male Scientists* (Cardiff: UK Resource Centre for Women in Science, Engineering and Technology (UKRC) and Cardiff University, 2008).
53 Christina Berndt, "Warten auf Methusalem", *Süddeutsche Zeitung*, February 27, 2001.
54 Berndt, "Warten auf Methusalem".
55 Ute Frevert, *"Mann und Weib, und Weib und Mann": Geschlechter-Differenzen in der Moderne* (München: C.H. Beck, 1995).
56 Petra Pansegrau, "Stereotypes and Images of Scientists in Fiction Films", in *Science Images and Popular Images of Science: Routledge Studies in Science, Technology and Society,* eds. Peter Weingart and Bernd Hüppauf (London: Routledge, 2007), 33–51.
57 Bahnsen, Ulrich, "Tanz der Primadonnen", *Die Zeit*, February 22, 2001.
58 Ulrich, "Tanz der Primadonnen".
59 Ulrich, "Tanz der Primadonnen".
60 Seeßlen, "Mad Scientist".
61 Kathrin Werner, "Sähen, spritzen und vielleicht ernten", *Süddeutsche Zeitung*, September 15, 2016.
62 Susanne Kinnebrock, "Puzzling Gender Differently? A Comparative Study of Newspaper Coverage in Austria, Germany, and Switzerland", *Interactions: Studies in Communication & Culture* 2, no. 3 (2012): 197–208; Chrstoph Neuberger, "Journalistische Objektivität: Vorschlag für einen pragmatischen Theorierahmen", *Medien & Kommunikationswissenschaft* 65, no. 2 (2017): 406–431.
63 Stanford M. Lyman, *The Seven Deadly Sins: Society and Evil.* Revised and expanded edition (Lanham: General Hall, 1989), 1–4.

64 Lyman, *The Seven Deadly Sins*, 212–231.
65 Lyman, *The Seven Deadly Sins*, 184–211.
66 Mathias Kepplinger, *Medien und Skandale* (Wiesbaden: Spinger VS, 2018); Jens Bergmann and Bernhard Pörksen, "Einleitung: Die Macht der öffentlichen Empörung", in *Skandal! Die Macht der öffentlichen Empörung*, eds. Jens Bergmann and Bernhard Pörksen (Köln: Halem, 2009), 7–33.
67 Ulrich, "Tanz der Primadonnen".
68 Jenny Kitzinger, "Questioning the Sci-Fi Alibi: A Critique of How 'Science Fiction Fears' Are Used to Explain Away Public Concerns About Risk", *Journal of Risk Research* 13, no. 1 (2010): 73–86; Helen Haste, "Myths, Monsters, and Morality: Understanding 'Antiscience' and the Media Message", *Interdisciplinary Science Reviews* 22, no. 2 (1997): 114–120.
69 Andrew Tudor, *Monsters and Mad Scientists: A Cultural History of the Horror Movie* (Oxford, UK: B. Blackwell, 1989); Andrew Tudor, "Seeing the Worst Side of Science", *Nature* 340 (1989): 598–592.
70 Julia Klöckner, "Bedenke das Ende", Frankfurter Allgemeine Zeitung, October 13, 2007.
71 Anke Richter, "Herz zu Herz, Hirn zu Hirn", Die Zeit, December 28, 2000.
72 Haynes, *From Madman*, 15.

References

Abbott, Horace Porter. *The Cambridge Introduction to Narrative*. Cambridge: Univ. Press, 2002.

Avraamidou, Lucy, and Jonathan Osborne. "The Role of Narrative in Communicating Science". *International Journal of Science Education* 31, no. 2 (2009): 1683–1707.

Back, Kurt W. "Frankenstein and Brave New World: Two Cautionary Myths on the Boundaries of Science". *History of European Ideas* 20, no. 1–3 (April 1995): 327–332.

Bauer, Martin W., and Jan M. Gutteling. "Issue Salience and Media Framing over 30 Years". In *Genomics & Society: Legal, Ethical and Social Aspects*, edited by Georg Gaskell, and Martin W. Bauer, 113–130. London: Routledge, 2006.

Bergmann, Jens, and Bernhard Pörksen. "Einleitung: Die Macht der öffentlichen Empörung". In *Skandal! Die Macht der öffentlichen Empörung*, edited by Jens Bergmann, and Bernhard Pörksen, 7–33. Köln: Halem, 2009.

Bilandzic, Helena. "Narrativer Journalismus, narrative Wirkungen". In *Medien und Journalismus im 21. Jahrhundert*, edited by Nina Springer, Johannes Raabe, Hannes Haas, and Wolfgang Eichhorn, 467–487. Konstanz: UVK, 2012.

Bilandzic, Helena, Susanne Kinnebrock, and Lena Klingler. "Evidencing Practices of Science Journalism in the Newspaper Coverage of Genomic Research". *Manuscript under Review*, 2023

Bird, S. Elizabeth, and Robert W. Dardenne. "Rethinking News as Myth and Storytelling". In *The Handbook of Journalism Studies*, edited by Karin Wahl-Jorgensen, and Thomas Hanitzsch, 205–217. New York: Routledge, 2009.

Blöbaum, Bernd. "Wissenschaftsjournalismus". In *Forschungsfeld Wissenschaftskommunikation*, edited by Heinz Bonfadelli, Birte Fähnrich, Corinna Lüthje, Jutta Milde, Markus Rhomberg, and Mike Schäfer, 221–238. Wiesbaden: Springer VS, 2017.

Bockelmann, Julia. "Wissenschaftsberichterstattung im SPIEGEL: Eine Inhaltsanalyse im Zeitverlauf". In *Molekulare Medizin und Medien: Zur Darstellung und Wirkung eines kontroversen Wissenschaftsthemas*, edited by Georg Ruhrmann, Jutta Milde, and Arne Freya Zillich, 41–69. Wiesbaden: VS Verlag für Sozialwissenschaften, 2011.

Bonfadelli, Heinz. "Fokus Grüne Gentechnik: Analyse des Medienvermittelten Diskurses". In *Biotechnologie-Kommunikation: Kontroversen, Analysen, Aktivitäten (Acatech DISCUSSION)*, edited by Marc-Denis Weitze, Alfred Pühler, Wolfgang M. Heckl, Bernd Müller-Röber, Ortwin Renn, Peter Weingart, and Günther Wess, 205–252. Berlin: Springer, 2012.

Brennen, J. Scott. "Magnetologists on the Beat: The Epistemology of Science Journalism Reconsidered". *Communication Theory* 28, no. 4 (2018): 424–443.

Bruner, Jerome. "The Narrative Construction of Reality". *Critical Inquiry* 18, no. 1 (1991): 1–21.

Dahlstrom, Michael F. "Using Narratives and Storytelling to Communicate Science with Nonexpert Audiences". *Proceedings of the National Academy of Sciences of the United States of America* 111, no. 4 (2014): 13614–13620.

Dudo, Anthony, Dominique Brossard, James Shanahan, Dietram A. Scheufele, Michael Morgan, and Nancy Signorielli. "Science on Television in the 21st Century: Recent Trends in Portrayals and Their Contributions to Public Attitudes Toward Science". *Communication Research* 38, no. 6 (2011): 754–777.

Dunwoody, Sharon. "Science Journalism: Prospects in the Digital Age". In *Routledge Handbook of Public Communication of Science and Technology*, edited by Massimiano Bucchi, and Brian Trench, 27–39. New York: Routledge, 2014.

Ehlers, Sarah, and Karin Zachmann. "Wissen und Begründen: Evidenz als umkämpfte Ressource in der Wissensgesellschaft", In *Wissen und Begründen: Evidenz als umkämpfte Ressource in der Wissensgesellschaft*, edited by Karin Zachmann, and Sarah Ehlers, 9–29. Baden-Baden: Nomos, 2019.

Eilders, Christiane. "News Factors and News Decisions: Theoretical and Methodological Advances in Germany". *Communications: The European Journal of Communication Research* 31, no. 1 (2006): 5–24.

Fludernik, Monika. *Towards a 'Natural' Narratology*, 2nd edition. London: Routledge, 2010.

Frevert, Ute. *"Mann und Weib, und Weib und Mann": Geschlechter-Differenzen in der Moderne*. München: C.H. Beck, 1995.

Godler, Yigal, and Zvi Reich. "Journalistic Evidence: Cross-Verification as a Constituent of Mediated Knowledge". *Journalism: Theory, Practice & Criticism* 18, no. 5 (2017): 558–574.

Gottschall, Jonathan. *The Storytelling Animal: How Stories Make Us Human*. New York: Houghton Mifflin, 2012.

Guenther, Lars, and Georg Ruhrmann. "Science Journalists' Selection Criteria and Depiction of Nanotechnology in German Media". *Journal of Science Communication* 12, no. 3 (2013): 1–17.

Hampel, Jürgen. "Die Darstellung der Gentechnik in den Medien". In *Biotechnologie-Kommunikation: Kontroversen, Analysen, Aktivitäten (Acatech DISCUSSION)*, edited by Marc-Denis Weitze, Alfred Pühler, Wolfgang M. Heckl, Bernd Müller-Röber, Ortwin Renn, Peter Weingart, and Günther Wess, 253–285. Berlin: Springer, 2012.

Haste, Helen. "Myths, Monsters, and Morality: Understanding 'Antiscience' and the Media Message". *Interdisciplinary Science Reviews* 22, no. 2 (1997): 114–120.

Haynes, Rosalynn D. *From Faust to Strangelove: Representations of the Scientist in Western Literature*. Baltimore: Johns Hopkins Univ. Press, 1994.

Haynes, Rosalynn D. *From Madman to Crimefighter: The Scientist in Western Culture*. Baltimore: John Hopkins Univ. Press, 2017.

Hinnant, Amanda, and Maria E. Len-Rios. "Tacit Understandings of Health Literacy Interview and Survey Research With Health Journalists". *Science Communication* 31, no. 1 (2009): 84–115.

Kepplinger, Mathias. *Medien und Skandale*. Wiesbaden: Spinger VS, 2018.

Kinnebrock, Susanne. "Puzzling Gender Differently? A Comparative Study of Newspaper Coverage in Austria, Germany, and Switzerland". *Interactions: Studies in Communication & Culture* 2, no. 3 (2012): 197–208.

Kinnebrock, Susanne, Helena Bilandzic, and Magdalena Klingler. "Erzählen und Analysieren". In *Wissen und Begründen: Evidenz als umkämpfte Ressource in der Wissensgesellschaft*, edited by Karin Zachmann, and Sarah Ehlers, 137–165. Baden-Baden: Nomos, 2019.

Kitzinger, Jenny. "Questioning the Sci-Fi Alibi: A Critique of How 'Science Fiction Fears' Are Used to Explain Away Public Concerns About Risk". *Journal of Risk Research* 13, no. 1 (2010): 73–86.

Kitzinger, Jenny, Mwenya Diana Chimba, Andy Williams, Joan Haran, and Tammy Boyce. *Gender, Stereotypes and Expertise in the Press: How Newspapers Represent Female and Male Scientists*. Cardiff: UK Resource Centre for Women in Science, Engineering and Technology (UKRC) and Cardiff University, 2008, http://cf.ac.uk/jomec/resources/Kitzinger_Report_2.

Kohring, Matthias, Alexander Görke, and Georg Ruhrmann. "Das Bild der Gentechnik in den internationalen Medien: Eine Inhaltsanalyse meinungsführender Zeitschriften". In *Gentechnik in der Öffentlichkeit: Wahrnehmung und Bewertung einer umstrittenen Technologie*, edited by Jürgen Hampel, 292–316. Frankfurt am Main: Campus, 2001.

Labov, William, and Joshua Waletzky. "Erzählanalyse: Mündliche Versionen persönlicher Erfahrung". In *Literaturwissenschaft und Linguistik*: Vol. 2, edited by Jens Ihwe, 78–126. Frankfurt am Main: Fischer, 1973.

Leon, Bienvenido. "Science Related Information in European Television: A Study of Prime-Time News". *Public Understanding of Science* 17, no. 4 (2008): 443–460.

Lewis, Celine, Mahrufa Choudhury, and Lyn S. Chitty. 'Hope for Safe Prenatal Gene Tests': A Content Analysis of How the UK Press Media are Reporting Advances in Non-Invasive Prenatal Testing". *Prenatal Diagnosis* 35, no. 5 (2015): 420–427.

Lippman, Walter. *Public Opinion*. New York: Harcourt, Brace & Co, 1922.

Lule, Jack. *Daily News, Eternal Stories: The Mythological Role of Journalism*. New York: Guilford Press, 2001.

Lyman, Stanford M. *The Seven Deadly Sins: Society and Evil*. Revised and expanded edition. Lanham: General Hall, 1989.

Maier, Michaela, Joachim Retzback, Isabella Glogger, and Karin Stengel. *Nachrichtenwerttheorie*. 2nd edition. Baden-Baden: Nomos, 2018.

Maier, Michaela, Tobias Rothmund, Andrea Retzbach, Lukas Otto, and John C. Besley. "Informal Learning Through Science Media Usage". *Educational Psychologist* 49, no. 2 (2014): 86–103.

Michelle, Carolyn. "'Human Clones Talk About Their Lives': Media Representations of Assisted Reproductive and Biogenetic Technologies". *Media Culture & Society* 29, no. 4 (2007): 639–663.

Neuberger, Christoph. "Journalistische Objektivität: Vorschlag für einen pragmatischen Theorierahmen". *Medien & Kommunikationswissenschaft* 65, no. 2 (2017): 406–431.

Nisbet, Matthew C., and Anthony Dudo. "Entertainment Media Portrayals and Their Effects on the Public Understanding of Science". In *Hollywood Chemistry: When Science Met Entertainment*, edited by Donna J. Nelson, Kevom R. Grazier, Jaime Paglia, and Sidney Perkowitz, 241–249. Washington: American Chemical Society, 2013.

Nisbet, Matthew C., and Bruce V. Lewenstein. "Biotechnology and the American Media: The Policy Process and the Elite Press, 1970–1999". *Science Communication* 23, no. 4 (2002): 351–391.

Nucci, Mary L., and Robert Kubey. "'We Begin Tonight With Fruits and Vegetables': Genetically Modified Food on the Evening News 1980–2003". *Science Communication* 29, no. 2 (2007): 147–176.

Pansegrau, Petra. "Stereotypes and Images of Scientists in Fiction Films". In *Science Images and Popular Images of Science: Routledge Studies in Science, Technology and Society*, edited by Peter Weingart, and Bernd Hüppauf, 33–51. London: Routledge, 2007.

Prince, Gerald. *Narratology: The Form and Functioning of Narrative*. Berlin: Walter de Gruyter.

Racine, Eric, Isabelle Gareau, Hubert Doucet, Danielle Laudy, Guy Jobin, and Pamela Schraedley-Desmond. "Hyped Biomedical Science or Uncritical Reporting? Press Coverage of Genomics (1992–2001) in Québec". *Social Science & Medicine* 62, no. 5 (2006): 1278–1290.

Schäfer, Mike S. *Wissenschaft in den Medien: Die Medialisierung naturwissenschaftlicher Themen*. Wiesbaden: Verlag für Sozialwissenschaften, 2007.

Schulz, Winfried. *Die Konstruktion von Realität in den Nachrichtenmedien: Analyse der aktuellen Berichterstattung*. 2. Auflage. Freiburg, München: Karl Alber, 1990.

Seeßlen, Georg, "Mad Scientist: Repräsentation des Wissenschaftlers im Film". *Gegenworte* 21, no. 3 (1999): 44–48.

Ten Eyck, Toby A., and Melissa Williment. "The National Media and Things Genetic – Coverage in the New York Times (1971–2001) and the Washington Post (1977–2001)". *Science Communication* 25, no. 2 (2003): 129–152.

Tudor, Andrew. *Monsters and Mad Scientists: A Cultural History of the Horror Movie*. Oxford, UK: B. Blackwell, 1989.

Tudor, Andrew. "Seeing the Worst Side of Science". *Nature* 340 (1989): 589–592.

Weitze, Marc-Denis, and Wolfgang M. Heckl. *Wissenschaftskommunikation: Schlüsselideen, Akteure, Fallbeispiele*. Berlin: Springer, 2016.

Wormer, Holger. "Vom Public Understanding of Science zum Public Understanding of Journalism". In *Forschungsfeld Wissenschaftskommunikation*, edited by Heinz Bonfadelli, Birte Fähnrich, Corinna Lüthje, Jutta Milde, Markus Rhomberg, and Mike Schäfer, 429–452. Wiesbaden: Springer VS, 2017.

Index

Printed in the United States
by Baker & Taylor Publisher Services